TECHNOLOGY OF FLUID POWER

Online Services

Delmar Online

To access a wide variety of Delmar products and services on the World Wide Web, point your browser to:

http://www.delmar.com
or email: info@delmar.com

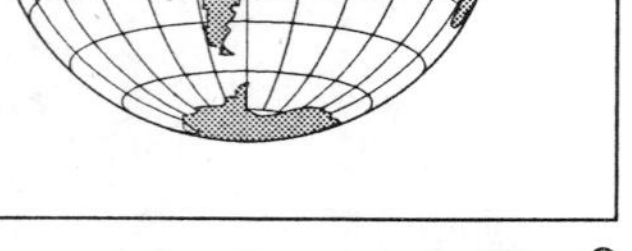

thomson.com

To access International Thomson Publishing's home site for information on more than 34 publishers and 20,000 products, point your browser to:

http://www.thomson.com
or email: findit@kiosk.thomson.com

A service of I(T)P®

TECHNOLOGY OF FLUID POWER

William W. Reeves

The Department of Industrial Technology
Ohio University

Delmar Publishers

an International Thomson Publishing company

Albany • Bonn • Boston • Cincinnati • Detroit • London • Madrid
Melbourne • Mexico City • New York • Pacific Grove • Paris • San Francisco
Singapore • Tokyo • Toronto • Washington

NOTICE TO THE READER

Delmar Staff
Publisher: Robert D. Lynch
Senior Administrative Editor: John Anderson
Production Manager: Larry Main
Art & Design Coordinator: Nicole Reamer
Editorial Assistant: John Fisher
Cover Design by Brucie Rosch

Printed in the United States of America
4 5 6 7 8 9 10 XXX 01 00 99

For more information, contact Delmar, 3 Columbia Circle, PO Box 15015, Albany, NY 12212-0515; or find us on the World Wide Web at http://www.delmar.com

International Division List

Japan:
Thomson Learning
Palaceside Building 5F
1-1-1 Hitotsubashi, Chiyoda-ku
Tokyo 100 0003 Japan
Tel: 813 5218 6544
Fax: 813 5218 6551

Australia/New Zealand
Nelson/Thomson Learning
102 Dodds Street
South Melbourne, Victoria 3205
Australia
Tel: 61 39 685 4111
Fax: 61 39 685 4199

UK/Europe/Middle East:
Thomson Learning
Berkshire House
168-173 High Holborn
London
WC1V 7AA United Kingdom
Tel: 44 171 497 1422
Fax: 44 171 497 1426

Latin America:
Thomson Learning
Seneca, 53
Colonia Polanco
11560 Mexico D.F. Mexico
Tel: 525-281-2906
Fax: 525-281-2656

Canada:
Nelson/Thomson Learning
1120 Birchmount Road
Scarborough, Ontario
Canada M1K 5G4
Tel: 416-752-9100
Fax: 416-752-8102

Asia:
Thomson Learning
60 Albert Street, #15-01
Albert Complex
Singapore 189969
Tel: 65 336 6411
Fax: 65 336 7411

Library of Congress Cataloging-In-Publication Data
Reeves, William W., 1950-
Fluid power : circuit and component design, operation, and analysis / William W. Reeves
p. cm.
Includes index.
ISBN 0-8273-6664-7
1. Fluid power technology. I. Title.
TJ840R385 1996 96-11155
621.2—dc20 CIP

Contents

To Barbara
for her love, patience, and understanding

To Jennifer
for giving me the opportunity to live long enough to write another book

And to Jill
because she always loved books more than anything else

Preface

This text is written primarily for beginning fluid power students in programs such as industrial technology, mechanical engineering technology, industrial engineering technology, mechanical engineering, and manufacturing engineering and in technical education programs such as industrial maintenance. It is devoted to providing students with fundamental concepts and basic skills necessary to understand and design a variety of fluid circuits. Emphasis is placed on exercises and assignments, with text to present concepts. The reader is instructed to apply these concepts through problem solving, color coding, schematic development, and component specification in order to develop a greater understanding of the practical applications of fluid power.

Various laws and theories have been devised to describe the nature and actions of fluids under pressure. Many of these are useful in predicting the operation of fluid circuitry under specific environments and conditions. Fluid circuits seldom operate within such limited parameters. Therefore, this text will minimize their usage and will incorporate only those laws and theories required to describe the major variables encountered in designing and understanding fluid circuitry.

The reader will be designing fluid circuitry based on practical applications and typical loads, cycles, or speed requirements. Although specific results will be expected, the variability of fluid components that control pressure and flow of fluids adequately compensates for most of the variance caused by conditional and environmental effects. Such variability is one of the major advantages of fluid power.

Throughout the text, emphasis is placed on developing an understanding of the construction of componentry such as pumps, valves, and actuators. With such understanding, the reader will be able to predict how circuits having fluid components will operate. The ability to design fluid circuitry will follow from the logical combination of components rather than rote memorization of standard circuits. The methods employed in this text have been field tested by the author for twenty years in the university classroom and laboratory environment; in industrial training seminars; in national and international presentations to peers; and through the multitude of college graduates who have used these techniques successfully in responsible positions as mechanical engineers, industrial engineers, manufacturing engineers, process engineers, plant engineers, and maintenance technicians and supervisors.

The reader who successfully completes this text should be able to "think fluid power." With further study and application of theories and laws, exposure, and hands-on experience, vocations such as those mentioned, as well as machine design and technical sales, are possible. Also, the casual or interested reader will find these concepts useful in developing an understanding of how fluid circuitry operates.

The author is indebted to the many technologists, scientists, educators, and engineers who have made the study of fluid power a possibility. Gratitude is expressed to Dr. Clyde M. Hackler, my mentor, who I am sure (being an educator myself) was never thanked often enough for the positive influence he had on many of our lives.

I would like to thank the many students who participated during the last ten years in special projects that resulted in many of the supplemental computer-aided instructional programs that are included with this book. Special recognition is given to Robert Parry, Scott Wagner, and Jody Paul, who contributed to those activities above and beyond the call of duty, and to my colleague Dinesh Dhamija, who provided much-needed assistance by structuring those programs to make them efficient and effective. Dan George spent many hours redrawing my initial concepts on a new CAD system. If the reader thinks that they are well done, Dan deserves the credit. If not, it was probably my fault for missing the concept.

I would like to acknowledge the contribution of the many industrial contacts who have provided needed information and many of the photographs that appear in this text. Although their companies are recognized and credited, theirs is the effort of the silent majority. So, thanks to Pete Everts, Larry Schrader, Michael Moore, Nancy Hendrix, and Larry Madden. Finally, I acknowledge the contribution of Ted Thieman, president of THT Enterprises in Dayton, Ohio. Ted has been a friend as well as an innovator in the use of hydraulics in manufacturing for many years. He kept bringing and sending me things long after I had any place to put them. Every time I began to challenge my sanity about trying to write another book, I realized that anything I had to live through would pale in comparison to what he has done. Ted, you were the inspiration. I hope the product meets with your approval.

In closing, to the informed reader, many of the illustrations that you will see in this text will be familiar, while others are quite unique. Those that are unique are special (at least to me). They are pictures of machinery, test stands, trainers, and other circuitry that we have designed and constructed over the years to serve specific educational, research-and-development, and industrial needs. They represent the actual applications of those things that appear in this text.

ACKNOWLEDGMENTS

The author and publisher wish to thank the following reviewers, who provided valuable feedback during the development of this book:

James A. Gray
School of Technology
Indiana State University
Terre Haute, Indiana

John Shelley
Drafting/CAD Dept.
School of Vocational Technology
Wellsville, New York

Jim Brumagin
New River Community College
Dublin, Virginia

Jim Steele
Automated Manufacturing Instructor
Ivy Tech State College
Sellersburg, Indiana

George C. Agin
Industrial Technology
Washtenaw Community College
Ann Arbor, Michigan

1

Introduction to Fluid Power

People, in general, have little understanding of various fluid power devices that they use every day. They may even be intimidated by fluid machinery that moves tons of soil (Figure 1-1) or raises whole houses while amazed by the accuracy of control available in robots using fluid devices (Figure 1-2). However, fluid circuits are normally no more complicated than house wiring circuits. With a basic understanding of a few fluid laws and the ability to perform simple algebra, anyone can understand basic fluid power circuitry.

FIGURE 1-1

Mobile hydraulic equipment. *(Courtesy of Richard W. Vockroth. From* Industrial Hydraulics, *Albany, N.Y.: Delmar, 1994, p.11.)*

FIGURE 1-2

Hydraulic robots move heavy loads with power and precision. *(Courtesy GMFanuc Inc. From Malcolm,* Robotics—An Introduction, *Albany, N.Y.: Delmar, 1988, p. 329.)*

1.1 FLUID POWER IN CONTEXT

The industrial workplace environment typically operates using an interacting network of materials, processes, and control; that is, the production of finished goods (whether an automobile or a hydroelectric dam) requires the manipulation (process) of specified materials through the use of sometimes sophisticated (sometimes simple) methods and power (control). This text will focus on one specific area of control: power transmission and control using fluid power energy. The production, operation, and maintenance of fluid power systems certainly employ materials and processes. However, the focus of this study will not be the production of fluid power components, but rather, the application of these components in systems intended for fluid power use or output.

1.1.1 Energy Sources

Energy is the ability to do work or move forces through distances. In general, there are six accepted forms of energy available as sources from which to do work: heat, light, nuclear, chemical, electrical, and mechanical. Theoretically, any form of energy may be changed to any other form (see Figure 1-3).

FIGURE 1-3

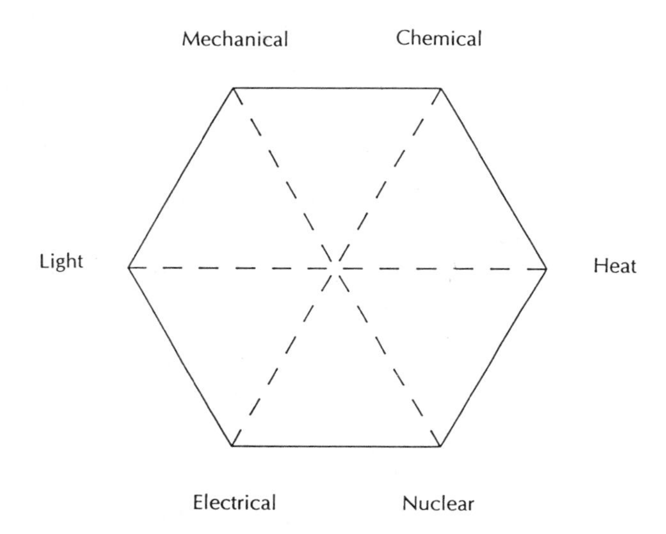

In a practical sense, this change is typically accompanied by unintentional changes to other forms that reduce the effectiveness of the change. For example, changing the chemical form of energy available in a storage battery into illumination from a flashlight is accompanied by a change, first to electrical energy for transmission purposes, then to light in the filament or ionized gas inside the bulb. Through this process, some of the potential energy has been lost as heat from the output lightbulb. This example cites some of the potential challenges faced in power and energy control and transmission. Fluid power is classified as a branch of mechanical energy in this context; and rather than being primarily concerned with producing light, heat, electrical, chemical, or nuclear output, fluid power focuses on the physical movement of objects from one location to another.

1.1.2 Fluid Power and Mechanical Energy

There are many similarities between the principles of mechanical energy and fluid power. Most fluid power components (valves, actuators, pumps, etc.) rely on the manipulation of six simple machines: the lever, pulley, wheel and axle, inclined plane, wedge, and screw. A fundamental understanding of the principles employed in these machines will be of benefit when examining the operation of fluid power components.

If fluid power is but a branch of mechanical energy, the question that arises is, "Why not just employ mechanical devices to transmit power from one location to another?" Most often, the answer is efficiency and flexibility. Consider the fact that most electrical energy used today originated in mechanical power such as falling water through a turbine, creating mechanical motion, and producing electron flow through electromagnetic fields. In turn, that electrical energy produced was then transferred over great distances to a final destination that may have been the rotation of an electrical motor (mechanical output) in a machine tool, fan, or the compressor unit in an air conditioner. Since the total system described here began as mechanical rotation of the turbine and ended as mechanical rotation of the output motor, a pulley and belt system could have been used to connect input and output devices. Obviously, this system would become very impractical and would have limited control capabilities. Similarly, the linear or rotary output evident in fluid power systems could also be achieved using mechanical machines. However, mechanical systems employ rigid components (made from solid materials) that inhibit sophisticated movement simultaneously through varying directions. The nature of fluids (gas or liquid) allows for a high degree of flexibility in movement and affords easy direction change with little alteration of the systems containing them. In addition, fluid power output (using gas and liquid) affords varying degrees of cushioning not readily available within a mechanical mechanism.

1.1.3 Fluid, Mechanical, or Electrical Power

Any operating industrial system is composed of a source of power (input) and an end-use device (output). Bridging these two is a system of transmission that affords varying degrees of control dependent upon the sophistication of devices located in the transmission section. Furthermore, components used in all three stages may be electrical, mechanical, or fluid devices. The selection of which type of device to use must be contingent upon the situation that exists and the type(s) of control required. To that end, modern industrial machinery may employ all three types of controlling devices. However, the output is nearly always mechanical. For example, the output of a hydrostatic (fluid power) transmission in an automobile is typically mechanical rotation of a gearing system; and the output of an electrical motor used in a machine lathe is mechanical rotation of a pulley or gear-head system. Most often, then, the question of whether to use electrical, fluid, or mechanical devices lies in the areas of input (source) and transmission or control. The input question is a very difficult one to tackle and is probably a matter of the perspective of where one starts. After all, the power supply of a typical fluid power system includes a drive device that is an electrical motor or a sophisticated mechanical mechanism such as an internal combustion engine. Obviously, to have a total understanding of the fluid power supply, one must understand the principles of mechanical or electrical power employed in these devices. This would require an understanding of the mechanics of turbine operation for electrical motor operation, or chemical energy for the explosions taking place in the internal combustion engine. Where did the turbine come from? How were the chemicals made? It appears that a never-ending regression of analysis would be needed to fully understand the question, "Just how does a fluid power supply work?" Actually, the answer probably lies in tracing everything back to our sun, which caused the light and heat that were transmitted through space, making the plants and animals grow, resulting (eventually) in chemicals and turbines, falling water, and all sorts of energy sources we have at our disposal. Just do not ask where the sun came from. Therefore, this study will be limited to include and begin with those components directly attached and related to the fluid power supply.

The big question, therefore, must be whether to use fluid, mechanical, or electrical transmission and controlling devices. The answer, of course, is yes. That is, use them all if needed and if they can be easily, practically, and effectively applied. Each has limitations.

It would be very difficult to stop one's car from rolling down a hill by piling electrons or oil in front of (or behind) the wheel, but a simple wedge can often achieve this goal. However, electrons can be moved very rapidly through a conductor if immediate response is required (e.g., electronic computers work much more rapidly than fluid computers or an abacus). At the same time, electrons are very small and afford little support for heavy loads. Mechanical devices with a high degree of rigidity do this type of job well, and fluid devices may perform similarly according to the degree of compressibility in the confined fluid (i.e., liquids are less compressible than gases). Consider the braking system in an automobile. Of course, it must be able to exert a high degree of and maintain a heavy force. Seldom, if ever, are electrical brakes employed in this type of application. Mechanical braking systems using cables, levers, and wedges afford good load-bearing capacity and may be employed in this type of application. However, the geometrical changes in direction from the brake pedal to the disk pad against the rotor would make for a mechanical nightmare. The flexibility of hydraulic hosing and the fluid within the hose allow for easy change of direction, while the virtually noncompressible liquid fluid allows for great load-bearing capacity. In brief, if one wants to transmit power over a long distance efficiently with nearly immediate response, electricity and electronics should be used. If one wants to maintain heavy load-bearing capacity, mechanical control should be used. If, however, one needs good response time with good load-bearing capacity, through medium distances with many changes in direction, fluid power should be considered.

The overcoming of long distances presents a cost to the modern industrial and consumer equipment manufacturer, as do load bearing and flexibility. Usually, the buyer "wants it all." This has led to the inclusion of all three (electrical, mechanical, and fluid) systems in many mechanisms, taking advantage of their individually unique characteristics in specific locations where applicable. Therefore, a thorough examination of the operation of modern equipment would require a total understanding of all three areas of power transmission and control. This would (by title) be beyond the scope of this text. Therefore, this text will concentrate on hydraulic, pneumatic, and fluidic component and system design, operation, and maintenance using those components. It will assume a basic understanding of the principles of mechanical power necessary to understand the design and operation of fluid power components (e.g., "What happens to forces when an inclined plane has a greater slope?" or "What happens to the tangential velocity of a wheel attached to an axle if the wheel becomes larger?"). Furthermore, it will limit the use of electrical power to consider only those components and parts of components directly attached to the operating fluid power system in a general fashion so as not to confuse the topical presentation. For example, the text will refer to some AC (alternating current) electrical motor specifications (such as horsepower) but will not dwell on the design and characteristics of various types of electrical motors to achieve that parameter resulting from fluid power considerations (nor would it examine the design and operation of internal combustion engines employed to provide similar characteristics to the fluid system). It will mention electrical solenoid applications to fluid components but will not examine solenoid construction or operation. It will not delve into the mysteries of electrical ladder logic—indirectly related to fluid power circuitry—that so often topically confuses other fluid power texts. Finally, it will not examine the world of electronic digital logic that is the basis of such things as programmable logic controllers sometimes used in combination with fluid systems. Logically, those who study such topics probably do not care to have their studies clouded with the design, operation, and characteristics of fluid components that may be attached to those types of components and systems. Rather, it considers these types of topics to be peripheral to one's understanding of fluid power; and they would be better left to studies in their own environment. Advanced training and education would then be wise to completely package these individual studies into a culminating survey of power transmission and control.

1.2 TRADITIONAL FLUID POWER STUDY

Traditionally, the study of fluid power has been a two-phase approach. The technical approach has emphasized the development of understanding of fluid power component design and operation. The engineering approach has emphasized the development of understanding of fluid power circuit design and operation. This book will synthesize both approaches in what may best be described as a technological approach. Technicians, technologists, and engineers all need to develop a better understanding of the total concept of fluid power. Such abilities will enable them to better communicate ideas that will result in improved design, maintenance, and operation of circuitry.

Fluid devices rely on the manipulation and control of various liquids and gases. The initial chapters in this book use hydraulic fluids, primarily oils, as the media for power transmission. Chapters 10 and 12 include circuit and component design variations necessary to understand power transmission using gases, primarily air, as the media for power transmission.

1.3 DESIGN OF THE BOOK

Most technical texts present a body of information, then require interaction with the reader in a summary at the end of each chapter. Most manuals rely heavily on illustrations and problem solving, with little verbal explanation. This book interlaces text, illustrations, and problems, where appropriate, in order of presentation. It is believed that this interaction results in a reduction of the loss of retention by the reader. However, the casual reader will find that the text can be read without interruption for interactive activities by simply skipping over those instructions. Typically, each chapter ends with a design problem that summarizes knowledge gained through practical demonstration of understanding, as well as review questions and suggested activities.

Some graphics in this book designated as "Figures" are primarily used to describe overall design and operation of hydraulic circuits and components. These figures are the line drawing types of illustrations. They do not describe specific sizing or construction. For this reason, some details may be eliminated for sake of clarity. However, important problems and assignments are frequently included in the "line-drawing" figures, as well as the text. Each line-drawing figure should be thought of as a logical extension of the text, as well as a means of clarifying text information. Therefore, *line-drawing figures in each chapter should be given close attention*. In addition, photographic illustrations are included to focus attention to "real-world" designs, situations, and applications.

1.4 CONVENTIONS USED IN THE BOOK

Many different methods exist for presentation of information. Although all methods may be correct, the reader must be informed about which methods will be incorporated. Consistency must also be maintained. Many of these conventions exist in designations.

Fluid power incorporates the use of distances, areas, and volumes to describe many components. Areas are typically measured in square inches in the U.S. customary system or square centimeters in the SI (metric) system and may be designated verbally, conventionally, or exponentially. This text will use exponential designations where appropriate and conventional designations when commonly employed. Areas may, therefore, be designated **(sq. in.)** or **{*cm²*}**. Similarly, volumetric measurement will be designated **(cu. in.)** or **{*cm³*}**. Distances such as feet or inches have been conventionally designated by the single and double prime symbols, respectively. In this book, linear distances will be designated verbally—for example, inch **(in.)** or foot **(ft.)** or *centimeter* **{*cm*}** or *meter* **{*m*}**. Forces or loads will be designated as pounds **(lbs.)**, *Newtons* **{*N*}**, or *kilograms* **{*kg*}**, rather than #. Ratings will be made verbally, such as gallons/minute {*liters / minute*}—or conventionally **(gpm)** **{*lpm*}**—and revolutions/minute **(rpm)** and should be read gallons per minute {*liters per minute*} and revolutions per minute, respectively. The text will primarily focus on U.S. customary system designations but will supply SI (metric) units in braces and with designations appearing in italics:

VOLume **(gal.)** = VOLume **{*cm³*}** * 3,785

Other conventionally used designations will be presented throughout the text and will be explained at those points.

Conventions are also used to describe mathematical functions. In this book, the following symbols will be used:

Symbol	Function
+	Addition
–	Subtraction
$\sqrt{\ }$	Square root
*	Multiplication
—— or /	Division
x^2	Square

Note that dual symbols will be used for division. In most cases, the line division symbol will be incorporated. However, in more complex formulation, both symbols will be used. In these cases, the division procedure using the slash (/) symbol should be performed first, followed by the major division procedure described by the line symbol. To clarify these situations, consider the following formulas:

$$\text{DIAmeter (in.)} = 2\sqrt{\frac{[\text{DELivery (gal./min.)} * 0.3208]/\text{VELocity (ft./sec.)}}{3.14}}$$

or

$$\text{DIAmeter } \{mm\} = 20\sqrt{\frac{[\text{DELivery } \{l/min.\} * 16.67]/\text{VELocity } \{cm/sec.\}}{3.14}}$$

First, notice that the designations such as gal./min. or *l/min* are separated from the values for delivery by parentheses (or braces when using the SI designations). This also affords a natural separation between values. Also note that the first letters in each value are capitalized. After repeated use of the values, the first letters will be used alone to describe these values. Finally, note that values are grouped through the use of brackets. The mathematical operation involved within these brackets should be performed first.

The mathematical operations should be performed as follows:

1. Multiply the DELivery value by the constant 0.3208 or *16.67*.
2. Divide the value achieved in Step 1 by the VELocity value.
3. Divide the value achieved in Step 2 by the constant 3.14.
4. Take the square root of the value achieved in Step 3.
5. Multiply the value achieved in Step 4 by the constant 2 or *20*.

This illustrates what may be called the worst-case scenario. In most cases, the formulas used to describe fluid circuitry are, by comparison, relatively simple and straightforward.

A final convention used in this book involves the method of describing fluid condition within a circuit or component. Some excellent fluid power texts illustrate component and circuit operation through the use of assembly drawings using various codes to describe fluid condition. Borrowing from this successful technique, this book will also use assembly-type drawings; and the condition of the fluid will be designated using color coding as presented in Figure 1-4. It is recommended that the reader use these codes (when given in illustrations) as guides for coloring the drawings to reinforce the learning of tracing fluid conditions within the circuit or component.

FIGURE 1-4

Illustration Coding

Condition	Color	Code
Pressurized fluid	Red	R R
Exhaust fluid	Blue	B B
Supply fluid	Green	G G
Volume- or flow-controlled fluid	Yellow	Y Y
Fluid under reduced pressure	Orange	O O
Fluid drainage or leakage	Brown	N N
Inactive fluid	White	
Fluid under intensified pressure	Violet	V V

1.5 ANCILLARY MATERIALS

Indices are included in this textbook. They may be used as guides for selection of components and designs for component illustrations. In addition, there is a complete answer section that presents suggested answers to problems that appear throughout each chapter in the text and figures. Also, there are suggested answers for each end-of-chapter design problem. Note that in many cases these should be considered as suggested answers that result from assumptions for selected sizes of components, based on given parameters. These suggested answers assume that design procedures presented in the text are followed. These answers also assume a limitation of not more than 1,200 pounds per square inch **(psi)** *(8,274 kiloPascals* {***kPa***}) of pressure existing within the circuit, except in special circumstances. Although in practical applications this limitation should not be considered an industrial standard, it does reflect a range of limitations imposed by typical control that limits maximum system pressure. That is, many versions of pressure controls operate between 250 and 1,200 psi {*1,724 and 8,274 kPa*}. The same manufacturer will produce other series of pressure controls having a lower limit at this upper limit and have an upper limit greatly exceeding this value. However, it is probably good design practice to use this limitation, for it represents the lower limit of operation. Also, special cases requiring pressure in excess of this limit will be able to be used.

A computer disk is included with this textbook. It contains computer-aided instructional (CAI) programs corresponding to the end-of-chapter design problems. It will operate on any MS-DOS based system, Version 4.0 or higher, having a 3.5-inch disk drive. The reader should install the programs on a hard drive using the installation program. The programs on this disk are auto-execute type programs, which cause a menu to appear when the disk drive is accessed from DOS. Following the directions that appear on the menu, any executable file may be accessed or the reader may exit back to DOS. Upon selection of any design problem from the menu, the reader will be guided through a step-by-step input, question-and-answer format, allowing the reader to test his or her learning ability or review design procedures for the specification of each component within the circuit. It will give immediate feedback on whether the selected answers are correct and give the appropriate method for derivation of that answer. Finally, it will give a summary of how well the reader did in answering the design questions and will allow him or her to review the correct or selected specifications for all components on-screen, using a hard-copy printout, or both.

In addition, beginning with Chapter 5, variations on the end-of-chapter design problems using a grid of potential parameters (loads, sizes, capacity requirements, etc.) appear. This will allow the reader to practice his or her abilities to design and specify components for those circuits using other possible combinations.

1.6 SUGGESTED ACTIVITIES

1. Take a walking tour of your shops, laboratories, or plant and make a listing of all equipment that employs fluid power components.
2. Conduct a survey of all fluid power devices found in your household or residence. (Do not forget the basement, garage, and carport.)
3. Visit a commercial facility (shopping center, restaurant, or mall) and attempt to identify any obvious fluid power equipment used in these facilities.
4. Take a trip to a construction site and identify any fluid power equipment used at the site.
5. From your observations and visitations, would you say there is more wide-scale use of fluid power in the shop, laboratory, or plant; in the household or residence; in the operating commercial facility; or on the construction site? Why?
6. Perform the same observations as in Activities 1 to 5, this time trying to identify mechanical systems and electrical systems found in those environments. Why do you think those systems were employed in those applications rather than fluid power?
7. Go to your local or school library and make a bibliographical listing of all books that appear in the card catalog or database system under the following subject categories: fluid power, hydraulics, pneumatics, and fluidics. While there, make a similar bibliographical listing of all

published article listings in the *Reader's Guide* or in the CD-ROM database for a recent year of periodical literature dealing with the same subject categories. Since you have identified all these sources, you might as well check out a book and read it or review some articles from magazines or journals.

1.7 REVIEW QUESTIONS

1. Define *energy*.
2. List six different forms (sources) of energy.
3. Why is fluid power considered to be a branch of mechanical energy?
4. List six simple machines of mechanical energy.
5. Compare and contrast input, control, and output in a mechanical mechanism, electrical circuit, and fluid power circuit.

2

Fluid Power and Hydraulic Principles

The general term used to describe the use of a fluid to transfer power is **fluid power**. The ancient Babylonians, Phoenicians, and Egyptians used oars and sails to transmit power using fluids such as water and air. The ancient Greeks developed a crude water-pumping mechanism more than 2,000 years ago. The Archimedean pump had a spiral screw located inside a watertight canister that raised water from inlet to outlet through rotation of the screw. This design is similar to the auger used today to transport particulate solids and viscous liquids (agricultural grains, polymers in the screw-type plastics injection-molding machines, etc.). Many cultures have used waterwheels for raising water from one level to another for irrigation, and the ancient Chinese used elaborate water clocks to keep record of time.

2.1 FLUID POWER DEVELOPMENT

In more recent times, fluids (water or air) were used to drive devices such as grainery mills, windmill water pumps, and paddlewheel sawmills. In each case, the fluid was only partially controlled and simply served as a prime mover for other mechanical devices.

Unpressurized and partially controlled fluid systems are known as **hydrodynamic**. Even today, uses of such systems are evident in forced-air furnaces, ships' propellers, and the automatic transmissions in automobiles.

2.2 THE MODERN FLUID ERA

2.2.1 Pascal's Law

In the seventeenth century, **Blaise Pascal**, a French Jesuit priest, expounded many theories that changed the ways we live today. Pascal was responsible for the markings of syllables (diacritical markings) that define pronunciation in the dictionary. He was also responsible for the development of logic systems on which some computer languages are based. Perhaps his most important discovery resulted in what today is known as **Pascal's Law**. Pascal's Law states: Pressure exerted on a confined liquid (1) is transmitted undiminished, (2) is transmitted in all directions, (3) acts with equal force on equal areas, and (4) acts at right angles to those areas.

The application of Pascal's Law has resulted in a modern era of fluid power. This era emphasizes the use of fluid, not as a prime mover, but as a mechanism for transferral of power through a closed fluid system. These fluid systems are known as **hydrostatic**. They are evident in common devices such as log splitters and the braking system on automobiles.

To demonstrate the effect created using Pascal's Law, consider a large container. Connected to the top of this container is a small neck having a cross-sectional area of 1 sq. in. {*6.5 cm*^{2}}. This container is filled with a liquid to a level extending into the neck section (see Figure 2-1).

FIGURE 2-1

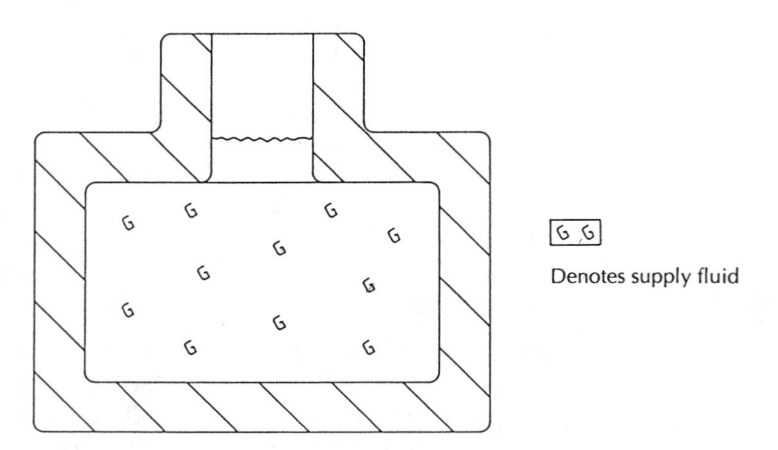

A piston is fitted into the neck of the container and allowed to slide down until it rests on the fluid. When a force of 10 lbs. {*44.5 N*} is placed on this piston, the result is a pressure of 10 psi {*68.9 kPa*} being exerted throughout the container. Remember, Pascal's Law states that pressure is transmitted undiminished and in all directions. Thus, pressure equals force divided by unit area **(P = F/A)** (see Figure 2-2).

If the bottom of the container has an area of 5 sq. in. {*32.3 cm²*}, the resulting force or load on the bottom surface will be 50 lbs. {*222.3 N*}. Remember, Pascal's Law states that pressure will act with equal force on equal areas. Thus, force equals pressure times area **(F = P * A)** (see Figure 2-3).

FIGURE 2-2

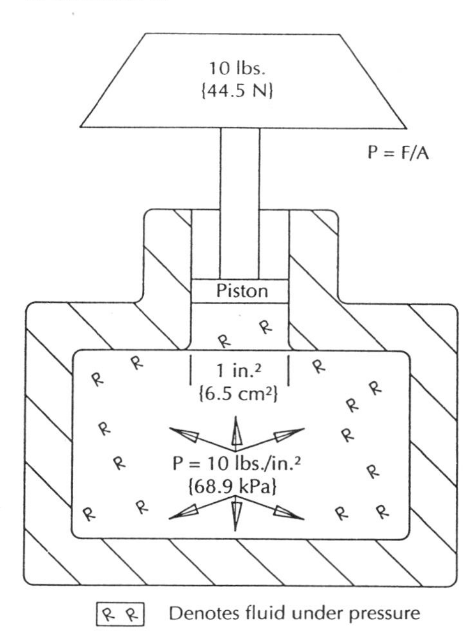

FIGURE 2-3

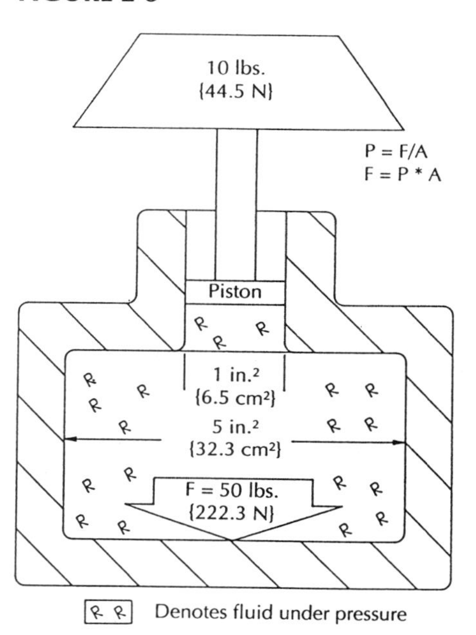

The multiplication of force in this manner is of little practical use, however, unless one is in the business of forcing the bottom or sides out of containers. However, the application of this principle is necessary to apply fluid power to do work.

The multiplication of force demonstrated in the previous section using the closed container can be related to the operation of a simple first-class lever. The balance point on the seesaw is called the **fulcrum**, the beam extending from the fulcrum to the output is the **load arm**, and the beam extending to the input is called the **lever arm**. If the load arm and the lever arm are equal in length and the load applied to the load arm and the force applied to the lever arm are equal, the lever is said to be *in balance* (see Figure 2-4).

FIGURE 2-4

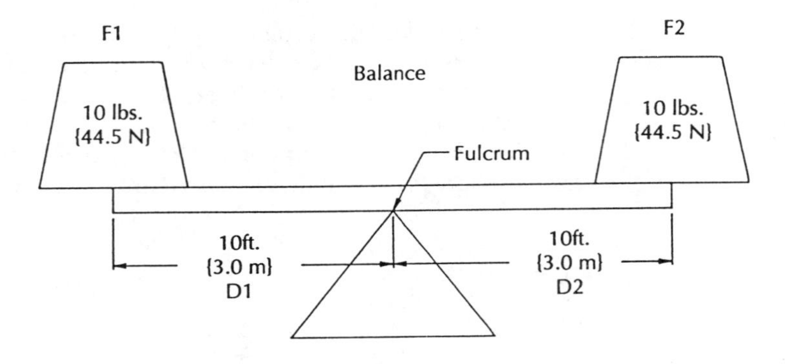

Problem 2-1

Cross-products

F1 (lbs.)/D2 (ft.) = F2 (lbs.)/D1 (ft.)

{F1 (N)/ D2 (m) = F2 (N)/D1 (m)}

Express this equation if F1 is an unknown force.

F1 (lbs.) = __________ F1 {*N*} = __________

The reason that the lever in Figure 2-4 is in balance is because the force and distance on either side of the fulcrum are equal. In the lever, balance exists if force **(F1)** is to distance **(D2)** as force **(F2)** is to distance **(D1)**. A more appropriate way of viewing balance in the lever is by calculating the cross-products of force and distance. Force times distance is defined as **work** (see Figure 2-5).

FIGURE 2-5

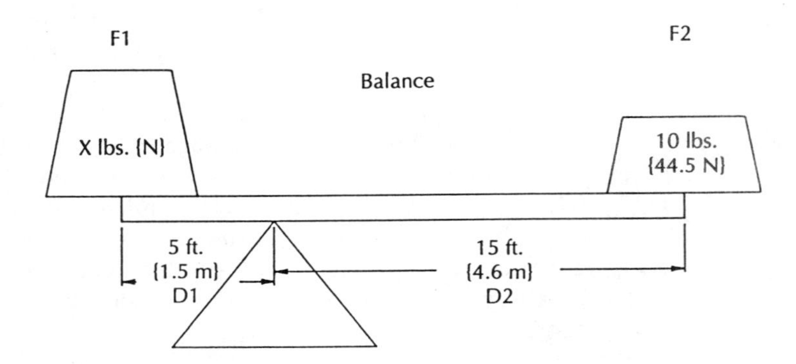

Problem 2-2

In Figure 2-5, if the distance from the fulcrum to a force 2 of 10 lbs. {44.5 N} is 15 ft. {4.6 m} and the distance from the fulcrum to a force 1 is 5 ft. {1.5 m}, what force 1 (X) would be needed to balance the lever?

F1 (lbs.) = __________ F1 {*N*} = __________

The concept of the lever can also be applied to a fluid system. First, consider a cylinder similar to the container presented earlier filled with oil. Into this cylinder is fitted a piston having a 2-sq. in. {*12.9-cm²*} cross-sectional area. This piston is loaded with a 30-lb. weight {*133.3 N*} (see Figure 2-6).

FIGURE 2-6

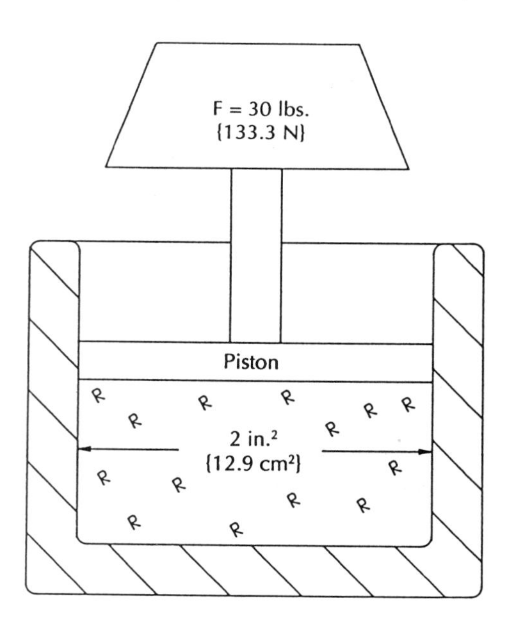

Problem 2-3

What will the resulting pressure be in this container?

P (lbs./in.²) = __________ P {*kPa*} = __________

The cylinder can be connected to another cylinder. The second cylinder has a cross-sectional area of 2 sq. in. {*12.9 cm²*}. A pipe or tube is used to connect the two cylinders. Oil is placed into the system to fill the voids in both cylinders and the pipe (see Figure 2-7). This fluid system is said to be in hydraulic balance.

FIGURE 2-7

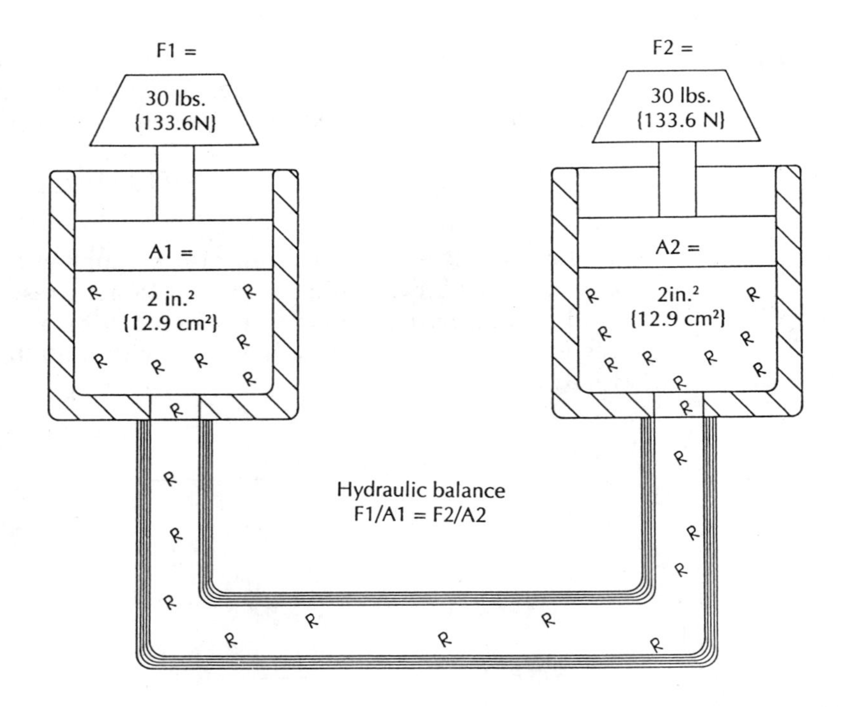

Problem 2-4 Express the equation in Figure 2-7 if F2 is an unknown force.

F2 (lbs.) = ________ F2 {N} = ________

If the second cylinder is increased in size to accept a piston having a cross-sectional area of 4 sq. in. {*25.8 cm²*}, a larger load can be balanced by the fluid system. Using the formula derived in Figure 2-7, the exact load that can be balanced may be calculated (see Figure 2-8).

Another more practical method of solving the problem in Figure 2-8 is to use Pascal's Law. Recall that the pressure developed by the 30-lb. {*133.6 N*} load exerted on the 2-sq. in. {*12.9 cm²*} piston was calculated to be 15 psi {*102.5 kPa*}. Since Pascal's Law states that pressure is transmitted undiminished in all directions, then every location within the closed fluid system will be acted upon by this pressure. That includes the first cylinder, the pipe, and the second cylinder.

Consider the fact that, using these principles, a person weighing only 150 lbs. {*666.4 N*} could stand on the 2-sq. in. {*12.9-cm²*} piston and balance an automobile weighing 3,000 lbs. {*13.3 kN*} if the second or output cylinder had a cross-sectional area of 40 sq. in. {*258.1 cm²*}. That output cylinder would have a diameter of slightly more than 7 in. {*18 cm*}. It could easily be grasped and carried by a person (see Figure 2-9).

FIGURE 2-8

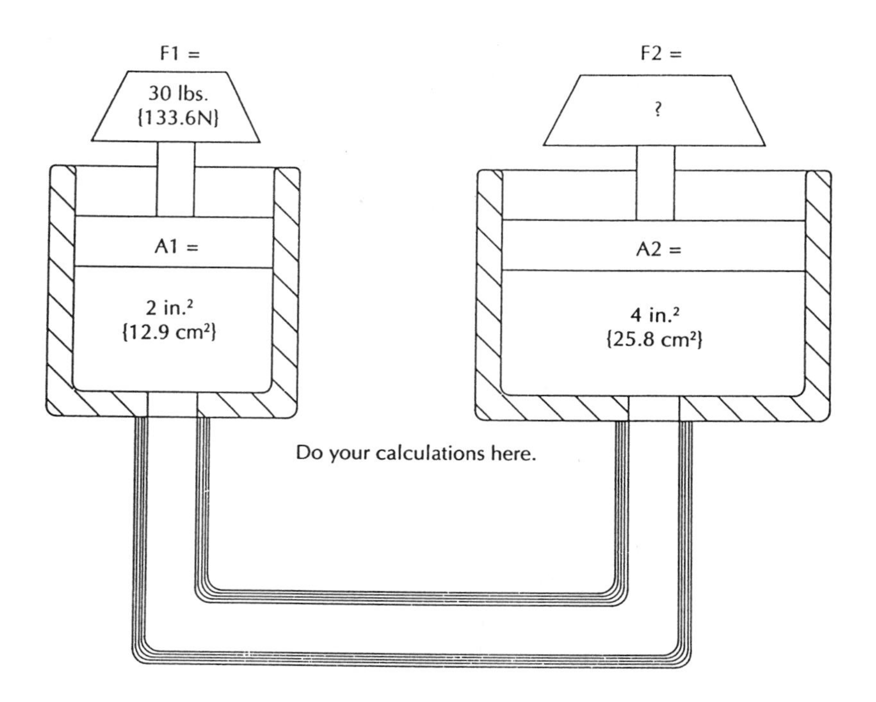

Problem 2-5

Do these calculations for Figure 2-8.

F2 (lbs.) = ________ F2 {*N*} = ________

Problem 2-6

Using the formula derived from Pascal's law to solve for force, calculate the force exerted on the 4-sq. in. {*25.8-cm²*} piston under the pressure existing in the system.

$F = P * A$

F = ________ (lbs.) F = ________ {*N*}

Check this answer with the answer for Problem 2-5. They should be the same.

Apparently, the answer to the problem of balancing any load is simply to use a large enough output cylinder. To some extent, this is true. The Greek scientist **Archimedes** said that, with a long enough lever, he could raise the world. However, Archimedes did not mention that he would prob-

FIGURE 2-9

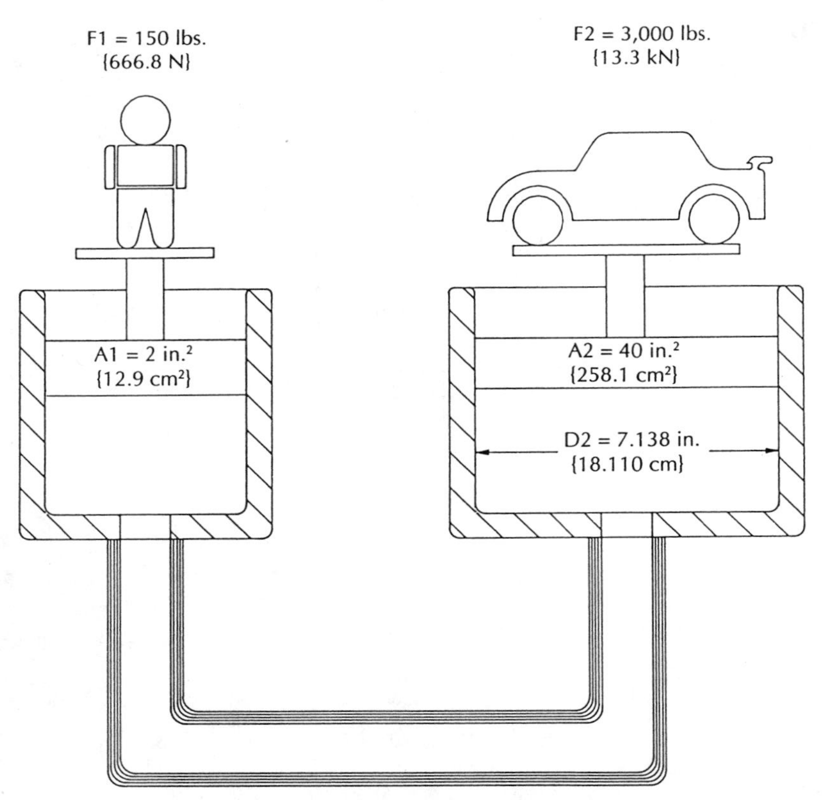

Assignment: Color-code the drawing.

ably need to be standing somewhere out of the solar system—and the fulcrum for his lever would need to be the moon—in order to accomplish this feat. Furthermore, Archimedes would need to move his end of the lever arm through miles {*kilometers*} of space to raise the Earth only 1 ft. {*30.5 cm*}.

It is seldom practical just to balance an output load. More often, it is desirable to displace or move it. For example, the person balancing the automobile might wish to raise it to change a tire. In theory, all he or she would have to do is pick up a tool. The added weight of the tool would cause the cylinder on which he or she stands to begin moving downward, and the car would begin to rise. Of course, the fluid in the system would be displaced accordingly. In this case, the input or first cylinder, moving through 1 in. {*2.54 cm*} of stroke would displace 2 cu. in. {*32.8 cm³*} of fluid. The result of this 1-in. {*2.5-cm*} stroke would move the 40-sq. in. {*258.1-cm²*} cylinder only 0.05 in. {*0.13 cm*}, raising the car less than 1/16 in. {*1.3 mm*} (see Figure 2-10).

FIGURE 2-10

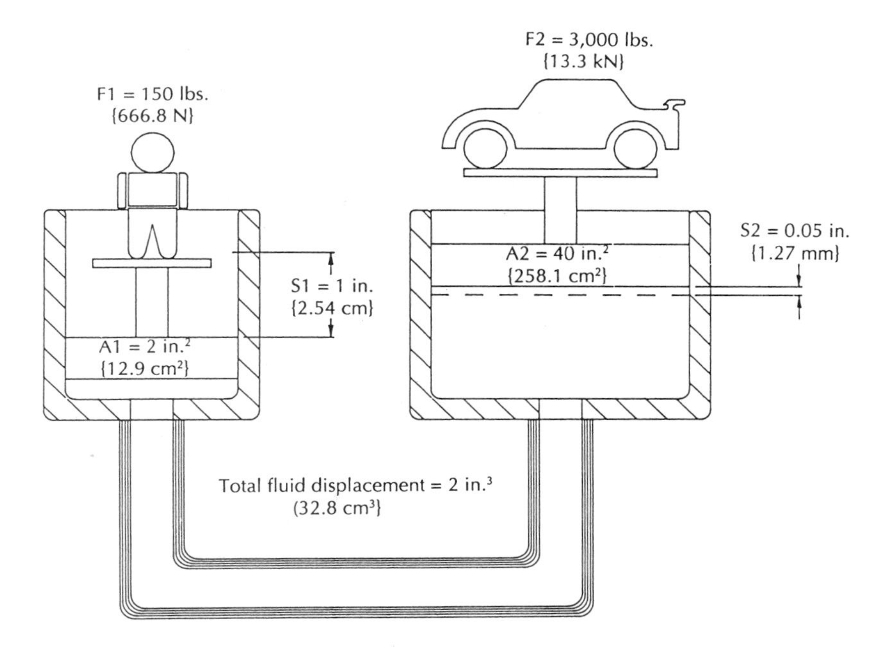

Problem 2-7

Cross-products displacement

S1 (in.)/A2 (in.^{2}) = S2 (in.)/A1 (in.^{2})

S1 {cm}/A2 {cm^2} = S2 {cm}/A1 {cm^2}

Express this equation if S1 is an unknown stroke.

S1 (in.) = ________ S1 {*cm*} = ________

Problem 2-8

How many inches {*centimeters*} of stroke in Cylinder 1 (Figure 2-10) would be necessary to raise the car 12 in. {*30.5 cm*}?

S1 = ________ inches

Even though the 2-sq. in. {*12.9 cm^2*} cylinder and the 40-sq. in. {*258.1 cm^2*} cylinder are small enough in diameter to be easily transported by the person, the length of the input cylinder necessary to extend the output cylinder through 12 in. {*30.5 cm*} of stroke would make this design impractical. The person would also need to transport a ladder nearly 20 ft. {*6 m*} long to be able to reach the platform on which he or she would begin his or her downward travel on the input cylinder.

Apparently, the fluid system operates well to multiply input force but poorly to transfer fluid and cause movement. Actually, this is not true. If the person were able to travel throughout the 20-ft. {*6-m*} stroke and "bottom out" the input cylinder, the output cylinder would "cap out" after 12 in. {*30.5 cm*} of stroke. If the person threw the tool into the car, the result would be a reversal of the previous operation. Only 12 in. {*30.5 cm*} of stroke of the output cylinder would raise the person through 20 ft. {*6 m*} of input cylinder stroke.

The Law of Conservation of Energy and Motion states that "*through normal procedures, energy cannot be created or destroyed but only changed in form.*" In a fluid system, energy is expressed as work **(W)**. Work may be calculated by multiplying force **(F)** times distance **(D)**:

W (lbs. ft.) {*J*} = **F** (lbs.) {*N*} * **D** (ft.) {*m*}

Problem 2-9

Calculate the work being performed to lower a person the distance of 20 ft. {*6.1 m*} if the person weighs 150 lbs. {*666.8 N*}:

W1 = ________ (lbs. ft.) {*kJ*}

Problem 2-10

Calculate the work being performed to raise the automobile a distance of 1 ft. {*30.5 cm*} if the automobile weighs 3,000 lbs. {*13.3 kN*}.

W2 = ________ (lbs. ft.) {*kJ*}

Applying the cross-products analysis of force and distance in the closed fluid system reinforces the Law of Conservation of Energy and Motion (see Figure 2-11).

Problem 2-11

Cross-products work (see Figure 2-11)

D1 (ft)/F2 (lbs.) = D2 (ft.)/F1 (lbs.) D1 {m}/F2 {N} = D2 {m}/F1 {N}

Express this equation if D1 is an unknown distance.

D1 (ft.) = __________ D1 {*m*} = ________

If the automobile needed to be raised a practical distance of 5 ft. {*1.5 m*} to work underneath it, more displacement of the input cylinder would be necessary (see Figure 2-12). As can be seen, raising the automobile 5 ft. {*1.5 m*} makes the job even more impractical. To make the automobile hoist more practical requires a method of allowing the input cylinder to repeatedly displace its fluid to the output cylinder without allowing the fluid to be redistributed from the output to the input cylinder.

FIGURE 2-11

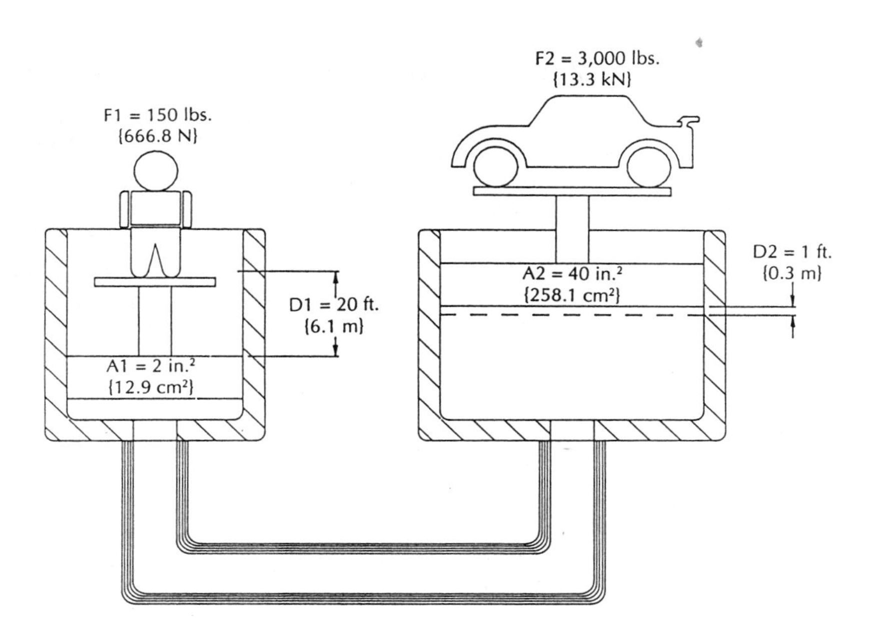

FIGURE 2-12

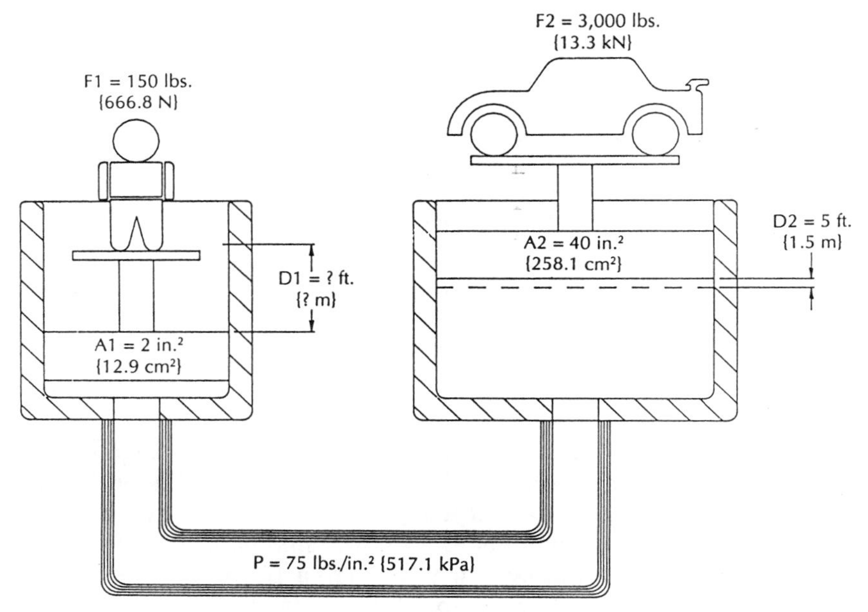

Problem 2-12 What distance (D1) must the input cylinder move to raise the automobile 5 ft. {1.5 m}?

D1 (ft.) = ________ D1 {*m*} = ________

2.2.2 Bramah's Jack

The first practical application of Pascal's Law was performed by the English industrialist **Joseph Bramah** during the early nineteenth century. Bramah devised a system that allowed the fluid pumped from the input cylinder to flow to the output cylinder while not allowing reversal of flow. This jack demonstrated the first application of control in a fluid system. In this case, the **direction** of flow was controlled. The directional control allowed flow to occur only one way. This type of directional control is commonly known as a **check valve** (see Figure 2-13).

FIGURE 2-13

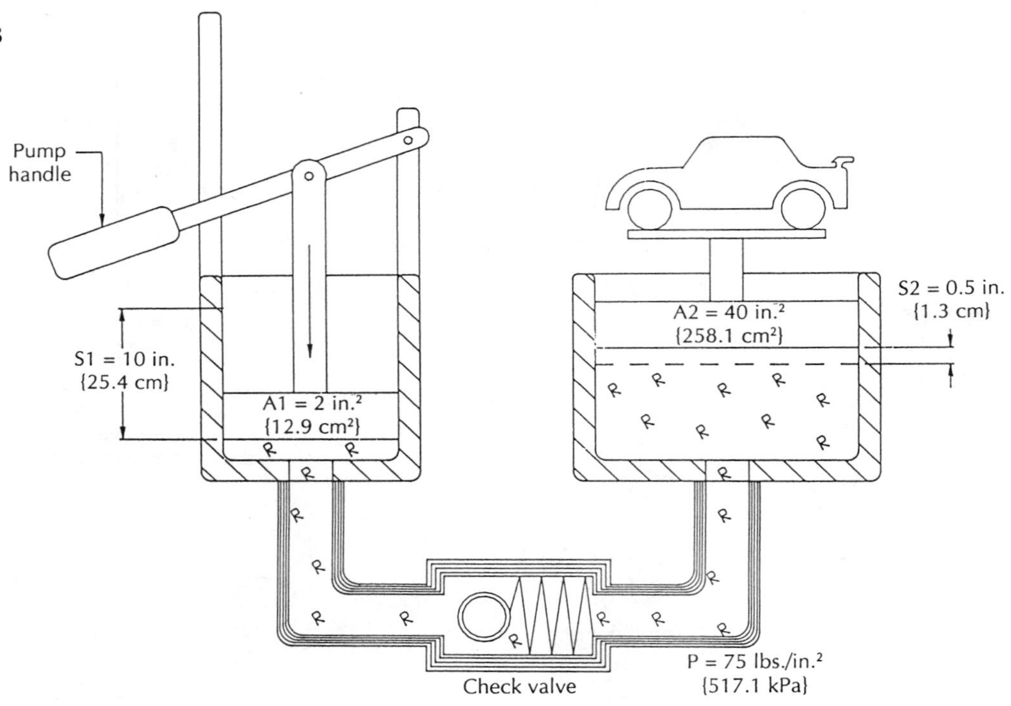

Any flow from the pump will cause the ball in the check valve to unseat and allow free flow to the output cylinder. Any fluid attempting to escape from the output cylinder back to the pump will cause the ball to be seated and will not allow flow. This condition exists if, after advancing the pump (input cylinder) through a 10-in. {*25.4-cm*} stroke and raising the output cylinder 0.5 in. {*1.3 cm*}, it is desirable to return or advance the pump piston to the top of its stroke by pulling up on the handle. Retracting the pump piston will create a vacuum or low-pressure area to exist in the pump chamber. A vacuum is simply a volume that exists under lower pressure than the surrounding environment. There are two primary ways to produce a vacuum: (1) removing material from the container without changing its volume or (2) increasing the volume of the container without adding any more material (air, oil, etc.). In this case, we have increased the volume without adding material by raising the piston through its stroke. The fluid in the output cylinder under high pressure attempts to return to the low-pressure (vacuum) area in the pump. However, this does not occur because the check valve blocks flow in that direction (see Figure 2-14).

FIGURE 2-14

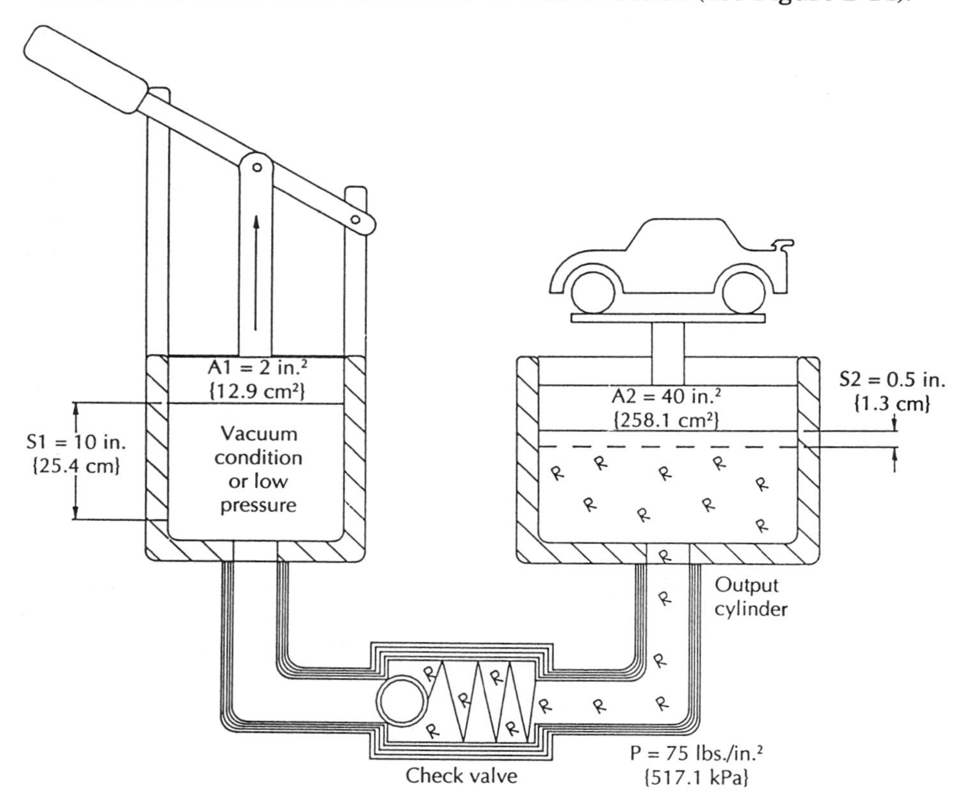

If the pump is connected to a tank or reservoir containing supply fluid while the vacuum condition exists, the higher atmospheric pressure will force the fluid in the reservoir up through the pipe and fill the pump. This creates another full charge of 20 cu. in. {*327.7 cm³*} of fluid that could be transferred to the output cylinder. Remember that throughout this pump-filling cycle, the output cylinder has remained stationary. It was previously extended through 0.5 in. {*1.3 cm*} of stroke (see Figure 2-15).

FIGURE 2-15

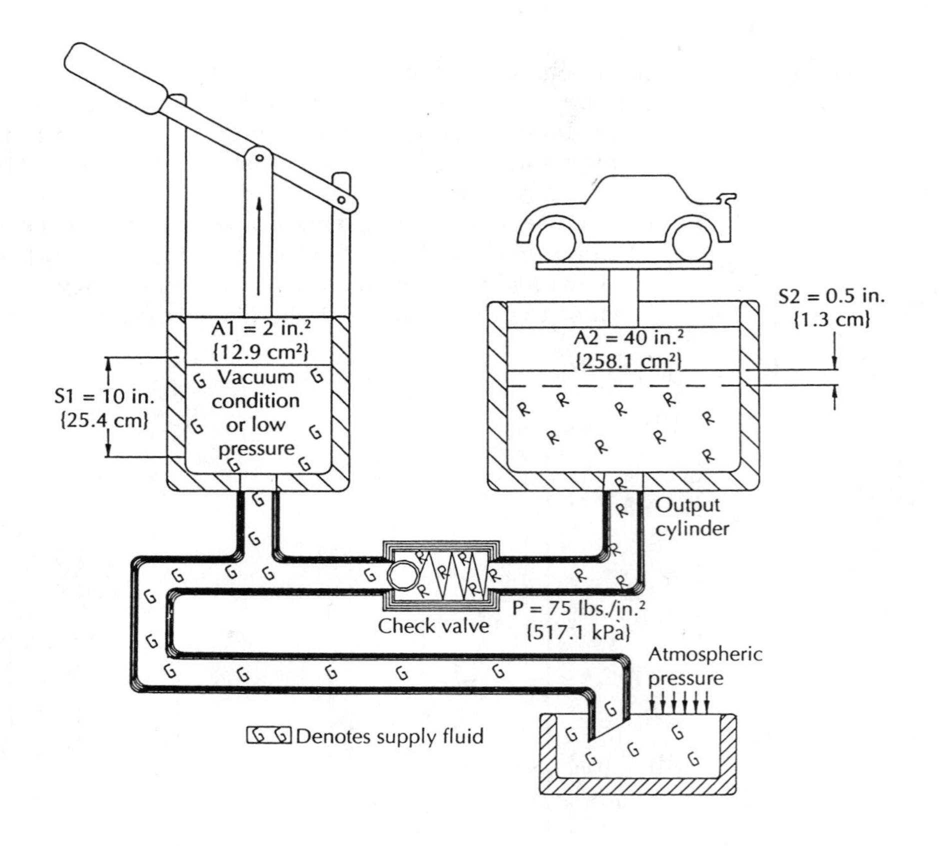

Of course, if the pump piston is forced downward at this time, the fluid simply flows back to the tank, taking the *path of least resistance*. Once again, the direction of flow needs to be controlled to allow flow from the tank to the pump but exclude flow from the pump to the tank (see Figure 2-16).

Each time the 2-sq. in. {*12.9-cm^2*} pump piston goes through the process of charging and displacing the charge **(cycle)**, the automobile will be raised 0.5 in. {*1.3 cm*}.

Problem 2-13

In Figure 2-16, how many cycles of the pump must be made to raise the automobile 12 in. {*30.5 cm*}?

________ Cycles

Finally, to fully understand the operation of Bramah's jack, another operational principle of fluid systems must be presented. Not only should

FIGURE 2-16

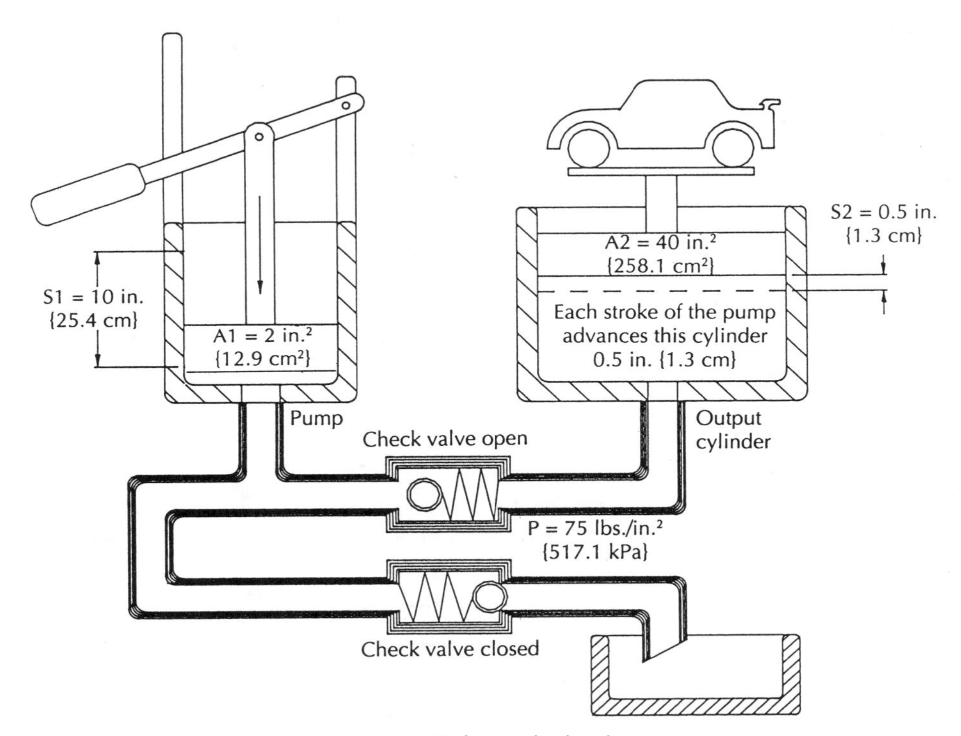

Assignment: Color-code the drawing.

a fluid system be able to transmit force and displace loads, but it should also be able to perform this operation repeatedly. The jack, as developed to this point, has made no arrangement to return the automobile to its normal position. Again, it is necessary to control the flow so that, on demand, it can be directed to the output cylinder or back to the tank from the output cylinder. To accomplish this task, a **two-position, one-way, directional control**—called a **gate valve**—may be used (see Figure 2-17).

With the gate valve in the open position, the weight of the automobile forces the fluid from the output cylinder and returns it to the tank. This action retracts the output cylinder's piston and lowers the automobile until the output cylinder's piston "bottoms out." Rotating the gate valve into the closed position allows the pumping and lifting operations to begin again. Fluid transferred from the pump to the output cylinder is blocked at the gate valve and not allowed to return to the tank. This results in the output cylinder's piston once again being extended (see Figure 2-18).

FIGURE 2-17

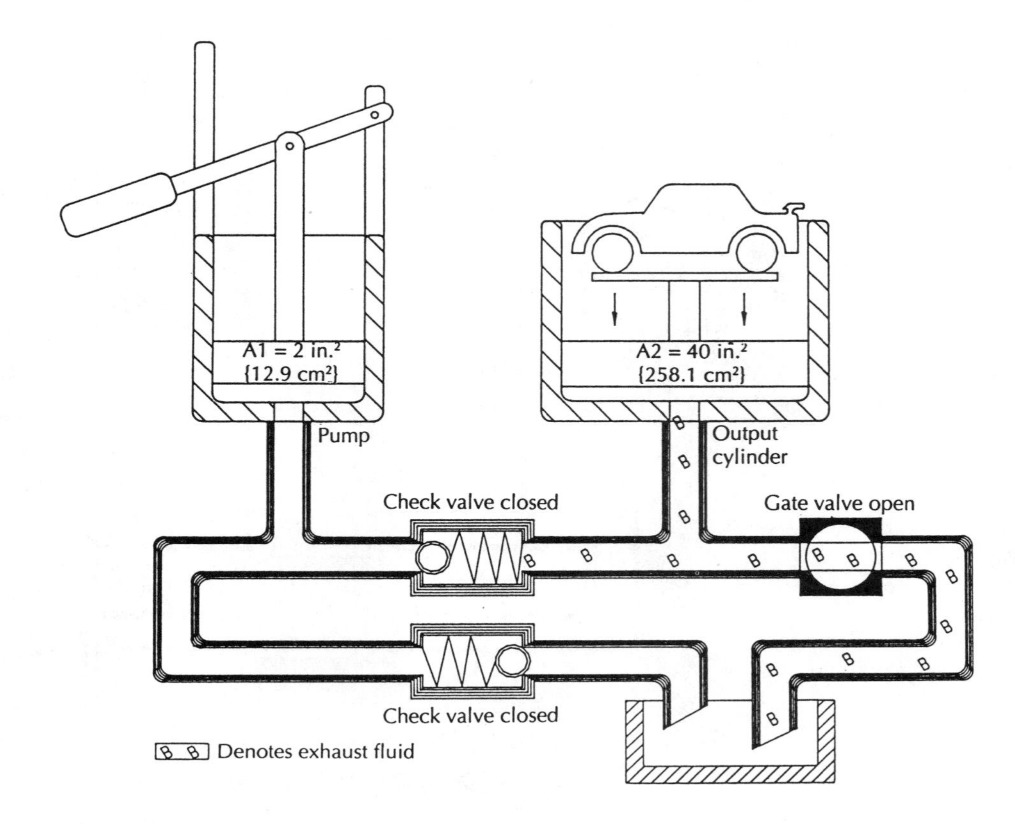

2.3 HYDRAULICS DEFINED

Prior to the close of the nineteenth century, many obstacles had to be overcome before Pascal's Law and Bramah's circuitry could have widespread application. The invention and subsequent refinement of the engine lathe greatly advanced the potential for production of fluid power components. Very accurately machined mating parts were required to prevent leakage in the fluid system. Steel, the material most commonly used to produce fluid devices, did not achieve widespread usage until the invention of refining furnaces during the 1860s. One of the major problems to be overcome existed in the fluid itself. Bramah's system operated using water as the fluid. The radical changes in the physical nature of water within a relatively narrow range of temperatures, and the rusting of many steel components within the circuitry using water, emphasized the need for a better fluid. Until the latter part of the nineteenth century, no fluid existed that could be easily obtained and would retain its physical characteristics over a broad range of temperatures while, at the same time, retarding rusting. The discovery of crude oil and its refinement resulted in such a fluid. Fluid power systems using oil as the medium for

FIGURE 2-18

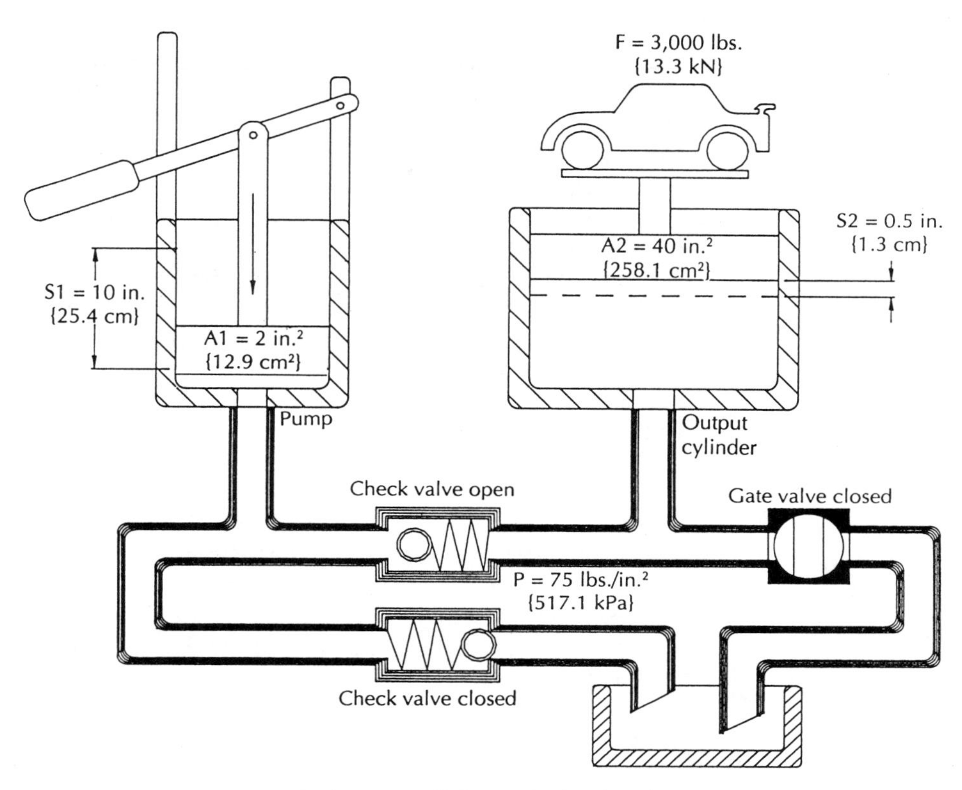

Assignment: Color-code the drawing.

transmission of forces are known as hydraulic. Other fluid power systems that use air (pneumatic) or water (fluidic) are available today. However, the combination of a virtually noncompressible fluid (oil) having advantages over water has resulted in more widespread use of hydraulics than any other fluid power system. The definitions of various fluid systems are summarized in Table 2-1.

TABLE 2-1

Fluid Power System Definitions

Fluid Power	That phase of physics that deals with the nature of confined fluids operating under pressure
Hydraulics	The phase of fluid power that deals with oil and/or water as the primary medium for power transmission
Pneumatics	The phase of fluid power that deals with air as the primary medium for power transmission
Fluidics	The phase of fluid power that deals with waters as the primary medium for power transmission and also a large body of knowledge related to the application of digital logic in fluid devices

2.4 HYDRAULICS REFINED

Other refinements to Bramah's basic circuit have been developed through the years. Although most of these have been in the area of control within the circuit, many have also taken place in the power supply itself. Few industrial hydraulic circuits rely on manual operation of a pump to create flow within a circuit. It would be quite tiresome for a person to spend eight hours per day pumping fluid to raise automobiles or to bend metal. The coupling of the hydraulic pump with propulsion devices such as electrical motors and internal combustion engines has relieved an operator from this tedious task. These devices deliver very high power output in a relatively small package. Their consistency in operational speed has fostered the design of many types of circuits that must perform operations repeatedly at a constant rate.

2.4.1 ANSI Symbols

Many diverse applications of hydraulic circuitry have developed since the nineteenth century. The development of these circuits were dependent upon skilled craftsmen building and constructing customized components. A multitude of component designs developed. Many divergent designs performed identical functions and operated similarly. For example, a component as simple as a check valve might have operated using a ball held against its seat by a spring, a facing valve seated with a spring, a poppet seated with a spring, or a piston seated with a spring. The engineering drawings that were produced to specify the manufacture of the various designs were radically different. However, all designs had the same function—to allow flow in only one direction. The valves, for the most part, were interchangeable if the plumbing used to connect them to the circuitry could be adapted.

It became apparent to industrialists that a common language was needed to describe the design of a circuit based on its function rather than its component design. The **Joint Industrial Commission** (JIC) first designed symbols that could be used to describe fluid power components based on their functions. Later, refinements to these symbols were made by the **National Fluid Powers Association**. The result was a set of standardized symbols that describe both the function and the operation of a hydraulic component using **American National Standards Institute** (ANSI) symbols. Since that time a major restructuring of ANSI symbols has been accomplished by the **International Standards Organization** (ISO); however, to date, ISO standards have not been formally adopted by ANSI and, therefore, are not currently in force. All three systems—JIC, ANSI, and ISO—have similarities. Acquiring a knowledge of the design and application of any set of standards will help you to understand all systems. A schematic representation of a hydraulic circuit is produced by combining symbols in a logical, ordered arrangement. These schematic representations are the universal language of fluid power. Using ANSI schematic drawings, the designer, the engineer, the machinist, the circuit fabricator, and the maintenance personnel can more easily communicate the intended operation of a hydraulic circuit. Figure 2-19 shows an ANSI schematic of Bramah's jack. Figure 2-20 illustrates an actual hand pump power supply.

Notice that the symbols used in the schematic shown in Figure 2-19 have a certain visual logic in themselves. The symbol for the cylinder looks like the outside casing of the cylinder with the piston and rod encased. The symbols for check valves look like a ball that may be forced against an angular seat. The symbol for the tank (reservoir) looks like a sectioned tank showing its sides and bottom, with the top open to the atmosphere.

The general symbol for a valve is a square. The two-position, one-way, directional control valve (gate valve) shows two squares adjacent to one

FIGURE 2-19

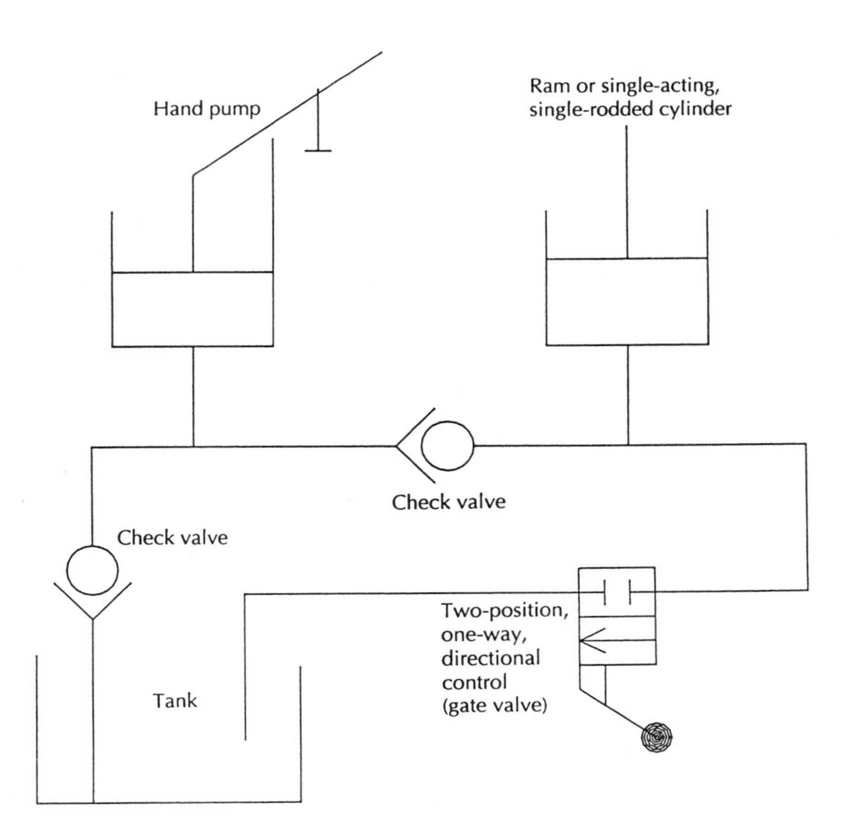

FIGURE 2-20

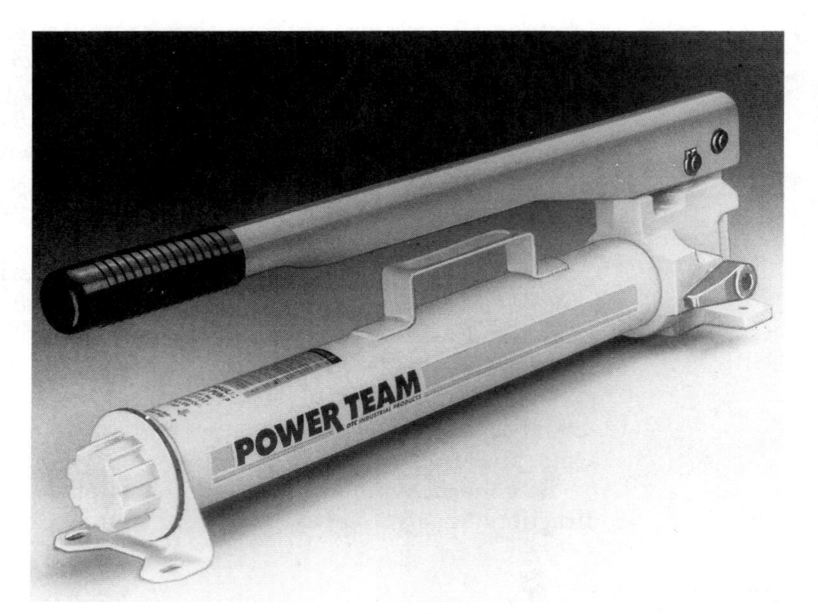

Portable hydraulic jack. *(Courtesy Power Team Division, SPX Corporation. From Vockroth,* Industrial Hydraulics, *Albany, N.Y.: Delmar, 1994, p. 8.)*

another. These two squares represent the two positions into which the valve may be shifted. Inside one of the squares is a line with an arrow on the end. This represents the position that allows fluid to flow through the valve from its inlet (where the blunt end contacts the side of the valve) to its outlet (where the arrow end contacts the opposite side of the valve). This position is referred to as the open position of the valve. The other position in this directional control shows a continuation of both the inlet and outlet ports into the valve. Each is terminated inside the valve so that the directional control blocks (excludes) flow through the valve in that position. This position is referred to as the closed position. Finally, the valve is controlled by an activator adjacent to its square. In this case, the open and closed positions are both controlled by shifting a lever.

Conductors (lines) that guide the fluid from one location to another are symbolized by straight lines. These conductor lines should always be drawn either horizontally or vertically.

The pump symbol used in this schematic is not a common ANSI symbol. Rather, nearly all hydraulic circuitry relies on a rotary-driven pumping device that performs the same type of operation as the single-piston pump used in Bramah's jack. Of course, in keeping with the graphic logic of schematic drawings, the pump would be shown as a modified circle to

represent that rotary design. Other ANSI symbols will appear throughout this book and will be discussed in detail at those points. However, a synopsis of typical ANSI symbols appears in Appendix C for those readers who are interested.

[] [] [] 2.5 PRACTICAL DESIGN PROBLEM

Figure 2-21 illustrates an application of Bramah's jack and Pascal's Law. In this situation, the jack is being used to compress plastic powder into a preformed tablet to be used in a later compression-molding operation. To compact the plastic powder requires 4,000 lbs. {*17.8 kN*} of force. The **pump** has a stroke of 8 in. {*20.3 cm*}. The **ram** or output cylinder has a cross-sectional area of 20 sq. in. {*129.0 cm²*}. Also, the preform mold must be able to open 4 in. so that the operator can pour the plastic powder into the cavity. Then the output cylinder must move through a full 4 in. {*10.6 cm*} of stroke and apply the 4,000-lb. {*17.8-kN*} force at the end of that stroke. Study Figure 2-21; then answer the following questions:

FIGURE 2-21

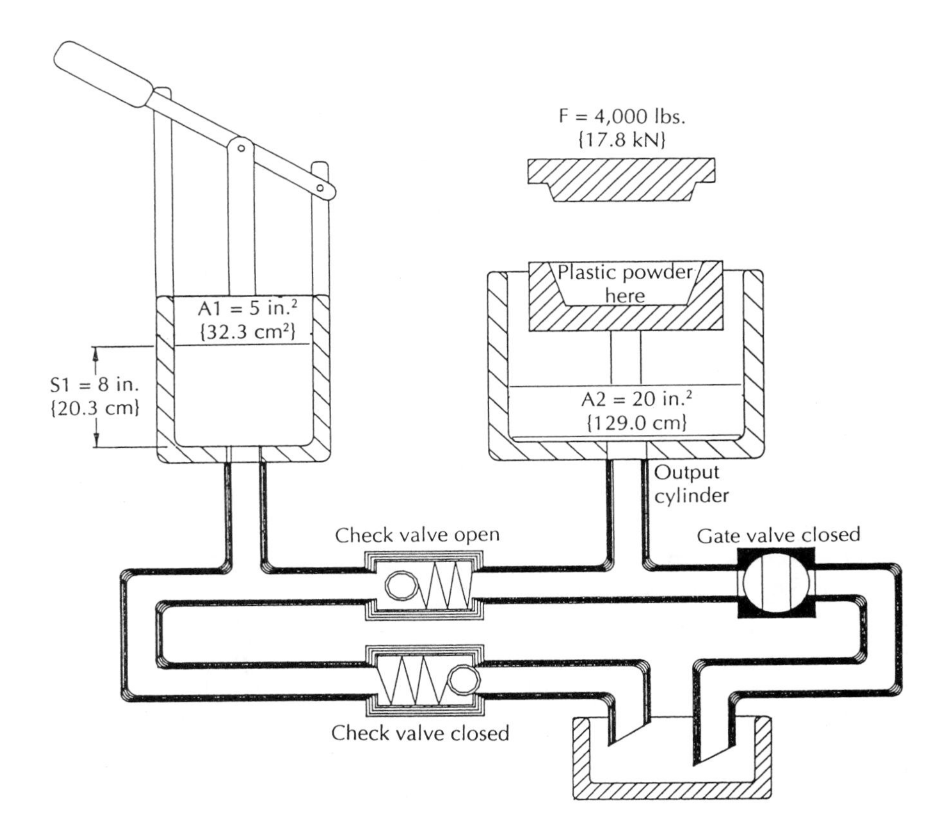

1. How much oil will be displaced in one full stroke of the pump?

   DISplacement (cu. in.) {*cm*^{3}} = __________

2. How far will the ram advance for each full stroke of the pump?

   Distance (in.) {*cm*} = __________

3. How many cycles of the pump must be made to fully extend the ram?

   Cycles = __________

4. If the machine operator, working at maximum rate, can fully extend and retract (cycle) the pump in ten seconds, how many seconds will it take for the ram to fully extend?

   Time {(sec)} = __________

5. What pressure will develop in the circuit while applying the 4,000-lb. {*17.8-kN*} force?

   Pressure (psi) {*kPa*} = _________

2.6 SUGGESTED ACTIVITIES

1. Visit your local or school library. Locate and read a biography or reference book (encyclopedia) entry for Blaise Pascal.
2. Make a survey of devices in your home, office, workplace, laboratory, and the like that are used to move a fluid (gas or liquid) from one location to another. Are these devices hydrodynamic or hydrostatic? Why?
3. Design yourself a first-class lever. Apply various loads to the load arm and pull down on the lever arm, using a force gauge (such as a fish-weighing spring gauge) to determine the force needed to overcome the load when the fulcrum is moved from one location to another. Measure and compare the distances moved at the ends of the load and lever arms when moving these loads.

4. Simulate the pumping action of Bramah's jack using a drinking straw and some water. Draw water into the straw; then place your finger over one end of the straw and transfer that quantity of water to a measuring cup. What would the displacement of fluid be per cycle for your straw pump? Measure the inside diameter of your straw and its length. Calculate the displacement of the inside of the straw. Express the volumetric efficiency of your straw pump by dividing the measured volume of transferred water by the calculated displacement.
5. If you have a small hydraulic jack and a pressure gauge, tap a hole in the lower casing of the jack and attach the pressure gauge (you can always remove it later and insert a pipe plug). Weigh a known mass and place the mass on the output rod of the jack. Advance the ram and record the pressure registered on the gauge. Calculate the cross-sectional area of the pump piston using Pascal's Law, the pressure registered, and the weight of the mass. Measure the distance that the ram advances for each pump cycle. Using this distance and the previously calculated piston cross-sectional area, determine the displacement of the jack's pump.

REVIEW QUESTIONS

1. Describe a hydrodynamic operation and provide four examples of hydrodynamic fluid systems.
2. State Pascal's Law.
3. Differentiate between hydrodynamic and hydrostatic operation.
4. Describe Pascal's Law using three equation variations defining the relationships among force, pressure, and area.
5. Define *work*.
6. Describe the operation of a check valve.
7. Verbally trace fluid flow in Bramah's jack during both extension and retraction of the output cylinder.
8. Contrast hydraulics, pneumatics, and fluidics.
9. Produce the schematic symbology for each of the following:
   a. check valve
   b. single-acting, single-end rod cylinder
   c. two-position, one-way, lever-activated, directional-control valve

3

Hydraulic Fluids and Fluid Supply

Hydraulic systems are unique from other power transmission systems because of the nature of hydraulic fluids. The ability of hydraulic fluid to change shape allows greater design flexibility than is possible with mechanical systems such as levers, screws, and wedges. At the same time, the fact that hydraulic fluid is virtually noncompressible allows for more precise control than may be achieved in pneumatic circuitry using air. This chapter describes some of those unique characteristics of hydraulic fluid. Some are beneficial, while others may be considered to be detrimental. However, they all present unique opportunities and challenges in hydraulic applications.

3.1 FLUIDS

Oil is the second-most plentiful liquid on Earth, surpassed only by water. Although oil, unlike water, is a nonreplenishable resource, it is highly abundant; and with worldwide commerce and transportation, it has become nearly universally available. Conservation measures, such as more efficient internal combustion engines and the development of synthetic lubricants, have greatly prolonged the future supply of oil. So much experimentation has gone into the development of modern-day oil that, in many ways, it no longer resembles the petroleum-based substance taken from the ground and refined for use. Therefore, more appropriately, the oil used in modern hydraulic circuitry should be referred to as **hydraulic fluid**.

3.1.1 Liquids as Fluids

In general, there are three accepted states of matter: **solid, liquid,** and **gas**. These states of matter may be described by their natures when placed in a closed container (see Figure 3-1). When placed into a closed container, a *solid retains its shape and occupies its own volume*. When placed into a closed container, a *gas takes the shape of the container and disperses to fill the container*. When placed into a closed container, the *liquid takes the shape of the container and occupies its own volume*.

Although relatively simplistic, the conception of these three states of matter is vital to the development of an understanding of why hydraulic systems operate as they do. Each of these states of matter describes materials composed of small building blocks called **molecules**. The ability to transmit a force from one point to another is affected by the relative closeness **(density)** of these molecules within the mass of the material. In a solid, the molecules are packed closely together. The result is that any force applied to one end of the solid will be transmitted, undiminished, in a straight line to the other end. In a gas, the molecules are separated by relatively large spaces. Any force applied to a molecule may or may not cause movement of another molecule. To ensure the transmission of force from one molecule to another requires a reduction of the volume in which the molecules are contained. Gases are, therefore, said to be **compressible**. In a liquid, the molecules are loosely packed together. Any force applied to any molecule will cause movement of that molecule, and it will impact upon another molecule. That molecule will impact upon

FIGURE 3-1

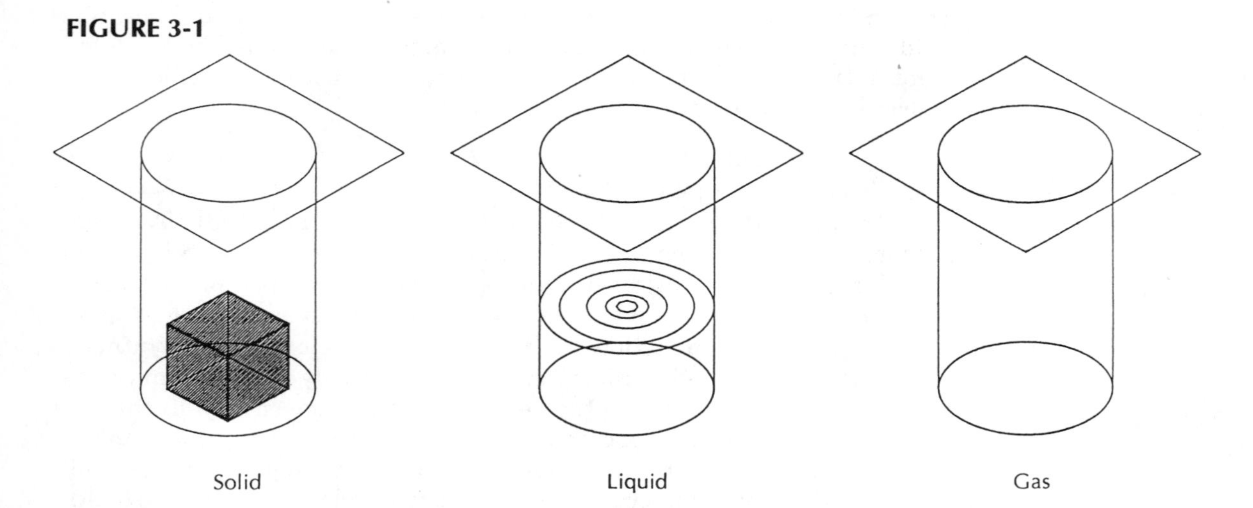

another and so on. The result is that the force will be transmitted in all directions. Consider these reactions to be similar to those of marbles placed in a frame (see Figure 3-2). Consider your finger to be the force pushing on each frame and visualize the effect on the marbles inside the frame.

If the force applied to one end of a material is needed to be transmitted in a straight line, the solid material would work best. The effect of outputting that force to the opposite end of the material would be immediate and undiminished. However, the solid materials, such as those used in mechanical power transmission systems, have great deficiencies when more complicated manipulation of the output force is required.

3.1.2 Fluids in Systems

Consider the situation that would exist if it were desirable to have a force exerted on a material to be transmitted to cause movement at multiple locations. If the material was contained within a sleeve such as a cross-tube, it could be guided to transmit that force from one of the ports to any or all of the other three locations (see Figure 3-3).

If the material inside the cross-tube in Figure 3-3 were a solid and a force was applied at any piston, movement could only occur in a straight line away from that force. The force applied at that piston would need to be greater than the combined forces on the other three pistons to cause movement. Then, the movement would be very slight. The amount of clearance between the solid material and the cross-tube would limit that movement in any direction (see Figure 3-4).

FIGURE 3-2

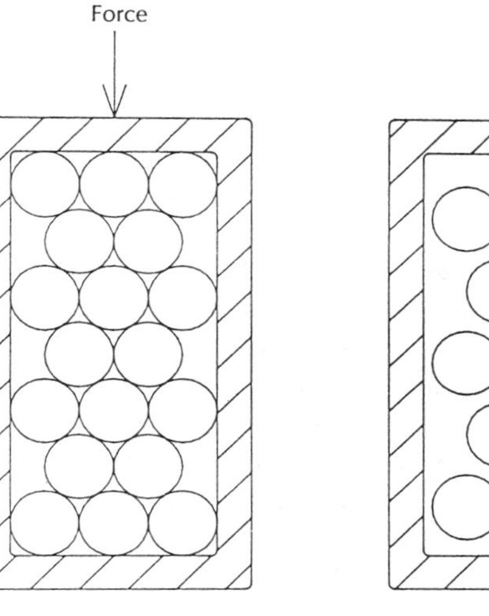

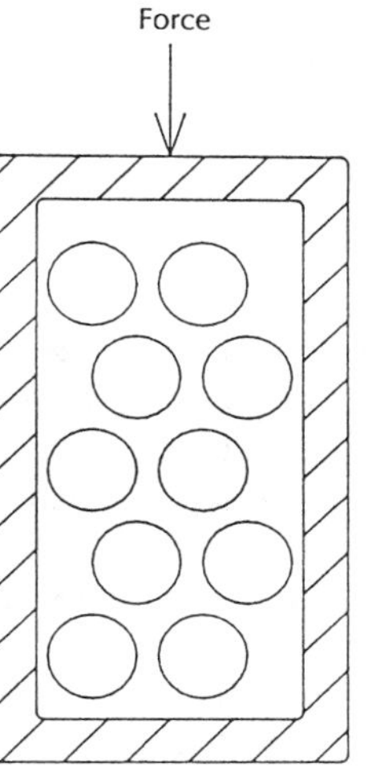

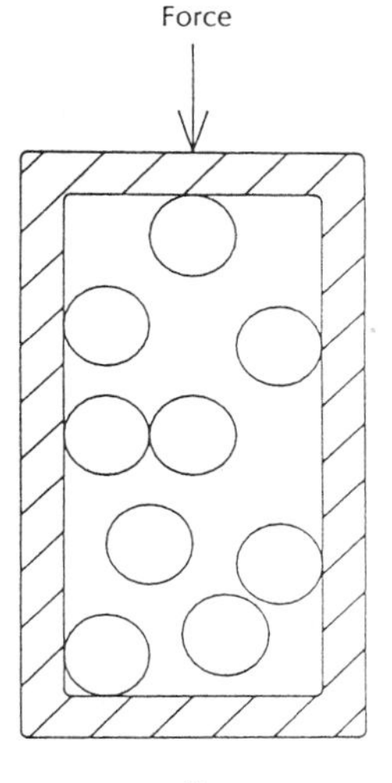

FIGURE 3-3

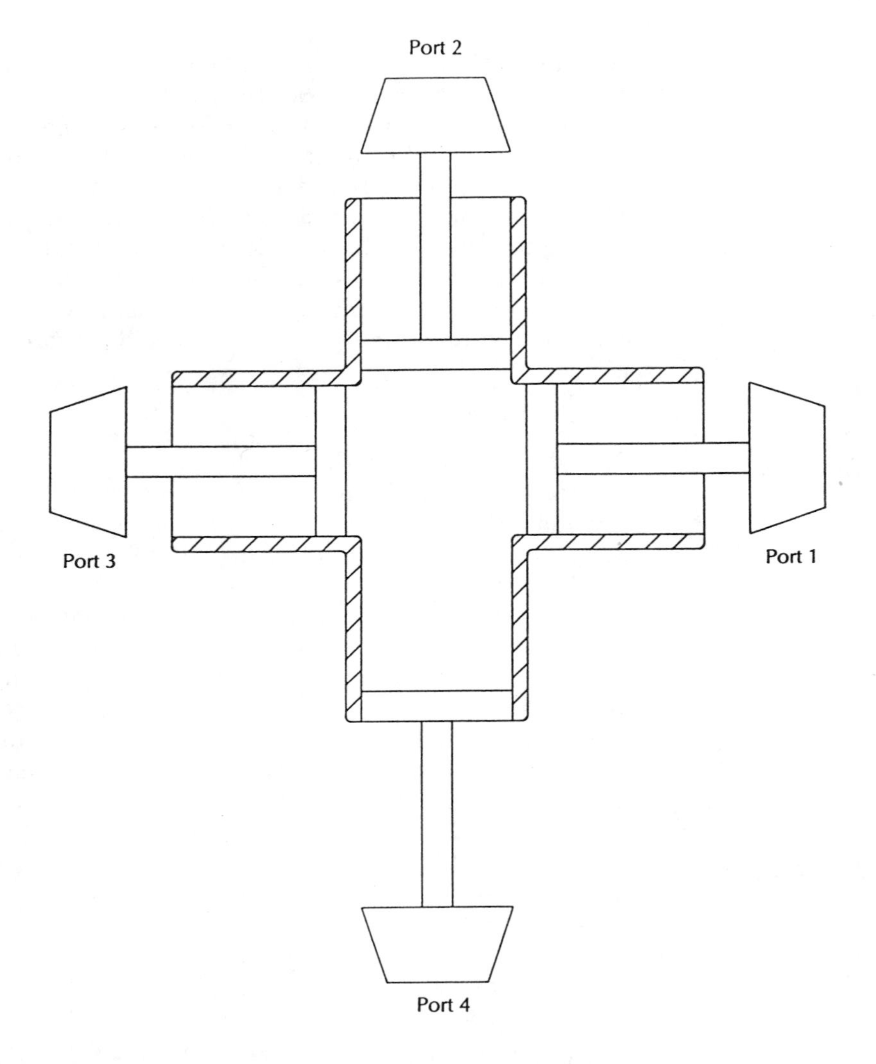

However, if the material inside the cross-tube is a liquid and force is applied on any piston, movement can occur radially from the center of the cross-tube. In this case, the cross-sectional area of all four pistons is identical. If the loads on three of the pistons are the same and a force slightly greater than the combined three loads is applied on the fourth piston, toward the center, the other three pistons will begin moving away, linearly, from the center (assuming that all four pistons are the same size). The movement will only be limited by the amount of movement of the fourth piston. In this case, each of the other three pistons will be displaced equally and to a distance equal to one-third the distance moved by the fourth piston (see Figure 3-5).

FIGURE 3-4

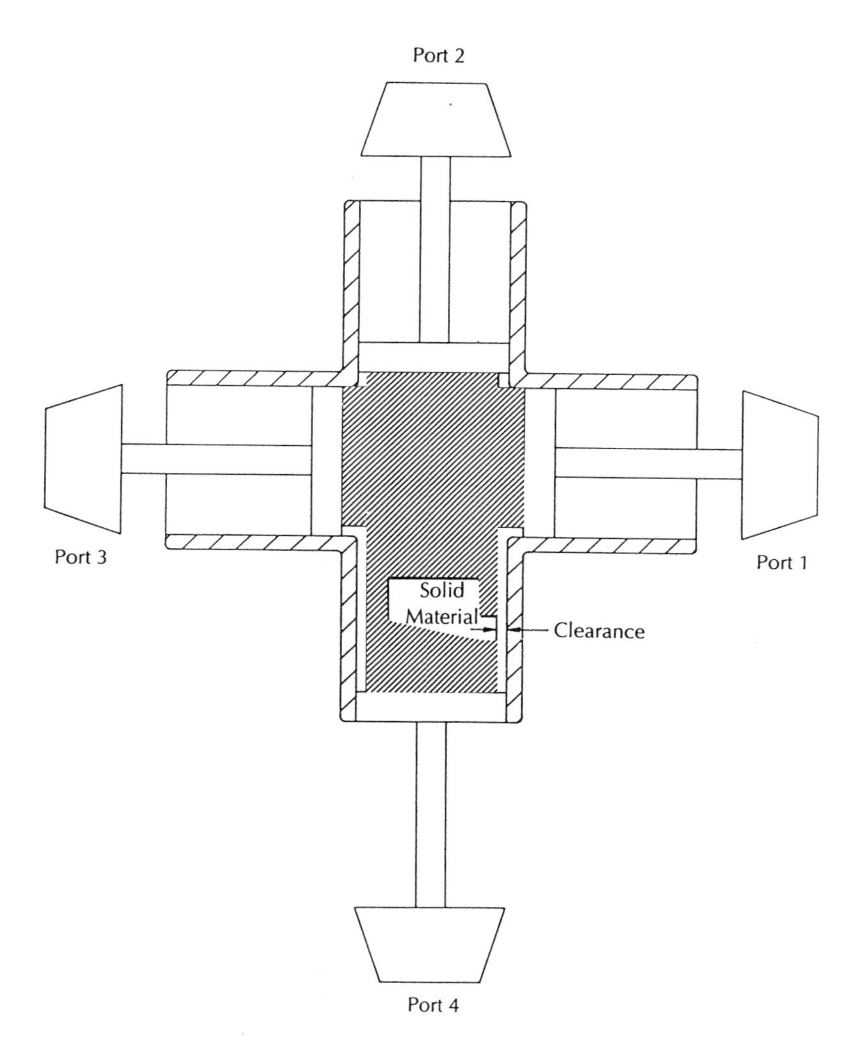

Radial movement can be attained in a solid mechanism (mechanical device) through the use of gears or cams. However, it should be obvious that the fluid mechanism results in a much simpler device. This simplicity is due to the nature of the fluid being able to conform to the shape of any container and its ability to transmit a force in all directions. Many books and other publications have been written on the subject of fluids and oils. As far as the study of hydraulics is concerned, the overwhelmingly important consideration is simply getting the fluid from one point to another within the system. There are, however, many factors that affect the efficiency of the movement of hydraulic fluid.

FIGURE 3-5

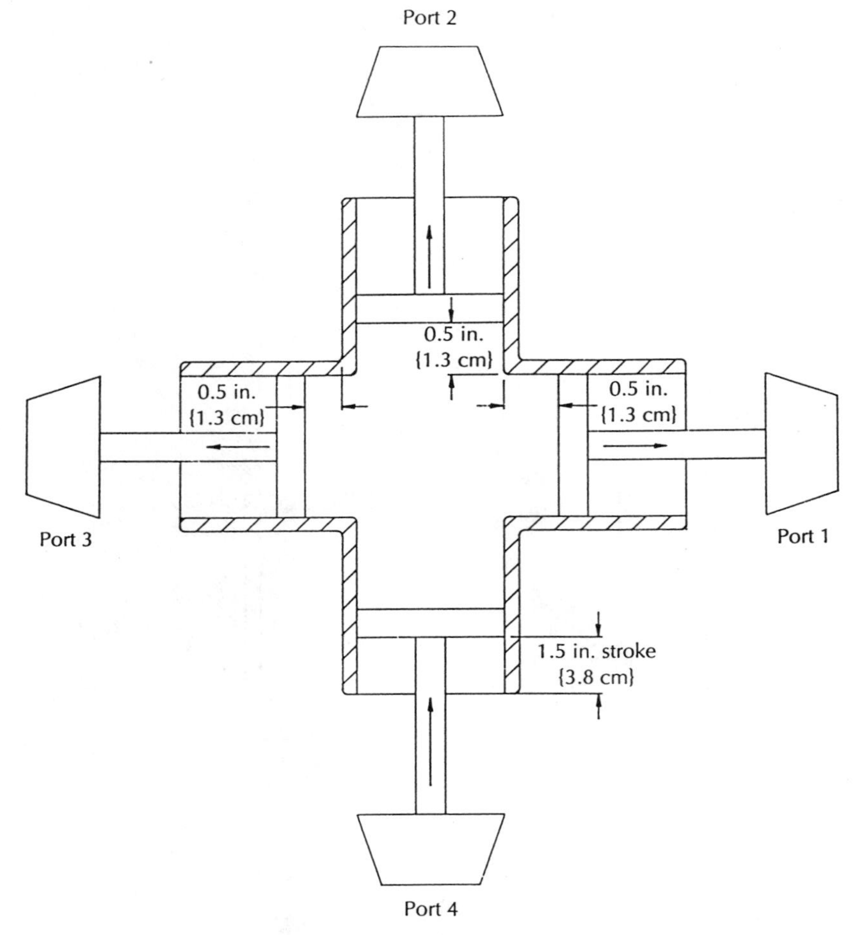

3.2 CHARACTERISTICS OF HYDRAULIC FLUIDS

The selection of which fluid to use in a hydraulic circuit was, for a long period of time, simply a matter of tradition. Few choices were available, and the consequences of changing from one to another type of fluid were minimal. Today, with the availability of **emulsion** (water and oil) fluids and various synthetic fluids as well as straight **(neat)** petroleum fluids, the circuit designer is able to specify certain fluids for use that will more readily operate within specified environments. These environments may include both very hot and very cold conditions. The hydraulic circuit will still be required to operate with precise control.

3.2.1 Fluid Density

The nature of materials and their responses to loads discussed in the previous section are largely determined by the weight per unit volume **(density)** of the material. In general, solids have greater densities than liquids, and liquids have greater densities than gases. However, there are wide variations in the densities of liquids. The result is that liquids having very high densities have characteristics similar to solids, while liquids having very low densities have characteristics similar to gases.

Since water is the most prevalent liquid on earth, its density has been established as the standard by which other liquids' and solids' densities are measured. The comparison of the density of a material to the density of water is referred to as **specific gravity {*apparent density*}**. The specific gravity of distilled water at 68 degrees Fahrenheit {*20 degrees Celsius*} is standardized to be 1.00. A cubic foot {*cubic centimeter*} of distilled water weighs 62.4 pounds {*1 gram*}. Therefore, the specific gravity of any material can be determined by dividing the density, in pounds per cubic foot, of that material by 62.4 {*or its mass in grams divided by its volume in cubic centimeters is its apparent density*}:

$$\text{Specific Gravity (S.G.)} = \frac{x\ \text{(lbs./cu. ft.)}}{62.4\ \text{(lbs./cu. ft.)}} \text{ or } \frac{x\ \{g\}}{y\ \{cm^3\}}$$

Problem 3-1

If a hydraulic fluid weighs 50 lbs. {*2,268 g*} per cu. ft. {*2,830 cm^3*}, what is its specific gravity?

S.G. = ________

Often, in the industrial workplace and in many industrial laboratories, the process of weighing a known quantity of hydraulic fluid and then calculating specific gravity is too time consuming. The **float-type hydrometer** (see Figure 3-6) allows for easy, direct reading of the specific gravity of a fluid, as long as the measurement falls within its range of calibration. The **hydrometer** is typically a sealed glass tube having two major sections: a bulb and a stem. The bulb, or enlarged end of the hydrometer, contains counterweights (typically small metallic spheres) at the tip and a large air- or gas-filled pocket that extends into the stem. The stem itself is also an air- or gas-filled pocket, having a graduated linear scale (like a small bench rule) contained within the stem or printed on the outside of the stem. Without the counterweights, the gas-filled bulb would float in nearly all liquid media. The counterweights offset this buoyancy

FIGURE 3-6

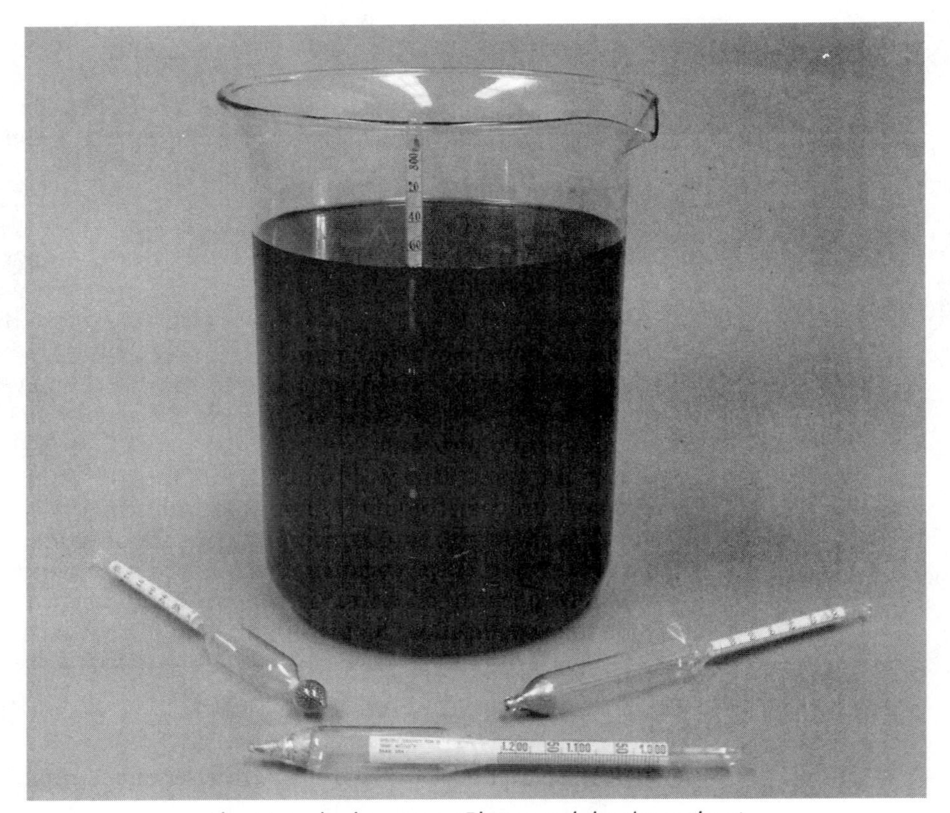

Float-type hydrometer. *(Photograph by the author.)*

and allow the hydrometer to stand up and somewhat sink (similar to the effect of a sinker on a bobber when fishing). The hydrometer will float higher (exposing more of the stem) in a dense liquid media and will become further immersed in a less dense liquid media. Graduations on the stem scale allow for a direct reading of the specific gravity of the liquid media.

One of the functions of a hydraulic fluid is to **seal** or isolate the mass of fluid to locations where it is intended to be from locations where it is not intended to be. In general, high-density fluids make superior sealing devices. This ability to seal will affect the efficiency of a pump because low-density fluids are more likely to slip through the mating parts of a pump and allow some of the fluid to return to the low-pressure side of the pump. Furthermore, low-density fluids are more likely to leak through mating parts on hydraulic output devices such as cylinders. The result of this leakage will be decreased speed or component response time because the effective volume supplied to that component within a period of time will be reduced (see Figure 3-7).

FIGURE 3-7

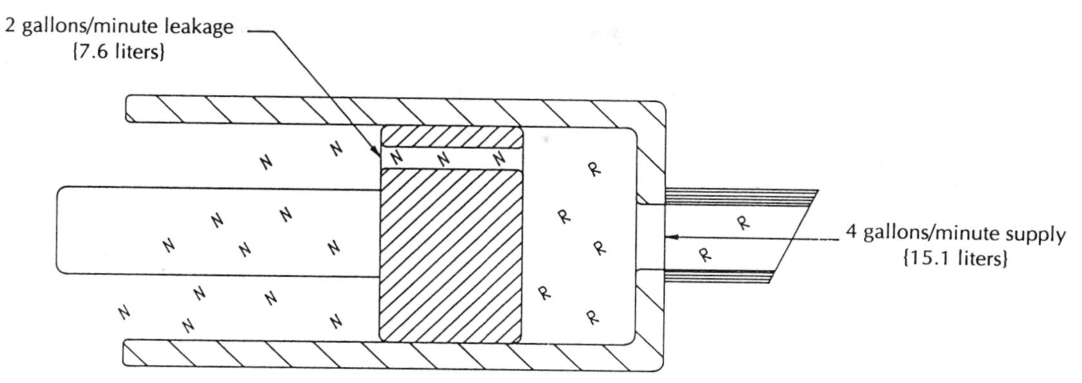

The result of this leakage will lessen the speed of the cylinder by one-half.

In general, denser liquids have greater resistance to flow **(viscosity)**. For example, consider pouring water from a bottle compared to pouring syrup from a bottle. The syrup has greater viscosity. However, viscosity is a much more involved concept than just a comparison of fluid densities.

3.2.2 Viscosity

Viscosity is the resistance to flow of a liquid inherent within the fluid itself. It is caused by the density of the liquid and the shape and size of the molecules making up the liquid mass. Because the molecules within the mass of a liquid impact upon and slide by one another, friction is created. If the fluid is dense, more impact occurs and greater resistance to flow is evident. However, dense fluid may have less resistance to flow if the size and shape of the moving molecules cause less resistance to movement. The molecules of hydraulic fluids are typically hydrocarbons formed into chainlike structures resembling a mass of spaghetti on a smaller liquid scale. However, not all chains are straight or simple wavy lines (linear). Some may have short chains emerging from the sides of the main chains (branched). The linear chain will typically have a higher density because the chains can pack closely together without branch interference. However, the linear chain will typically have better flow characteristics (less viscosity) because the side chains on the branched molecules inhibit the chains from sliding by one another (see Figure 3-8). In the low-viscosity fluid, when molecules meet, they have a tendency to slide or roll past one another, resulting in less friction than is apparent in the high-viscosity fluid having irregularly shaped molecules.

Viscosity is often exemplified by the rate of movement of a solid through a liquid. The movement of a pearl through thick shampoo relates the re-

FIGURE 3-8

Low-viscosity fluid

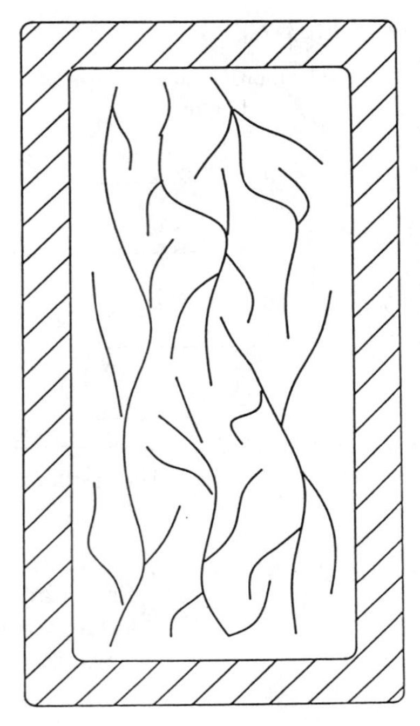

High-viscosity fluid

action of a very viscous fluid. This type of viscosity measurement is referred to as **static viscosity** and may be measured using the falling-ball viscosity test (see Figure 3-9). The more common, standard measurement of viscosity for hydraulic fluid is **dynamic viscosity**.

Dynamic viscosity may be measured using many standardized methods. One such method is by a testing procedure using the **Saybolt viscometer** (see Figure 3-10). The **American Society for Tests and Materials** (ASTM) standard D88 describes this procedure. Basically, this measurement is used to determine the ease with which a liquid flows through a standard opening under standard conditions. A small container or capillary tube is housed inside an oil bath (typically mineral oil). This capillary tube may be plugged at the bottom to block any flow through the standard orifice. Furthermore, the oil bath may be temperature controlled through the use of the thermostatically controlled electrical resistance heater transferring heat to the capillary tube through the oil bath. The hydraulic fluid is then placed into the capillary tube and allowed to fill up to the height of the integral lip. Any excess fluid will flow into the outer cup and subsequently will not be used in the operation. This creates what is called a **standard head** of oil. When the fluid in the capillary tube has

FIGURE 3-9

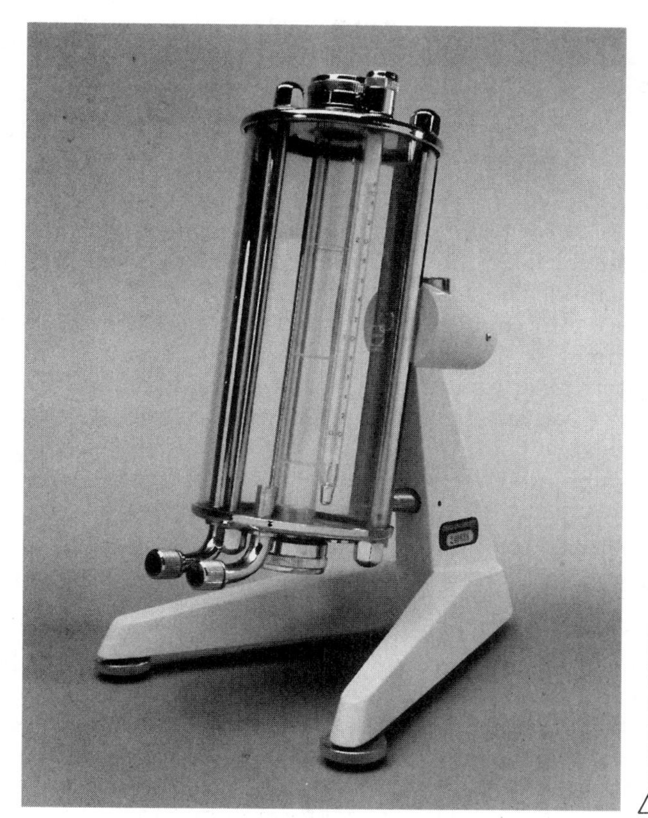

Falling-ball viscosimeter. *(Photograph by the author.)*

FIGURE 3-10

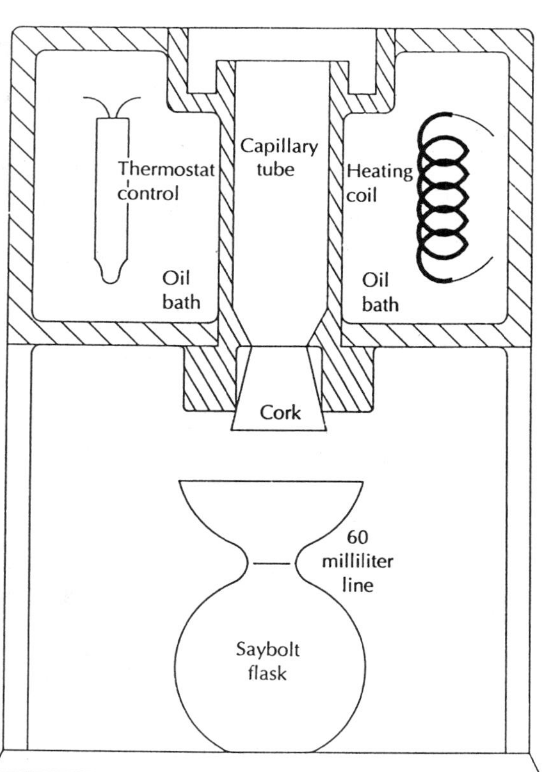

stabilized at a predetermined temperature, the cork is removed and the fluid is allowed to drain into the Saybolt flask. The time, in seconds, required to fill the Saybolt flask to a level of 60 ml is designated the **Saybolt Seconds Universal** standard for that fluid. Through the use of this apparatus and procedure, the relative viscosity of different fluids may be compared.

Viscosity is often considered to be the most important characteristic of a hydraulic fluid. Low-viscosity fluids flow more easily than high-viscosity fluids. However, low-viscosity fluids typically have lower densities; thus, they act poorly as seals and have somewhat slower response time. Low-viscosity fluids also leak through seals more easily. This loss of supply will necessitate an increase of the amount of supply necessary to perform an operation. The result is a reduction of efficiency and power in the hydraulic circuit. Also, the low-viscosity fluids do not separate moving parts as readily as high-viscosity fluids. This will cause more wear in components. The result will be increased friction generating increased heat, further reducing the viscosity of the oil and exaggerating this effect.

However, high-viscosity fluids result in similar problems within the hydraulic circuit. High-viscosity fluids require more force to move from one location to another. Loss of power in the hydraulic circuit using high-viscosity fluid will occur because more power is used to move the fluid. Although highly viscous fluids serve well as seals, their higher densities sometimes preclude them from penetrating closely mating parts. The result is loss of lubrication and increased friction and heat. Furthermore, the high-viscosity fluid will not flush the metallic chips created through wear as readily as a low-viscosity fluid. This may result in gouging and scoring of mating parts within a hydraulic component. Highly viscous fluids greatly affect the efficiency of pumps. Although pumps do not leak internally as much with high-viscosity fluid, the charging of the pump with fluid is more difficult. If the reservoir in the circuit is pressurized with atmospheric pressure only, a pressure of about 14.7 psi {*101 kPa*} will be available. Any reduction in this pressure created through moving a highly viscous fluid greatly reduces the efficiency of the pump through starvation.

3.2.3 Foam in the Fluid

Another important consideration in fluid stability is resistance to foaming. Gas bubbles, similar to those present when pouring cola or water, may be present when hydraulic fluid is agitated. If these bubbles become entrained in the fluid stream, they will cause gas pockets to form. Compression and expansion of the captured gases will result in sluggish response in components. Along with this loss of control, the entrapped gases, normally air, will accelerate oxidation (chemical formulations of ceramic particles and coatings) in hydraulic components. Hydraulic fluid suppliers can usually compound additives that retard foaming. These additives should be considered a requirement in hydraulic circuits operating with high fluid velocity or in circuitry that has a history of foaming. The telltale sign of excessive air pockets in the fluid stream is a pinging or hammering noise similar to the noise that you may have heard in your household water plumbing system when it gets air entrained in the lines.

3.2.4 Fluid Compatibility

Nearly all liquids act as solvents on some materials, and such is the case with hydraulic fluids. Commercially available hydraulic fluids, with the exception of water, rarely cause destructive chemical reactions with mild steels or brasses. Although the greatest masses of hydraulic components are made from these metals, other vital parts—such as gaskets and seals—are normally made from organic materials. Deterioration of gaskets and seals through chemical attack by hydraulic fluid will cause failure as certainly as rusting does. Therefore, care should be taken to ensure that hydraulic fluid and gasket and seal materials are compatible.

3.2.5 Viscosity Index

Finally, a hydraulic fluid should maintain its viscosity and character over a broad range of temperatures. As the hydraulic fluid becomes hotter, it will thin and flow more easily. By performing the Saybolt Seconds Universal test repeatedly on samples of hydraulic fluid at different temperatures, a graph of the reaction can be attained. The graph relates to the **viscosity index** of the fluid (see Figure 3-11).

Sample 1 in Figure 3-11 describes a fluid having a relatively high viscosity index (less slope), compared to the low viscosity index (more slope) of Sample 2. The hydraulic fluid in operating circuitry will heat through shear and friction of the fluid's molecules. *A hydraulic fluid having a low viscosity index will cause inconsistent speed in hydraulic output devices in normal operation.* Heat conditions external to the hydraulic circuit due to environmental conditions will also affect the viscosity index of a fluid. This should be considered when selecting hydraulic fluids for systems such as those operating in automobiles, aircraft, and other portable equipment. In general, the temperature of hydraulic oil should be maintained between 80 and 150 degrees Fahrenheit {*26.7 and 65.6 degrees Celsius*} in a circuit.

FIGURE 3-11

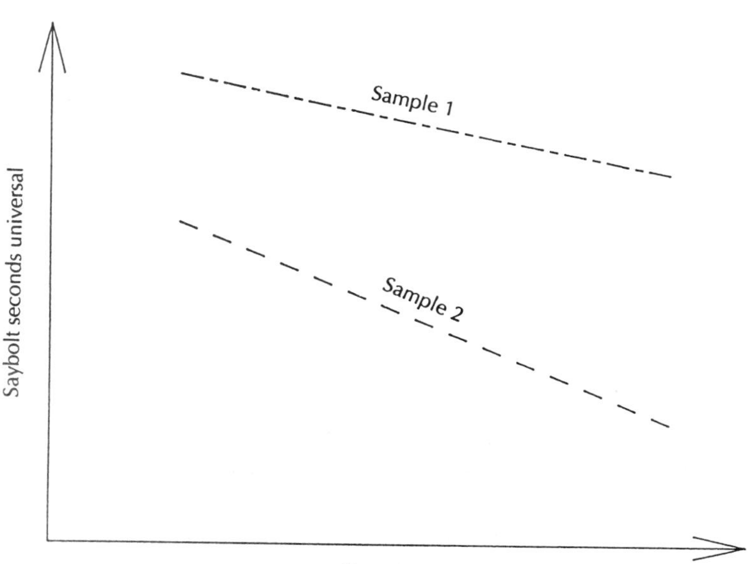

3.2.6 Other Fluid Characteristics

Density and viscosity account for a majority of the favorable (or unfavorable) characteristics of a hydraulic fluid. Just as surely, other unfavorable characteristics within the fluid can cause system inefficiencies or failure.

Probably the most important of these other characteristics is cleanliness. Ideally, the hydraulic fluid should remain free from particulate contamination. Methods employed in filtration are discussed in Chapter 4. In addition, thermal stability at extremely high and low temperatures is an important consideration and may be examined through procedures such as flash and fire point testing (ASTM D92, D93, or D1310) or pour point testing (ASTM D97), respectively. Vapor pressure testing (ASTM D323, D2879) may be used to determine or compare the character of hydraulic fluids in low-pressure (vacuum) conditions that affect pump operations.

Specialized operational environments demand special character and testing of hydraulic fluid. Medical equipment and biological applications of hydraulic circuitry would demand more rigorous examination for potential contaminants. For nearly any desired fluid character, there are devices and procedures available to maintain it and standardized tests to monitor their effectiveness. Those interested in such topics are referred to the ASTM and ANSI standards as well as hydraulic fluid manufacturers' data sheets.

3.3 TYPES OF HYDRAULIC FLUIDS

Basically, the choice of which type of fluid to use in a hydraulic circuit is a decision between water and oil. Water is denser and cheaper, while oil is a superior lubricant. Water promotes oxidation of metal surfaces. Oxidation of the fluid itself may occur. When long-chain hydrocarbon molecules break down, active chemical sites occur at the terminal ends of the molecules. These chemically reactive sites typically absorb oxygen. The result of fluid oxidation is acid and sludge formation in the oil. Water retards fire, while oil promotes it when exposed to flame or high temperature. Much development has been made to produce oils more like water and water more like oils. Synthetic substitutes have also been produced that can be used in extreme situations or in instances where a balance of oil–water characteristics are needed.

3.3.1 Oils

There are two types of **neat** oils that serve as the basis for most hydraulic fluids. These hydrocarbon oils are either **naphthenic** or **paraffinic**. Naphthenic oils are generally extracted from the offshore Gulf coast drilling operations in the United States. Naphthenic oils operate well in the higher ranges of permissible system temperature (up to 150 degrees Fahrenheit) {*65.6 degrees Celsius*} and have high resistance to foaming. However, their natural low viscosity index must be adjusted through the incorporation of additives to make them suitable for use in hydraulic circuitry.

Paraffinic oils are refined from high-grade Pennsylvania crude. Paraffinic oil has a naturally high viscosity index and is resistant to moisture and fluid oxidation. In general, the paraffinic oils rated at SAE (Society of Automotive Engineers) 10 weight or SSU (Saybolt Seconds Universal) 200 at 100 degrees Fahrenheit {*37.8 degrees Celsius*} are the superior neat hydraulic fluids.

3.3.2 Water

Water is a good alternative when attempting to lower the cost of fluid in the hydraulic system. Current movements toward the use of metallic components that resist chemical corrosion and rusting may signal a return to the use of water as the primary medium for hydraulic power transmission. Toady, water is seldom used in its natural state as a hydraulic fluid. Water is typically combined with oil or glycol and other additives for use as hydraulic fluids.

Water–oil emulsions are primarily oils with up to 40% water added in an emulsion (nonmiscible) form. When operating properly, these white oils have low viscosity and the fire retardancy of water while maintaining the lubricating properties of paraffinic oils. However, these properties are dramatically changed when the water separates or evaporates during normal operation. Therefore, the correct constituency must be maintained through constant monitoring.

Unlike oil, **glycol** is miscible in water and forms a solution. This allows for a wide variation of character in the fluid. Water–glycol solutions offer fire retardancy, good lubrication, and high density. However, their usage is limited to the lower range of permissible operating temperatures (up to 130 degrees Fahrenheit) {*54.4 degrees Celsius*}. Above this temperature, the glycol and water separate from solution. Water–glycol solutions also tend to retain higher levels of entrained air, causing foaming, while their higher densities may cause pump starvation. These pink oils may also act as solvents on many painted surfaces of hydraulic components; and, like water–oil emulsions, the ratio of water to glycol must be carefully monitored to maintain optimal character in the fluid.

3.3.3 Synthetic Fluids

Other fluids may be synthesized from petroleum-based substances to fit a very specific need. **Phosphate-ester** fluids are rated as fire resistant and may be used continually in circuits operating at 150 degrees Fahrenheit {*65.6 degrees Celsius*}. Even at this high temperature, phosphate-esters will maintain very good lubrication character and resistance to fluid oxidation. However, these fluids are strong solvents to many plastic and rubber materials used for seals and gaskets. Close attention to follow the fluid suppliers' suggestions for seal compatibility must be made to assure efficient operation in the hydraulic circuit using phosphate-ester fluid.

Chlorinated hydrocarbon–based and **silicone-based** oils may also be used in hydraulic circuitry. These fluids have low viscosities and elevated temperature resistance. However, their low densities typically result in more component wear, especially in high-speed circuitry.

3.3.4 Fluid Selection

Typically, hydraulic circuits are designed to operate using neat paraffinic oils. However, during operation it may be found that the circuit requires special character fluid to resist temperature, oxidation, or other influences. When changing from one fluid to another, seal compatibility, breathing capacity of the reservoir to remove entrained air, and necessary temperature controls should be matched to those suggested by the fluid manufacturer. The costs of unit volumes of fluid rises from a low of water to oil through phosphate-esters, silicones, and chlorinated hydrocarbons. However, the lowest-cost fluid is not always the worst selection; nor is the selection of the highest-cost fluid always the best. Control of the environment or conditions may be a more appropriate approach to solving fluid problems, rather than alteration of the fluid.

3.4 SUGGESTED ACTIVITIES

1. Take a sample of oil (or any liquid other than distilled water) and determine its specific gravity by accurately calculating the volume of a container to hold the liquid, the initial weight of the container, and the weight of the container and liquid after filling to the predetermined volume.
2. Determine the specific gravity of the sample used in Activity 1 using a float-type hydrometer.
3. Visit a local store that sells oil; examine the label on the oil containers; and record the trade name, SAE designation, and SSU designation (if given).
4. If you have a Saybolt viscometer available, perform some experiments to determine the SSU of some of the oils that you found in a local store. Do your results correspond with the designations supplied on the containers? If not, why do you think they are different?
5. If you do not have a Saybolt viscometer, use a large and small paper cup to design your own viscometer. Punch a small hole in the bottom of the small cup. Draw a line on the inside of the small cup to represent your standard head. Draw a line at a convenient location (perhaps 2 in.) {*5.1 cm*} up from the bottom of the large cup. Using rubber or plastic gloves, cover the outside

of the hole with your finger and pour a sample of oil into the small cup up to your standard head line. Holding the small cup over the large cup, remove your finger from the hole and measure the time that it takes to fill the large cup to your standard line. Repeat this experiment with another oil sample (or water) and compare your results.

6. Using the Saybolt viscometer or a homemade viscometer, a heat source, and a thermometer, heat your fluids and repeat the experiment. (Be careful. Do not get them so hot that you cannot handle them.) Compare and contrast your results from this set of observations with your previous observations by graphing lines that represent change in viscosity. What did you find out about the viscosity index of each fluid?
7. Repeat your viscosity index experiment; but this time, cool your fluids by placing them in a refrigerator for a period of time immediately before testing. Add these observations to your previous graph for viscosity index. Do the results appear to be consistent (linear projections of the graph lines) with your heated observations? If not, why do you think they differed?

3.5 REVIEW QUESTIONS

1. Discriminate among solids, liquids, and gases in terms of containment, density, and compressibility.
2. Contrast the characteristics of water with oil when used as hydraulic fluids.
3. Define *specific gravity*.
4. Describe the float-type hydrometer.
5. Discuss the effects of hydraulic fluid density on its ability to seal.
6. Define *viscosity*.
7. Why may some higher-density fluids have lower viscosity?
8. Describe the operation of the Saybolt viscometer.
9. Discuss the effects of high and low viscosity on the character of hydraulic fluid.
10. Why do hydraulic fluids incorporate antifoaming agents?
11. Discuss viscosity index in terms of fluid character and hydraulic application.
12. Describe the following tests in terms of the character of hydraulic fluid:
    a. flash and fire point
    b. pour point
    c. vapor pressure
13. Contrast two types of neat oil—naphthenic and paraffinic.
14. Describe emulsion hydraulic fluid.
15. Describe glycol hydraulic fluid.
16. Describe phosphate-ester hydraulic fluid.

4

Hydraulic Power Supply

Hydraulic circuits may be described as secondary power transmission devices. Advancements have been made in pump designs to the point that some pumps are capable of driving themselves. However, these pumps are useful only in limited applications as hydrodynamic water pumps. The supply of hydraulic fluid necessary to cause displacement or movement in output devices requires external prime movers such as electrical motors and internal combustion engines. Electrical motors and internal combustion engines both cause rotary, not reciprocal, output. Therefore, adaptations of the reciprocating pumping mechanism first designed by Joseph Bramah are necessary.

4.1 PUMPING CONCEPTS

4.1.1 Pressure Differential

The principle of pump operation requires the existence of a vacuum condition to charge the system, using atmospheric pressure to force supply fluid into the pump. Although atmospheric pressure varies from one geographical location to another, it is conventionally standardized to be 14.7 psi {*101.4 kPa*}. Very simply, this means that a column of air having a cross-sectional area of 1 sq. in. {*1* cm^2} that is as high as the Earth's atmosphere weighs 14.7 lbs. {*1.0 kg*}. Hydraulic fluids range in specific gravity from 0.75 to 1.5. Therefore, these fluids will weigh between 46.8 and 93.6 lbs./cu. ft. {*0.75 and 1.5* g/cm^3} (see Problem 3-1). Most of the hydraulic fluids currently in use have specific gravities between 0.8 and 0.9. Consequently, a hydraulic fluid having a specific gravity of 0.85 would weigh approximately 53 lbs./cu. ft. {*850 kg/m³*}. Each of the 1,728 cu. in. {*1,000,000 cm³*} of fluid in that 1 cu. ft. {*1 m³*} would weigh 0.031 lbs. {*0.85 g*}. Two cu. in. {*cm³*} of that fluid stacked on top of one another would weigh approximately 0.062 lbs. {*1.70 g*}; 3 cu. in. {*cm³*} stacked would weigh 0.093 lbs. {*2.55 g*}, and so on. The use of atmospheric pressure to move such a column of fluid from the tank to the pump may limit the efficiency of the hydraulic circuit.

The lever principle may be applied to determine the maximum distance a hydraulic fluid having a specific gravity of 0.85 may be raised using atmospheric pressure (see Figure 4-1).

FIGURE 4-1

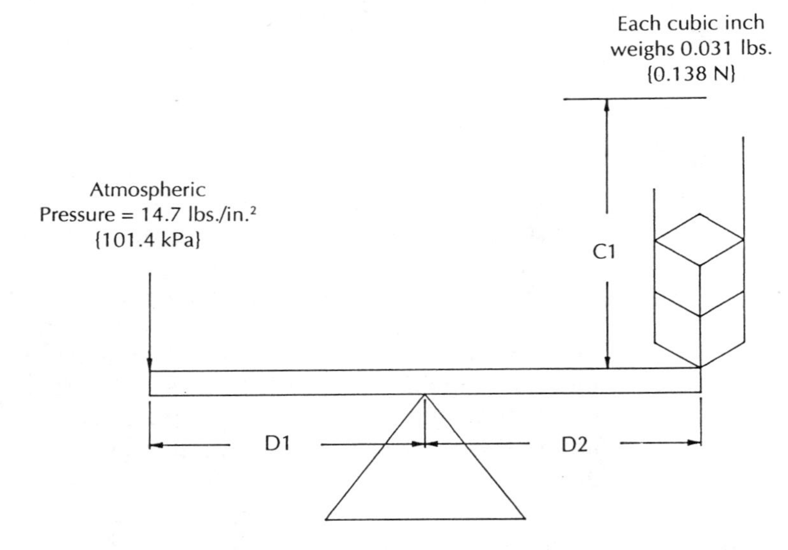

Problem 4-1

What is the maximum column height (C1) that may be balanced using atmospheric pressure?

C1 (in.) = ________

C1 {cm} = ________

Applying the findings from Figure 4-1 to a practical hydraulic circuit reveals that the vertical distance from the intake tube, inside the reservoir, to the pump inlet is limited by the weight of the fluid. Other factors will further affect the efficiency of pump charging, including the cross-sectional area of the intake tube and the velocity at which the fluid flows through the tube (see Figure 4-2).

In general, anything that would cause resistance to the flow of the fluid will decrease the efficiency of the pump-charging operation. Therefore, it

FIGURE 4-2

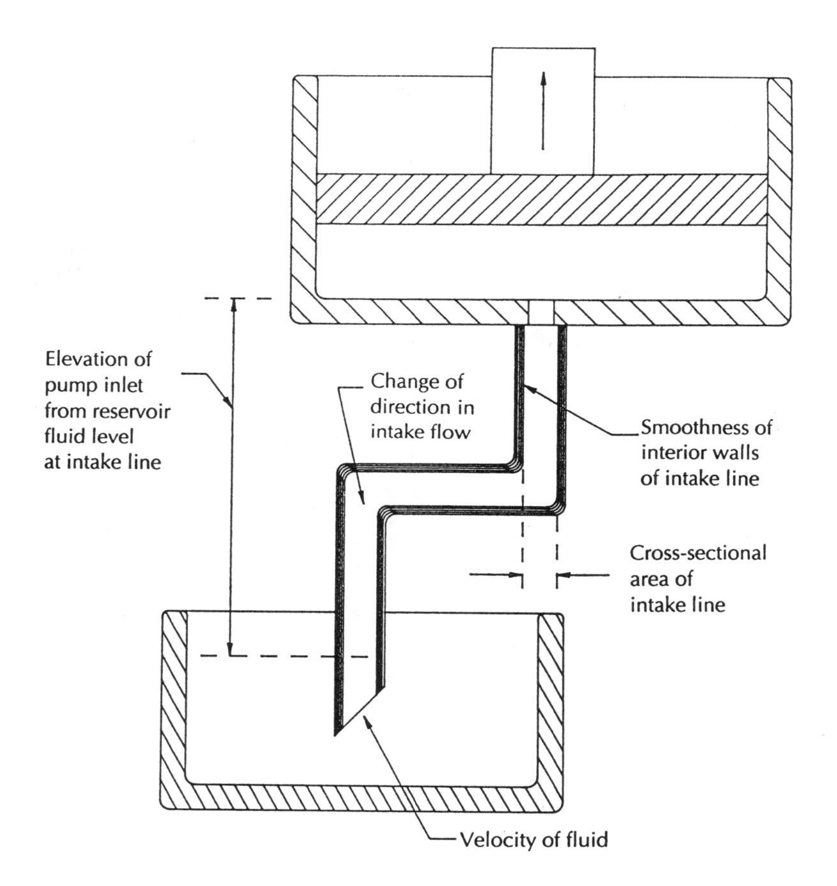

is best to (1) keep the pump inlet in close proximity to the reservoir, (2) eliminate bends or changes of direction of flow from the reservoir to the pump inlet, (3) use a conductor with extremely smooth interior surfaces, and (4) keep fluid velocity as low as possible within the inlet tube. These considerations are important because most industrial hydraulic circuits locate the pump above the reservoir, creating a condition known as a **negative head**. In most negative head situations, the limitation is about 18 in. {45.7 cm}.

Thus far in this chapter, the fluid has only been transferred from the reservoir to the pump. After this transferral, the fluid must be captured within the pumping mechanism, then expelled into the circuit. The operations necessary to cause charging, capturing, and expelling of the hydraulic fluid define the function of a pump. Often, the pump is incorrectly considered to be the component that creates pressure. However, *the function of a pump is simply to create flow. Resistance to that flow results in pressure*. However, nearly all component parts of the pump must withstand, at some time, that resulting pressure.

4.1.2 Pump Displacement

The pump will capture hydraulic fluid within the confines of its physical design. The capacity of the pump is expressed as the volume contained within the pump or discharged by the pump in one cycle. This rating of the pump is called its **displacement**. Since the vast majority of pumps used in industrial hydraulic circuits are driven by rotary devices, a cycle of a pump is designated as a revolution. *Displacement is, therefore, expressed in cubic inches per revolution {cubic centimeters per revolution}* (see Figure 4-3).

FIGURE 4-3

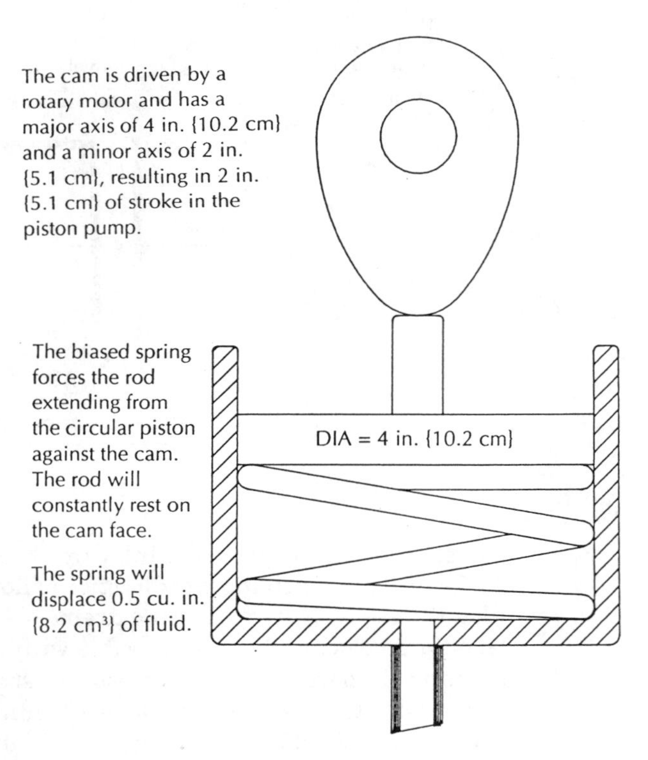

Problem 4-2 What is the displacement of this pump?

DIS (cu. in./rev.) = ________

DIS {cm^3/rev.} = ________

The geometrical calculation of displacement is theoretical. Theoretical displacement is often provided in manufacturers' specifications. Circuit pressure and the effectiveness of seals between the inlet and the outlet of the pump will reduce this theoretical displacement to an actual displacement that may be measured during operation. The relationship between the theoretical displacement and the actual displacement is the **volumetric efficiency**. Volumetric efficiency (V.E.) is expressed as a percentage:

$$\text{V.E. (\%)} = \frac{\text{Actual Displacement}}{\text{Theoretical Displacement}} * 100$$

Problem 4-3

The theoretical displacement of a pump is calculated to be 12 cu. in. {*196.7 cm*^{3}} per revolution, and the actual displacement of the pump is measured as 9.6 cu. in. {*157.3 cm*^{3}} per revolution. What is the volumetric efficiency of the pump?

V.E. (%) = __________

Caution should be taken when specifying the size (displacement) of a pump because its volumetric efficiency will vary with changes in system pressure, age of the pump, and changes in hydraulic fluid. Pump manufacturers will generally provide charts that describe the changes in actual displacement of pumps at various pressures. However, only experience and testing will reveal the effect of wear caused by age and changes in hydraulic fluid. In many cases, the displacement of the pump is fixed. However, it is possible to vary the displacement in some pump designs. For example, limiting or increasing the stroke of the piston in the pump in Figure 4-3 would vary its displacement. The result would be increased or decreased movement of fluid within a period of time. Typically, the amount of fluid movement in a given period of time is determined by the operational speed of the rotary drive or motor in combination with the displacement of the pump.

4.1.3 Delivery

The rate at which fluid is supplied in a hydraulic circuit is known as **delivery** and is rated as volume per unit time. In general, delivery is expressed as **gallons/minute (gal./min.), {*liters/minute (l/min.)*}**, or **gpm {*lpm*}**. There are 231 cu. in. {*1,000 cm*^{3}} in 1 gallon {*liter*} (see Figure 4-4). This factor, along with the operating speed of the motor drive and the displacement of the pump, will establish the pump's delivery:

$$\text{DELivery (gal./min.)\{l/min.\}} = \frac{\text{DIS(cu. in./rev.) \{cm}^3\text{/rev.\} * Motor speed (rev./min.)}}{\text{231 (cu. in./gal.) \{}1{,}000\ (cm^3/l)\}}$$

FIGURE 4-4

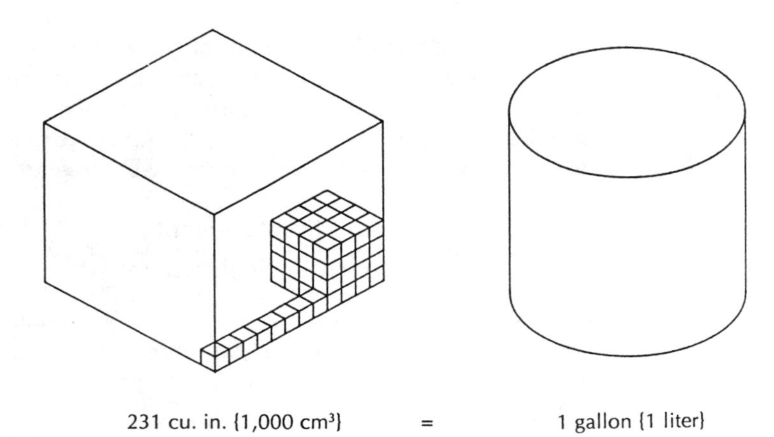

231 cu. in. {1,000 cm³} = 1 gallon {1 liter}

Problem 4-4 Continuing the development of the pump presented in Figure 4-3, calculate the theoretical delivery of the pump if the motor driving the cam is operating at 1,200 rev./min.

DEL (gal./min.) {l/min.} = ________

4.1.4 Volumetric Efficiency

Since pumps typically operate at speeds well above 1,000 rev./min., it is difficult to stop the pump after one revolution to measure its actual displacement. A more practical method of determining the volumetric efficiency of a pump is to compare its theoretical delivery to its actual delivery:

$$\text{V.E. (\%)} = \frac{\text{Actual Delivery (gpm) \{lpm\}}}{\text{Theoretical Delivery (gpm) \{lpm\}}} * 100$$

Problem 4-5 Using the theoretical delivery calculated for the pump in Problem 4-4, calculate its volumetric efficiency if its actual delivery were measured at 104 gal./min. {*393.7 l/min.*}.

V.E. (%)= ________

4.2 FORCE, DISPLACEMENT, AND TIME

Hydraulic circuitry would serve no useful purposes if there were no loads to be overcome, forces to be exerted, or objects to be moved. The loads and the necessary displacement of them define the parameters under which a hydraulic circuit operates. This section examines the factors that contribute to the determination of the power necessary to drive a hydraulic circuit. Although very thorough in examination, it serves as only an introduction to develop formulation for determining the horsepower or **wattage** required to do a job. Most commonly, people working with hydraulics use the result of this process (horsepower or power formulation, found in section 4.3).

4.2.1 Work

The movement of a load (force) results in work being performed. In Chapter 2, work was calculated by multiplying force times distance. Thus, work may be expressed as:

Work (ft. lbs.) {*N m*} = Force (lbs) {9.8 kg_f} * Distance (ft.) {*m*}, or

Work (in. lbs.) {*J*} = Force (lbs.) {*N*} * Distance (in.) {*m*}

In hydraulics it is often more appropriate to consider the operation of the circuit and its components to understand how formulas are derived. For example, in Bramah's jack, the distance **(D)** moved by the ram (output cylinder) may be defined as the quantity **(Q)** of fluid displaced by the cylinder divided by the cross-sectional area **(A)** of the cylinder's piston:

$$\text{Distance (in.)}\{cm\} = \frac{\text{Quantity (in.}^3\text{)}\ \{cm^3\}}{\text{Area (in.}^2\text{)}\ \{cm^2\}}$$

Furthermore, the force component in the circuit may be expressed using Pascal's Law:

Force (lbs.) {*N*} = Pressure (lbs./in.^{2}) {*Pa*} * Area (in.^{2}) {m^2}

If these two expressions are substituted into the work formula, a variation is achieved:

$$\text{Work (in. lbs.)}\ \{J\} = \text{P (lbs./in.}^2\text{)}\ \{Pa\} * \text{A (in.}^2\text{)}\ \{m^2\} * \frac{\text{Q (in.}^3\text{)}\ \{m^3\}}{\text{A (in.}^2\text{)}\ \{m^2\}}$$

Simplified, this equation becomes:

Work (in. lbs.) {J} = P (lbs./in.^{2}) {Pa} * Q (in.^{3}) {m^3}

In many cases this variation is the more appropriate form to use when determining the work being performed by a hydraulic circuit. The force component will often vary in a hydraulic circuit and may be difficult to measure. However, simple pressure gauges are often incorporated into hydraulic circuits. Measurement of the pressure in a circuit may be achieved by monitoring the pressure gauge.

Problem 4-6

A hydraulic jack is being used to raise a car transmission through a distance of 8 in. {*20.3 cm*}. The jack has a 2-in. {*5.1-cm*} diameter piston, and the pressure gauge on the jack reads 50 psi {*344.7 kPa*}. How much work is being performed during this operation?

Work (in. lbs.) {*J*} = __________

4.2.2 Power

Another important consideration in any device that performs work is the rate at which the work is performed. Nearly anyone could move a 1,000-lb. {*453.6-kg*} pile of gravel between two points that are 10 ft. {*3.1 m*} apart. If one person moved the pile in one hour and another moved it back in two hours, the first person may, incorrectly, be said to have done more work. In each case, 1,000 lbs. {*453.6 kg*} of gravel were moved a distance of 10 ft. {*3.1 m*}, or 10,000 ft. lbs. {*13.6 kJ*} of work were performed. However, the first person was more productive. Therefore, time of operation must be accounted for to determine productivity. *Work per unit time is referred to as power*:

Power (ft. lbs./min.) {*W*} = Work (ft. lbs.) {*0.167 N m*} / Time (min.), or

Power (in. lbs./sec.) {*W*} = Work (in. lbs.) {*J*}/Time (sec.)

Using the variation derived in the previous section, power may be defined as pressure multiplied by quantity divided by time:

$$\text{Power (in. lbs./sec.) } \{W\} = \frac{\text{Pressure (lbs./in.}^2)\{Pa\} * \text{Quantity (in.}^3) \{m^3\}}{\text{Time (sec.)}}$$

To make this formula even more practical, the quantity divided by time may be expressed as delivery:

$$\text{DELivery (gal./min.) \{l/min.\}} = \frac{\text{Quantity (in.}^3) \{cm^3\}/231 \text{ (in.}^3\text{/gal.) } \{1{,}000\ (cm^3/l)\}}{\text{Time (sec.)/60 (sec./min.)}}$$

Simplifying and solving the constant yields:

$$\text{DELivery (gal./min.) } \{l/min.\} * 3.85\ \{16.67\} = \frac{\text{Quantity (in.}^3) \{\text{cm}^3\}}{\text{Time (sec.)}}$$

Substituting this formula into the power equation yields:

$$\text{Power (in. lbs./sec.)} = \text{P (psi)} * \text{DEL (gpm)} * 3.85 \frac{\text{in.}^3}{\text{gal.}} \frac{\text{min.}}{\text{sec.}}, \text{ or}$$

$$\text{Power } \{W\} = \text{P } \{Pa\} * \text{DEL } \{lpm\} * 16.67 \frac{cm^3}{l} \frac{min.}{sec.}$$

Using a **flow gauge** in combination with a **pressure gauge**, one may determine the power consumed in a hydraulic circuit.

Problem 4-7

A hydraulic circuit is being used to compress cardboard in a recycling operation. The pump for this circuit has been measured to deliver 3 gal./min. {*11.4 l/min.*}, and the circuit is operating under 200 psi {*677.3 kPa*} pressure. How much power is being consumed during this operation?

Power (in. lbs./sec.) {*W*} = ________

4.2.3 Horsepower

James Watt, an English inventor, developed the first practical steam engine. While developing his engine, he decided that he would need to be able to compare the power produced by his invention to a more common working system. Through observation, Watt calculated that a draft horse working in a vertical shaft mine could raise 550 lbs. {*249.5 kg or 2.4 kN*} of coal through a distance of 1 ft. {*0.3 m*} in 1 sec. Thus, Watt described the standard unit of power as 1 horsepower:

1 hp = 550 (ft. lbs./sec.)	*1 hp = 746 {N m / sec.}*
33,000 (ft. lbs./min.)	*746 {J / sec.}*
6,600 (in. lbs./sec.)	*44,760 {N m / min.}*

(Note: The exercise of expressing power as horsepower in the SI standard is not appropriate and appears in this presentation for the sake of continuity.)

Conversion of the power formula may be achieved by dividing the equation by the horsepower standard:

$$\text{hp} = \frac{\text{P (psi)} * \text{DEL (gpm)} * 3.85 \frac{(\text{in.}^3)}{(\text{gal.})} \frac{(\text{min.})}{(\text{sec.})}}{6{,}600 \text{ (in. lbs./sec.)}}, \text{ or}$$

$$hp = \frac{P\ \{kPa\} * DEL\ \{lpm\} * 0.01667\ \frac{cm^3}{l}\ \frac{min.}{sec.}}{746\ \{W/hp\}}$$

Simplifying the equation and performing the division yields a dimensionless value:

hp = P (psi) * DEL (gpm) * 0.000583, or

hp = P (psi) * DEL (gpm)/1,715

Simplifying the equation in the SI standard yields the following:

*hp = P {kPa} * DEL {lpm} * 0.00002235, or*

*hp = P {kPa} * DEL {lpm} / 44,743*

The calculation of horsepower is a very practical exercise, since hydraulic circuits typically are driven by devices rated in horsepower units, such as electrical motors and internal combustion engines. Although hydraulic circuitry has many advantages to these other input devices, it must be remembered that *no more hydraulic horsepower may be developed than is applied to the circuit by the driving device*. In nearly all cases, the output of the hydraulic circuit will be reduced below the input value of the driving device. Since pressure is transmitted undiminished throughout the circuit, it will develop until it overcomes a load or another weaker point in the circuit. Therefore, there are no real pressure losses in the hydrostatic circuit resulting from the effect of pressure operating on something other than the output hydraulic device. The power loss from the input value of the electrical motor must be attributed to loss of delivery. This loss is expressed by the volumetric efficiency of the pump. Compensation for this loss in a circuit is made by dividing the necessary horsepower by the volumetric efficiency of the pump:

$$hp = \frac{P\ (psi)\ \{kPa\} * DEL\ (gpm)\ \{lpm\} * 0.000583\ \{0.000022\}}{V.E.\ (\%\ \text{expressed as a decimal value})}$$

This compensation will describe the horsepower needed to supply the extra delivery needed to operate a circuit under a specified pressure at a predetermined delivery (see Figure 4-5). Often, this formulation is shortened to express volumetric efficiency at about 80%:

hp = P (psi) {*kPa*} * DEL (gpm) {*lpm*} * 0.0007 {*0.00003*}

FIGURE 4-5

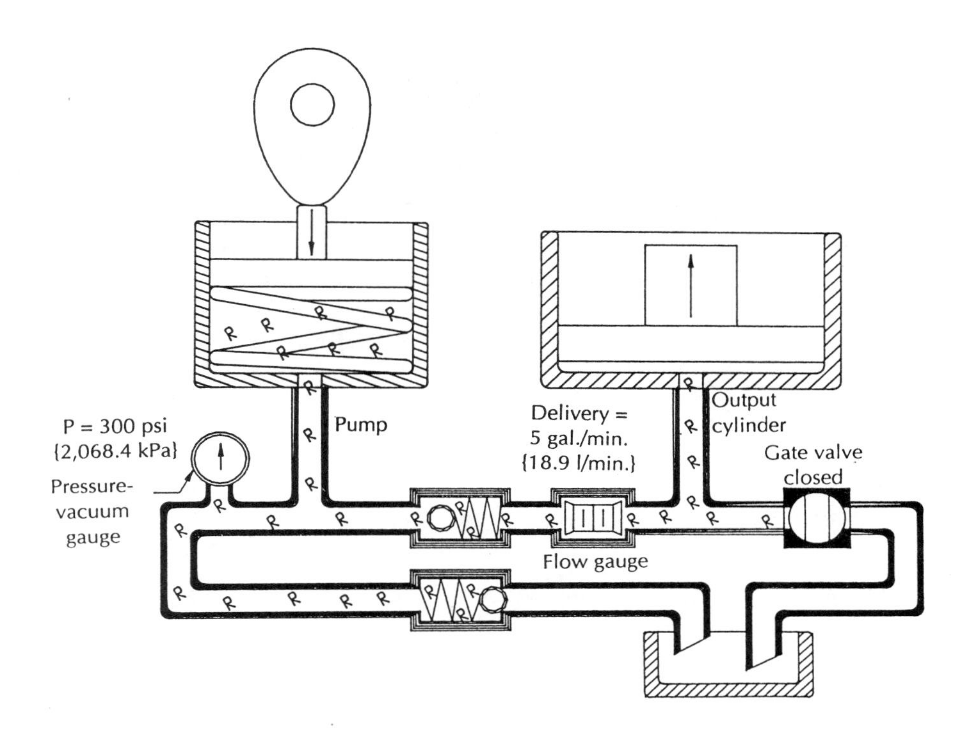

Problem 4-8

The hydraulic pump is supplying fluid at a rate of 5 gal./min. {18.9 l/min.}, and the pressure within the circuit measures 300 psi {2,068.4 kPa}. What horsepower is being consumed if the pump is rated at 80% volumetric efficiency?

hp = ________

4.3 HYDRAULIC PUMP DESIGNS

Unlike water pumps and blower units used to move air, industrial hydraulic pumps must be designed to withstand high pressures on the outlet side. In theory, the hydraulic pump will transfer all of the fluid that enters the inlet to the outlet. Pumps designed with closely mating parts and seals to assure this high-efficiency transferral are known as **positive-displacement** or **hydrostatic pumps**. Although many variations of pump designs exist, the more common ones are **gear, vane,** and **piston** types. In each case, *the pump is designed to convert the rotary motion created by the motor drive into streamlined flow of hydraulic fluid.*

4.3.1 Gear Pumps

The concept of the gear is nearly as ancient as that of the wheel. Undoubtedly, some person long ago discovered that cutting designs and chambers into the outside of the wheel produced much better traction. Through observation, it was learned that materials trapped within these chambers would be captured at one location and deposited at another. Many practical applications of this observation evolved, including lifts to raise water from one level to another for irrigation purposes. The result was a device that could transfer fluid, if the gear inlet location was immersed within the supply and the pressure at the outlet location was relatively low (see Figure 4-6).

FIGURE 4-6

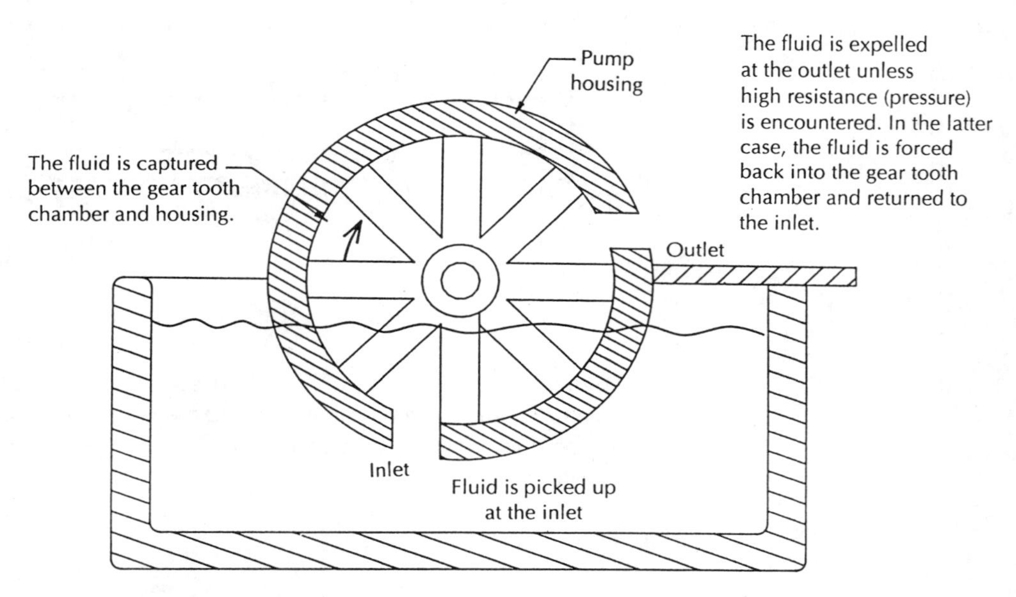

Obviously, this design is not practical in hydraulic circuits, where high pressure is necessary for power transmission. A method is needed to exclude flow from the system back to the tank. This is achieved by incorporating a more complicated gearing system in which one gear is used to drive another (see Figures 4-7 and 4-8).

The design presented in Figure 4-7 represents an **external gear** or **gear-on-gear pump**. Advancements in gear designs have been incorporated into hydraulic gear pumps. Rather than using straight sides on the gear faces, a gradual curve or involute is used to reduce the sliding friction of the meshing gears and increase the mechanical efficiency of the pumping mechanism. An extreme application of this design feature is illustrated in the **lobe pump** (see Figure 4-9).

FIGURE 4-7

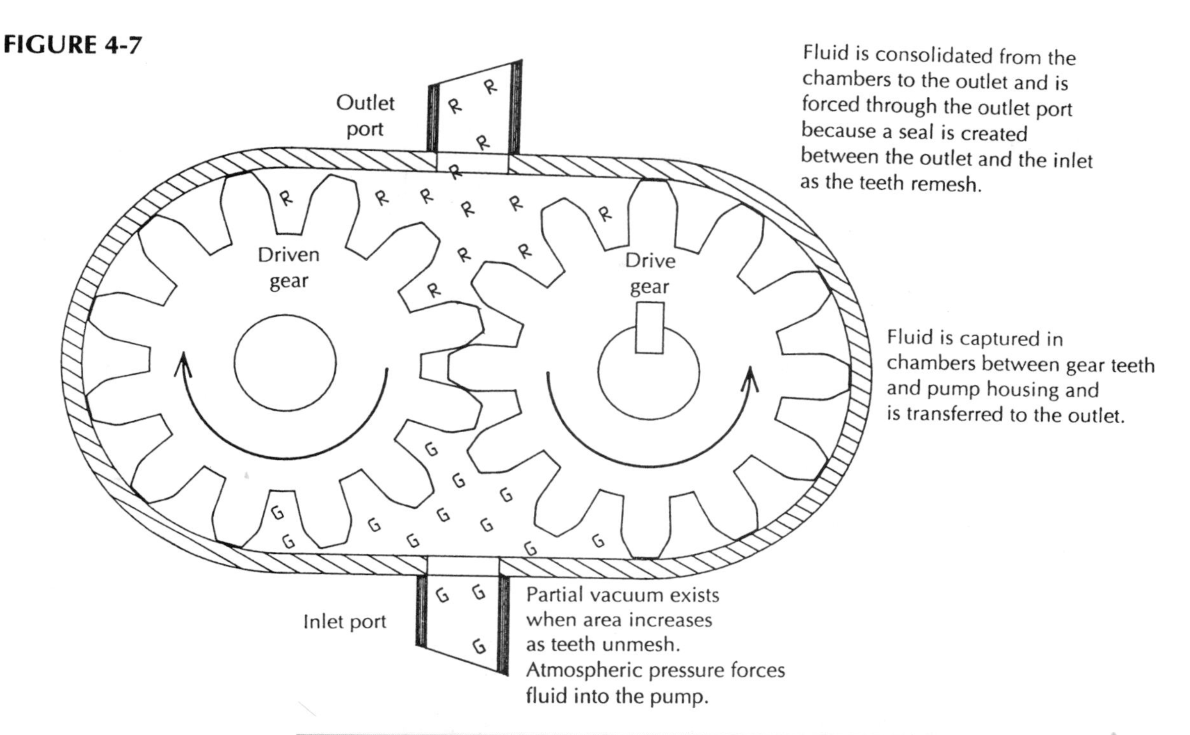

FIGURE 4-8

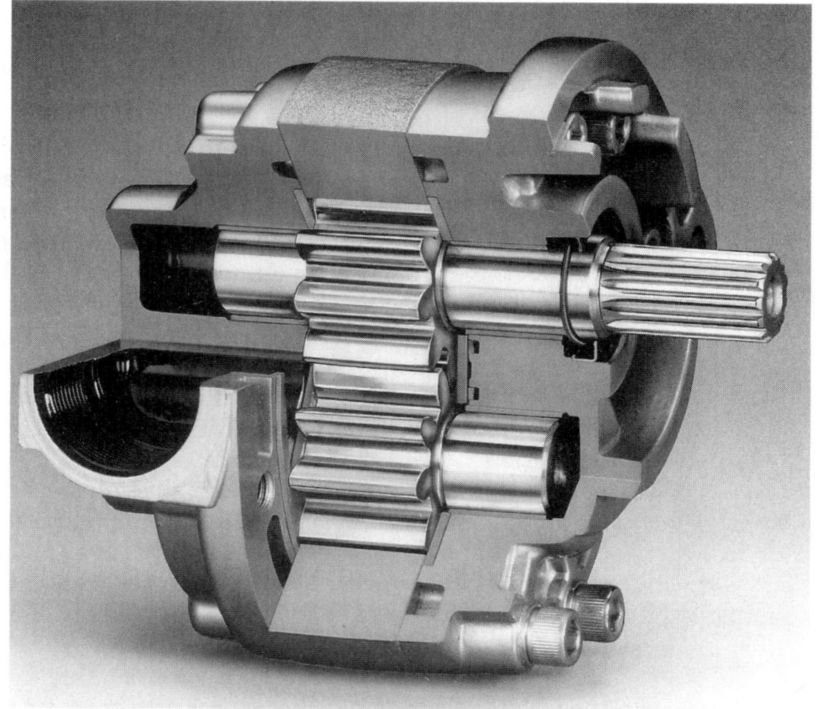

External or gear-on-gear pump. *(Courtesy Danfoss Fluid Power. From Vockroth,* Industrial Hydraulics, *Albany, N.Y.: Delmar, 1994, p. 115.)*

FIGURE 4-9

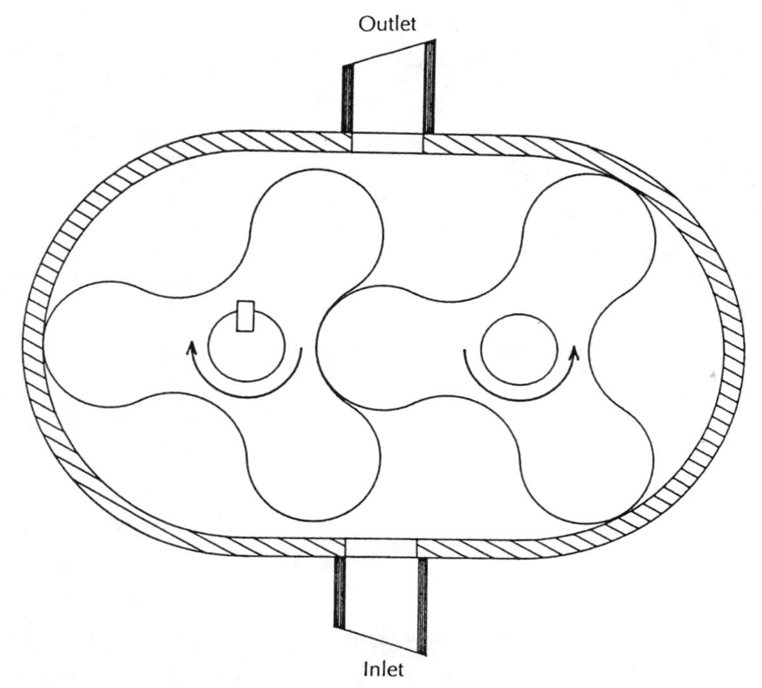

Assignment: Color-code the drawing.

The lobe pump operates on the same principle as the external gear pump. The increased size of the chambers in the lobe pump results in greater displacement than in an external gear pump of the same physical size. However, the lobe pump does not seal the inlet from the outlet as well as the external gear pump. The resulting slippage and leakage reduces the volumetric efficiency of the lobe pump. In some cases, the reduction in volumetric efficiency surpasses gain in mechanical efficiency through reduction of sliding friction between gears. The external gear pump remains the more popularly used design, except in some cases where high volume is required from a relatively small package.

Another variation on the external gear pump design is the **screw-type** pump. This design probably is the most ancient in concept, for it is based on the principles described by the Archemedian screw. In this design, two or three auger-type screws are mounted in such a manner that rotation of one screw will rotate all the others through intermeshing of the gears (screws). If one were to look into the end of the meshing screws, the appearance would be similar to the external gear arrangements. Thus, increasing areas appear where the screws unmesh, allowing for charging of these areas from atmospheric pressure operating upon hydraulic fluid in the reservoir and a partial vacuum existing at this inlet. However, unlike the external gear designs, flutes are machined along the face of each gear that direct the inlet fluid longitudinally along the face of each screw and deposit this charge at the opposite end of the screws into a chamber of decreasing volume at the outlet. Continual deposition of fluid into the outlet chamber will exceed its volumetric capacity and expel fluid through the outlet port.

A relatively small pump package may be achieved by using an **internal** or **gear-in-gear** design. In the internal gear design, a spur gear, similar to the one used in external gear pumps, is driven inside a mating ring gear block. The gear block is free to rotate within the pump housing. The spur gear and the ring gear block are located on different centers. The resulting eccentric arrangement allows for increasing and decreasing pumping chamber size that charges the pump inlet and expels the fluid from the outlet (see Figure 4-10). This pump design is also often referred to as a crescent pump. The crescent seal allows pressurized fluid to be directed to the edges of the gear mating surfaces. The decreasing area of the crescent seal in the areas toward the pump inlet side creates a dynamic seal that tends to force any fluid that might slip by the edges toward the pump outlet.

FIGURE 4-10

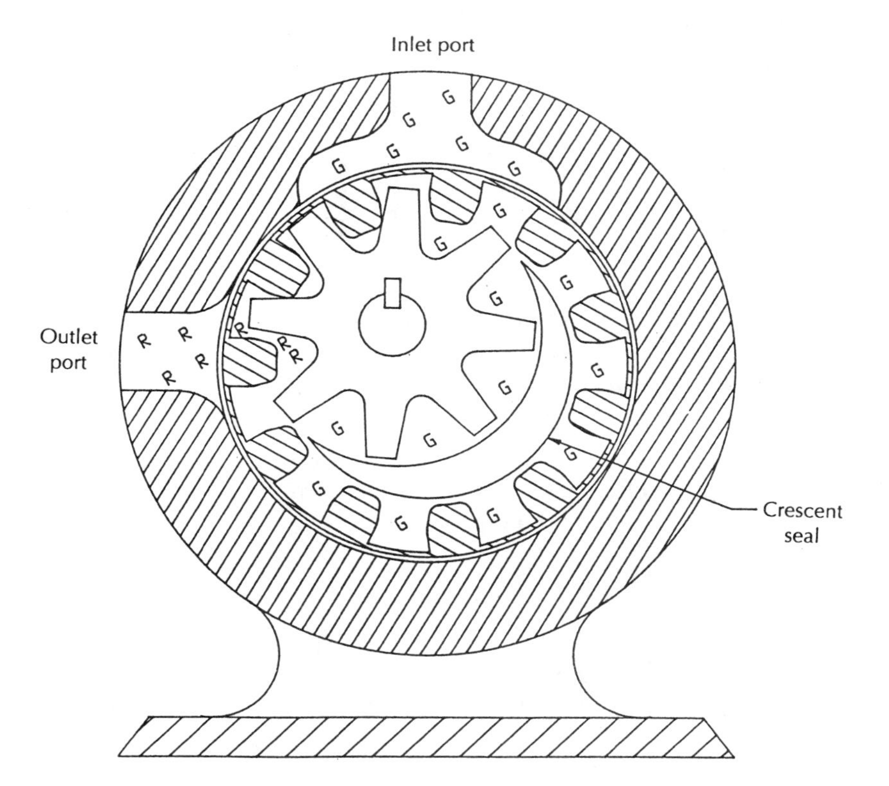

By incorporating advanced gear design features into the internal gear pump a more efficient pumping mechanism is achieved. The **gerotor pump** described in Figure 4-11 incorporates such design features (see also Figure 4-12).

FIGURE 4-11

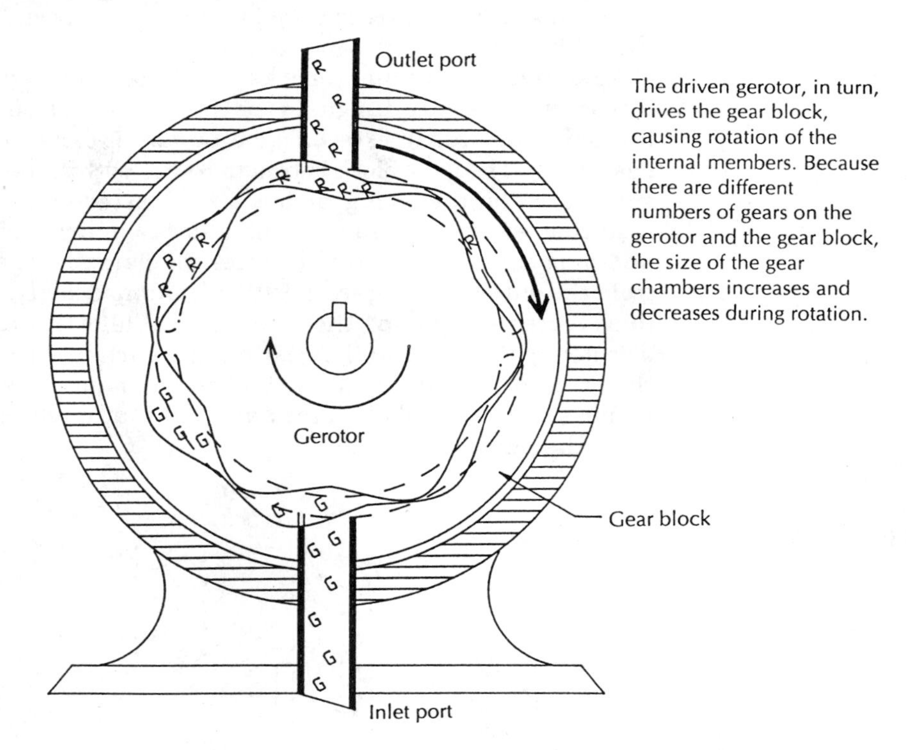

4.3.1.1 Gear pump displacement Displacement in the gear pump is fixed. It is a function of the physical size of the gear tooth chamber and the number of chambers used.

Problem 4-9

An external gear pump has thirty teeth on each gear, and each gear tooth chamber contains 0.5 cu. in. {*81.9 cm³*} of space. What is the displacement of the pump?

DIS (in.³/rev.) {cm³/rev.} = ________

Gear pumps range from low to high displacement. In general, internal gear pumps have low to medium displacement. External gear and screw

FIGURE 4-12

Gerotor or gear-in-gear pump. *(Photograph by the author.)*

pumps have medium displacement, and maximum displacement in a gear design is available in the lobe pump.

4.3.1.2 Gear pump characteristics While operating, gear pumps are generally noisy compared to other types of pumps. This characteristic may be attributed to the impact when gear teeth mesh. However, screw-type pumps are known to be very quiet operating mechanisms. The volumetric efficiency of gear pumps ranges between 70% and 80%. Increases in system pressure and pump age will exaggerate the noise created by the pump and will result in lower volumetric efficiency. Gear pumps are subject to wear on the drive bearing and shaft from side loading (see Figure 4-13). External gear pumps are suitable for use in circuitry operating at a maximum pressure of somewhere between 2,000 and 4,000 psi {*13,790 and 27,580 kPa*}. Internal gear pumps are typically rated somewhat lower, at a maximum level between 1,500 and 2,000 psi {*10,343 and 13,790 kPa*}.

Gear-type pumping mechanisms are popular because they generally cost less than other types of pumps. High-pressure operation will cause more wear on the gear pump bearings; and as the pump ages, sloppy operation will result. Still, gear pumps are often used, especially in applications where contamination or dirty oil is prevalent. Greater clearance between mating parts in the gear pump (compared to other pump types) accounts for this characteristic.

FIGURE 4-13

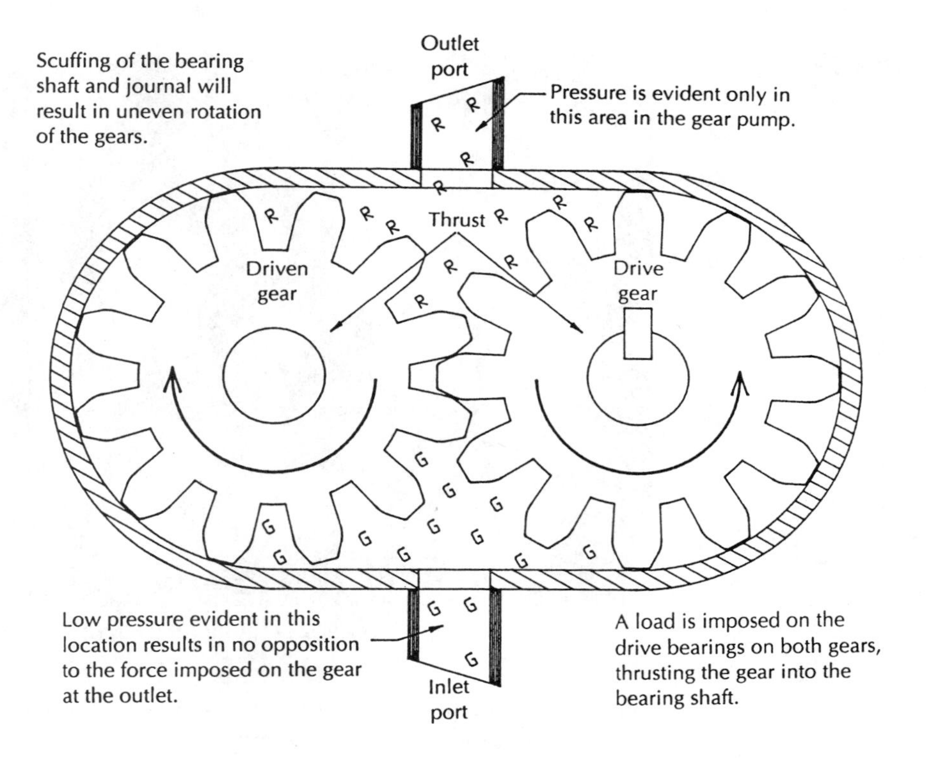

4.3.2 Vane Pumps

In the **unbalanced vane pump,** a driven rotor is located inside the housing. The rotor has slots machined around the outside edge. Into each slot is fitted a spline of metal called a vane. The vanes are sized to allow free movement inside the slots. Centrifugal force created through rapid rotation of the rotor forces the vanes to be thrown against the pump housing (see Figures 4-14 and 4-15).

The basic vane pump design may be varied to achieve design and operational characteristics not capable with gear and other types of pumps. One such adaptation will allow variation in the displacement of the vane pump. By altering the degree of eccentricity between the rotor and the reaction ring chamber housing, the difference between the size of the pumping chambers at inlet and outlet locations will vary. As the eccentricity increases, displacement becomes greater. As the rotor and reaction ring become concentric, the displacement of the pump decreases. Figure 4-16 describes the special arrangements needed in a **variable-displacement vane pump** (see also Figure 4-17).

When the centerline of the reaction ring is concentric with the centerline of the rotor, no flow occurs. Since the chambers would never enlarge in this situation, no vacuum condition would exist to charge the pump.

FIGURE 4-14

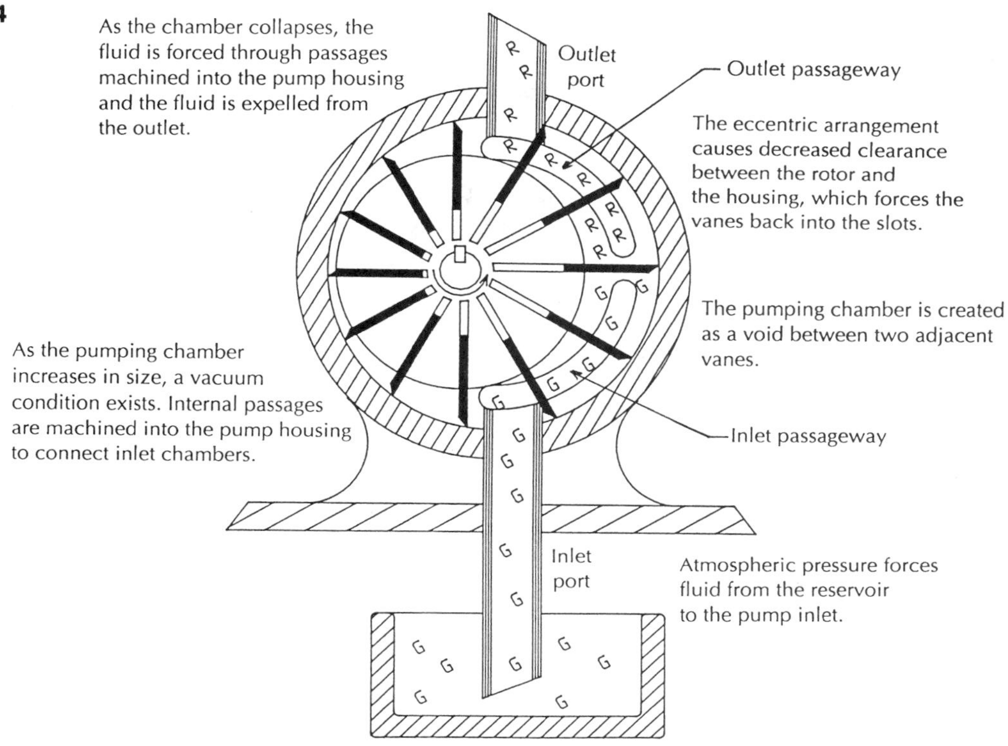

FIGURE 4-15

Vane pump. *(Photograph by the author.)*

FIGURE 4-16

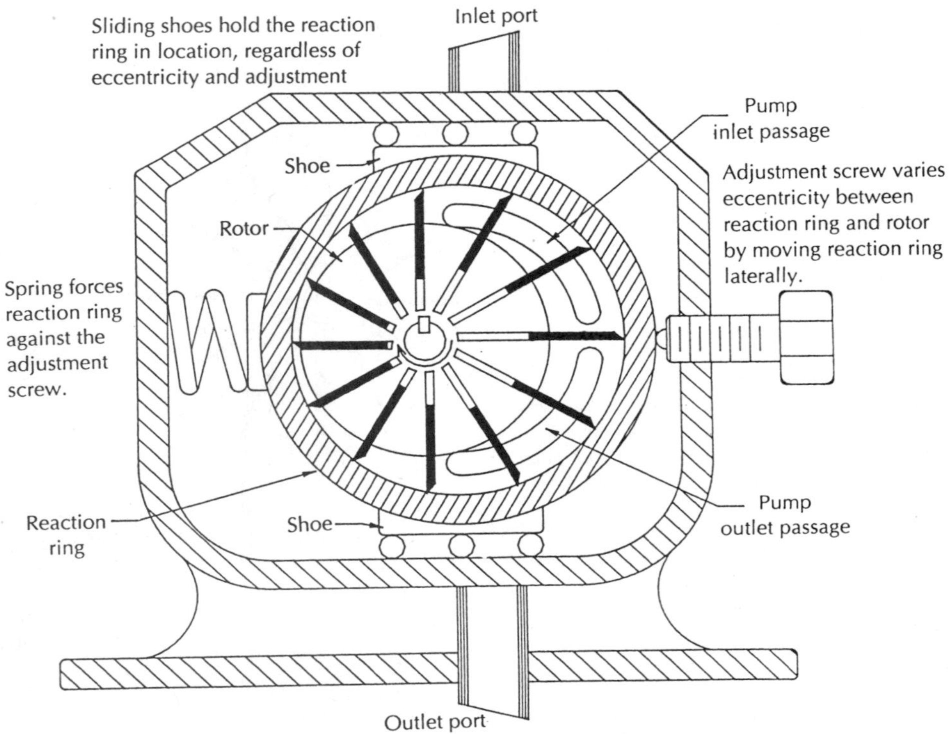

Assignment: Color-code the drawing.

FIGURE 4-17

Variable displacement vane pump. *(Photograph by the author.)*

Side loading is evident in the vane pumps illustrated in Figures 4-14 and 4-16. However, adaptations of the vane pump design may be made to eliminate side loading. In the **balanced vane pump**, the circular rotor is located concentrically within an elliptical pump housing (see Figures 4-18 and 4-19). Two inlets are located 180 degrees apart as the pump chambers increase in size toward the major axis of the ellipse. Also, two outlets are located 180 degrees apart as pump chambers decrease in size toward the minor axis of the ellipse. In both cases, the inlets and outlets are connected inside the pump casing to lead to one external inlet and one external outlet port each.

FIGURE 4-18

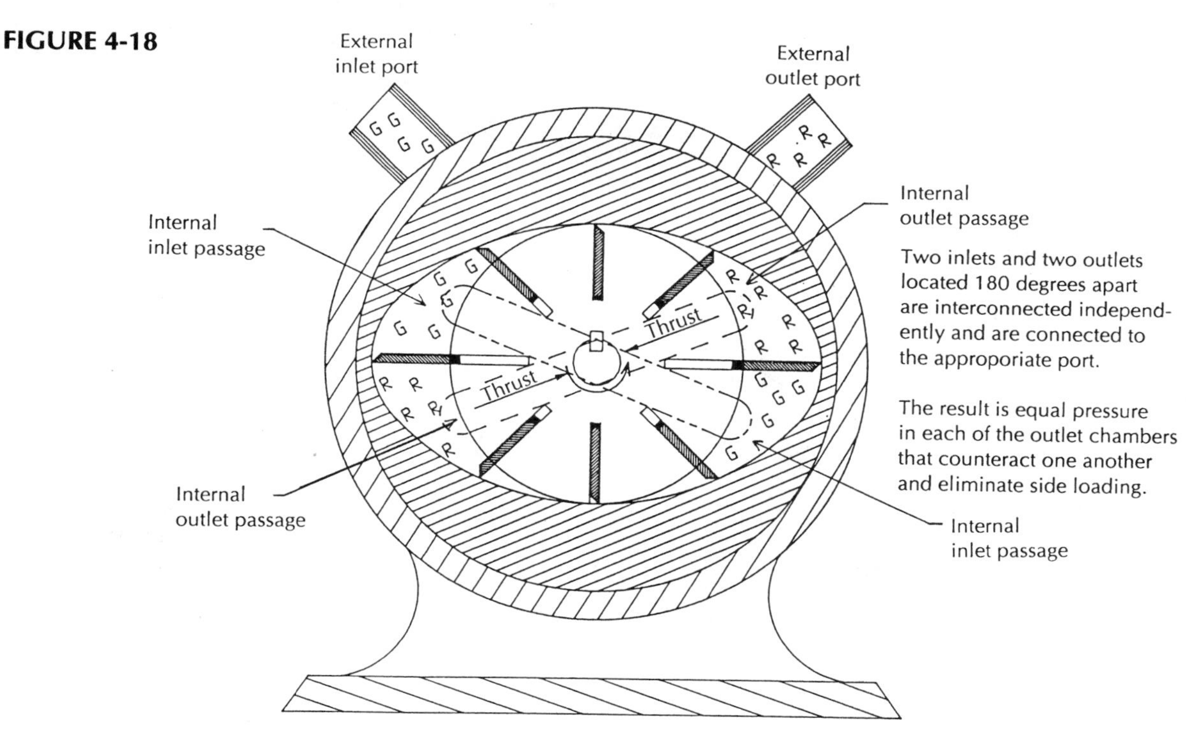

4.3.2.1 Vane pump displacement The volume contained within the pump chamber when the vane is at its apex of extension multiplied by the number of vane chambers in the pump is the theoretical displacement of the vane pump (see Figure 4-20). Of course, this is only true if the vanes, when retracted in their slots, are in very close proximity to the pump chamber or reaction ring. Otherwise, the difference between the maximum chamber size and the minimum chamber size multiplied by the number of vane chambers in the pump is the theoretical displacement.

FIGURE 4-19

Balanced vane pump. *(Photograph by the author.)*

FIGURE 4-20

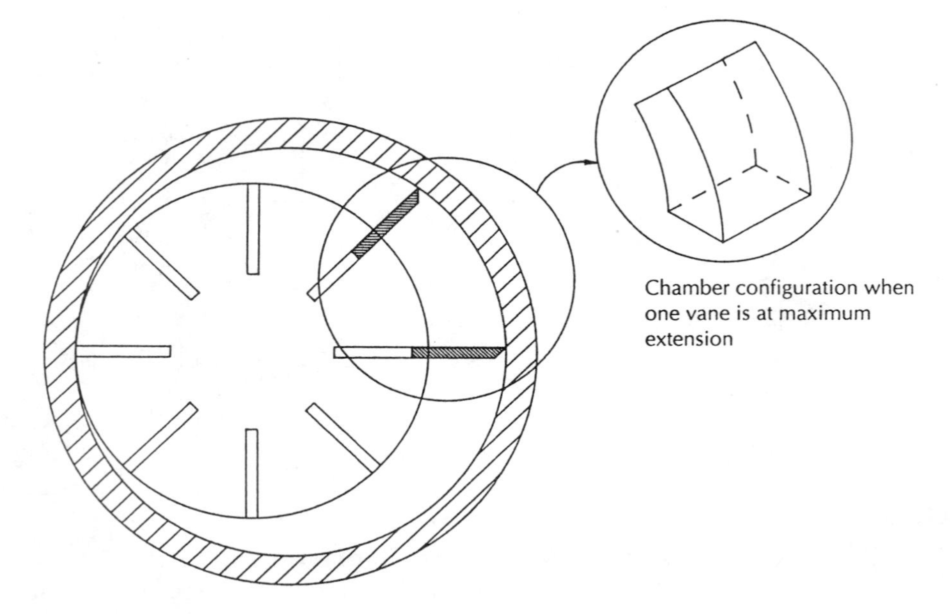

Problem 4-10

Calculate the theoretical displacement of the vane pump having eight vanes if maximum pump chamber size yields 0.90 cu. in. (2.3 cm^3).

DIS (cu. in./rev.) = ________

DIS {cm^3/rev.} = ________

Vane pumps typically vary in displacement from low to medium displacement. However, many vane pumps have no arrangement to extend the vanes other than centrifugal force and, therefore, may not operate at low drive speed. The necessity to drive these pumps at rather high speeds (up to 1,800 rpm) results in pump delivery typically rated in the medium to high range. Furthermore, the ability to vary the displacement of some vane pump designs allows for flexibility in circuit design not easily achieved when using the gear-type pumps.

4.3.2.2 Vane pump characteristics Vane pumps are generally noisy upon start-up. This characteristic may be caused by brief **cavitation** (vacuum voids in the fluid) and the physical impact of the vanes upon the chamber housing. Normally, cavitation decreases momentarily after start-up and a cushion of oil reduces the noise created by impacting vanes. However, the pressure created at the vane tips may tend to force the vanes back into the rotor slots. To overcome this characteristic, many vane pumps channel a supply of pressurized fluid into the rotor slots behind the vanes (see Figure 4-21).

FIGURE 4-21

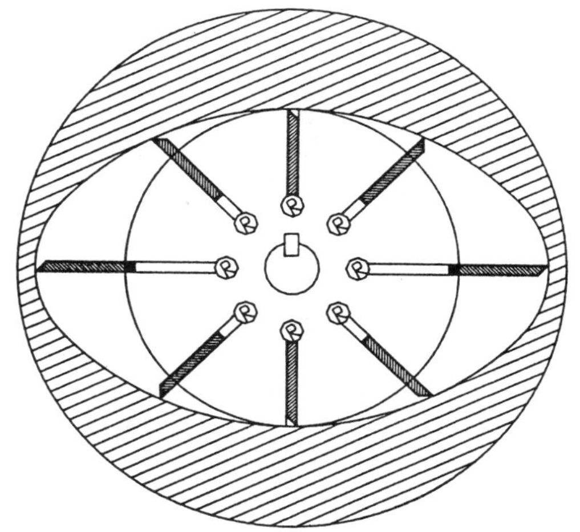

Pressurized fluid taken from the outlet and led through passages in the end plate of the pump is channeled behind the vanes to asssist vane tension and reduce vane impact on retraction.

Vane pumps are not highly tolerant of dirty or contaminated fluid. Abrasion caused by dirty fluid corrodes the vane tips and causes leakage of the fluid from the pressurized chamber back to the low-pressure inlet. This will greatly reduce the volumetric efficiency of the pump. Wear in the vane pump is also typically encountered on the chamber or reaction ring wall because of vane tracking. The vanes themselves are normally much longer than the clearance between the rotor and the chamber. Therefore, the vane pump will self-adjust for this normal wear by simply allowing the vane to extend a bit further out of the rotor slot. It should be noted that this normal wear will slightly increase both the chamber size and the theoretical displacement of the vane pump. However, leakage and slippage associated with increased age of the pump will, in general, more than compensate for this increase. Vane pumps typically have volumetric efficiencies in the range of 80% to 90%.

Vane pumps may be used in both high- and low-pressure operations. Typically, the unbalanced vane designs are used in low-pressure operations up to 1,000 psi {*6,895 kPa*}, while balanced vane and special vane design pumps may be used in circuits requiring 2,000 psi {*13,790 kPa*} or more.

4.3.3 Piston Pumps

Early pumping mechanisms, such as the one described in Bramah's jack, used a piston reciprocating in a bore. Although gear and vane pumps are the more modern designs, it would be incorrect to consider the **piston pump** an antiquated design. Piston pumps are extremely useful today, especially in high-pressure operations and in high-efficiency circuitry.

The major problem encountered in designing an efficient piston pump is to allow for high displacement while assuring streamlined flow to and from the pump. The single-piston pump, described earlier, will supply fluid to the output device in pulses, similar to the heart pumping blood. The resulting erratic movement of the output device might not be detrimental in a floor jack used to raise an automobile. However, erratic output device movement could be disastrous in a hydraulic circuit used to feed a piece of metal through a milling machine.

The piston in a piston pump typically advances in its bore faster during the middle of the stroke than at the beginning or end. A graph of the supply from one cylinder of a piston pump over the course of the elapsed time of the stroke would appear similar to part of a sine wave curve. By combining multiple cylinders and beginning their strokes out of phase (at different times), a rough approximation of streamlined flow may be achieved (see Figure 4-22).

To achieve this type of output, a piston pump may rely on a mechanical device such as a crankshaft to cause phased extension and retraction of the pistons. The crankshaft arrangement also allows for one drive device to cause pumping in all cylinders simultaneously (see Figure 4-23).

Further advancements in piston pump design have resulted in devices that afford greater streamlined flow by eliminating the necessity to rap-

FIGURE 4-22

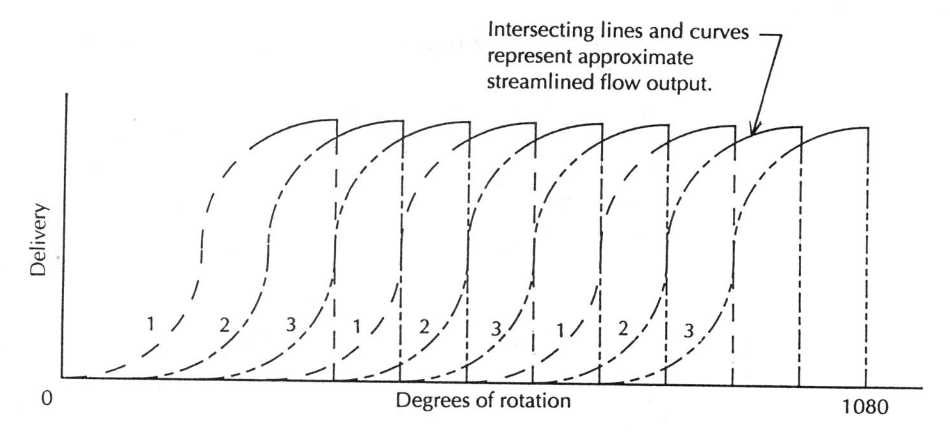

FIGURE 4-23

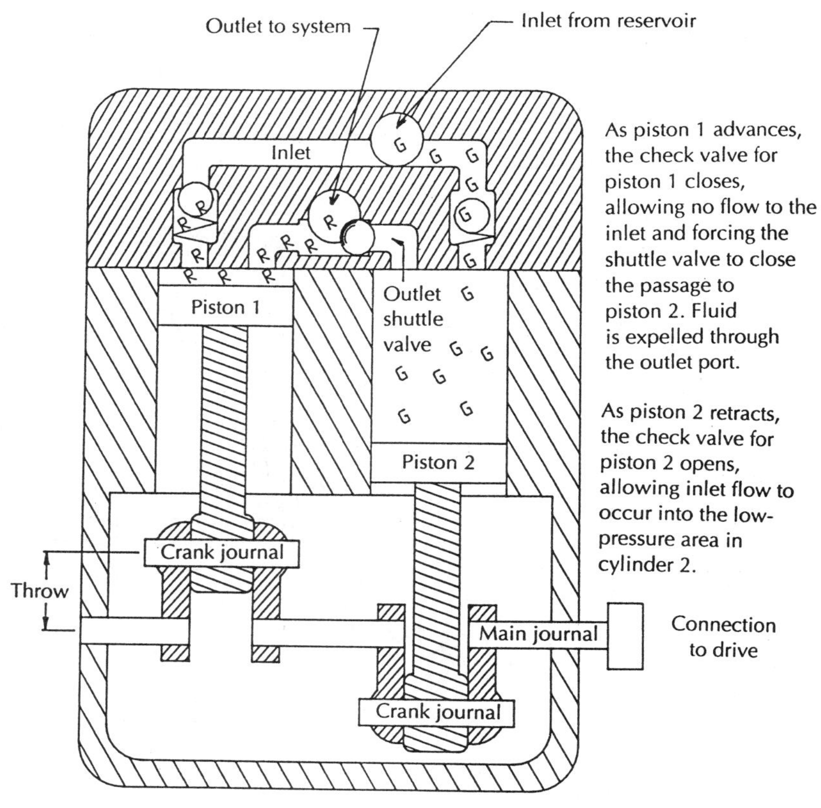

idly change piston direction at the end of each stroke. To this end, crankshaft designs, although typical in air compressors, are unusual in hydraulic application.

4.3.3.1 Radial piston pump The radial piston pump is similar in appearance to the unbalanced vane pump (see Figures 4-24 and 4-25). In the **radial piston pump**, an odd number of bored holes are equally spaced in a cylindrical chamber. A piston is fitted into each hole and is free to slide into the bore until a shoe on the end of the piston meets the perimeter of the cylinder chamber. The chamber is located eccentrically inside a pump housing and held by a stationary hub that contains inlet and outlet ports. The chamber is then rotated by the driving device, and centrifugal force throws the pistons away from the center hub until their shoes contact the inside of the pump housing. Because of the eccentric arrangement between the cylinder chamber and the inside of the pump housing, constantly varying degrees of clearance are evident throughout the rotation. Through 180 degrees of rotation, increasing clearance occurs; and the pistons are moving away from the hub. This allows for expansion of volume in the cylinder and charging through the inlet port. Through the other 180 degrees of rotation, the clearance between the cylinder chamber and the inside of the pump housing decreases. This causes the pistons to move back toward the hub and forces the fluid in the cylinders to be discharged through the outlet port.

FIGURE 4-24

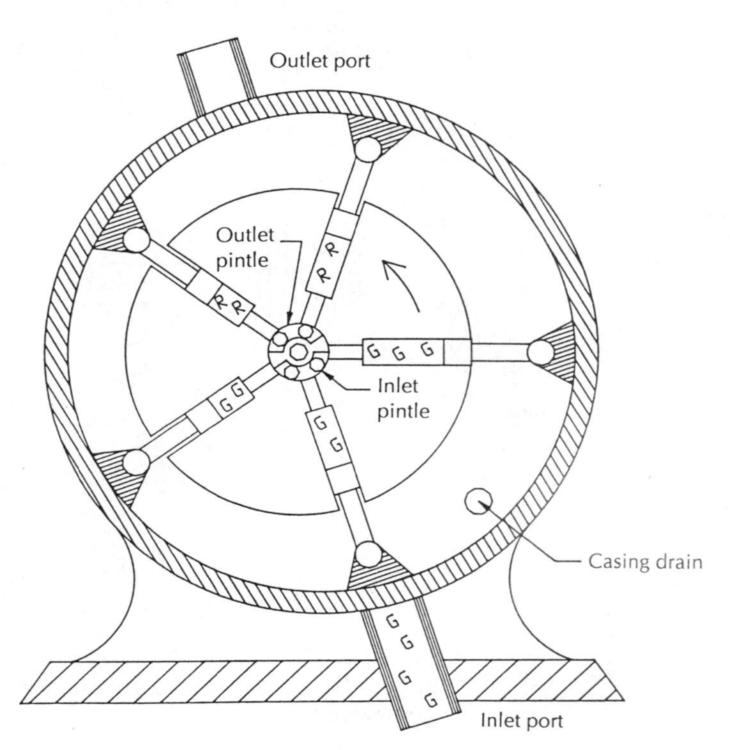

FIGURE 4-25

Radial piston pump. *(Courtesy Vickers, Inc.)*

Each piston has a small hole drilled through it to allow a small seepage of oil to lubricate the interface between the piston shoes and the inside of the pump housing. This necessitates the use of an external drain for the pump housing to ensure that excessive buildup of oil behind the piston will not cause back pressure in the pump. However, side loading is evident in the radial piston pump because of the existence of pressure on the outlet side and no pressure opposition at the inlet.

4.3.3.2 Axial piston pump Side loading is eliminated in the **axial piston pump** design (see Figure 4-26). In the **aligned-axis axial piston pump** (see Figure 4-27), the cylinder chamber rotates inside the pump housing on a roller bearing sleeve. Unlike the radial piston pump that has cylinders located around the periphery of the cylinder chamber, the axial piston pump has an odd number of cylinders bored through the face to the base of the cylinder chamber. The pistons are assembled into a **reaction yoke** that corresponds to the cylinder arrangement. A ball machined onto one end of each piston is fitted through the hole in the reaction yoke and held into this position by a mating socket in the **shoe**. The piston assembly is inserted into the base of the cylinder chamber, and the pistons are allowed to slide back and forth in unison within the

FIGURE 4-26

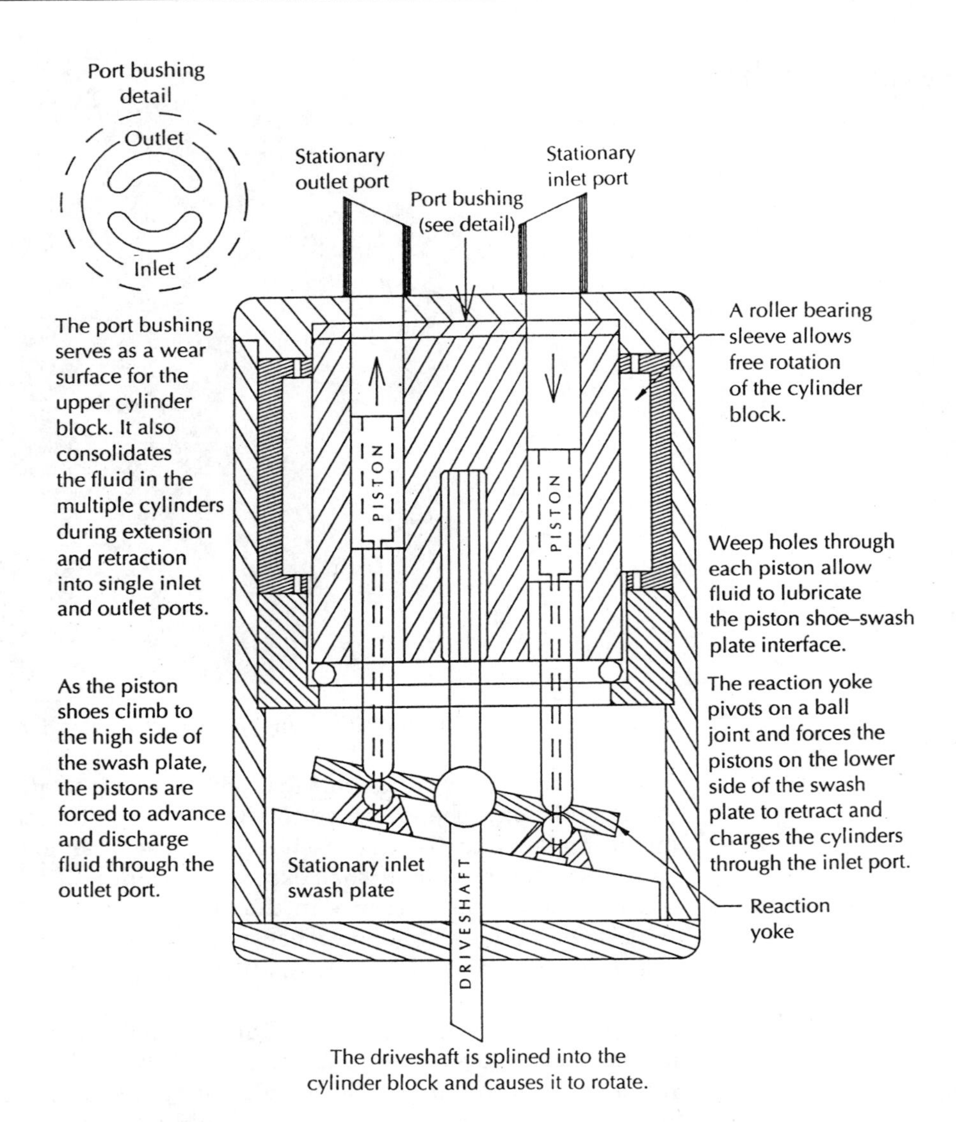

cylinders. The piston shoes rest on a beveled cylindrical block called a **swash plate** that is held in position by the base cap. The bevel results in varying clearance between the swash plate and the cylinder chamber. Through 180 degrees of rotation, as the clearance decreases, the pistons are forced toward the face of the cylinder chamber. At the same time, the reaction yoke forces the pistons on the opposite side of the pump to move away from the face of the cylinder chamber. Porting, in the aligned axial piston pump, is made through the face end cap. Two semicircular ring slots are machined into a port plate that is held stationary above the face

FIGURE 4-27

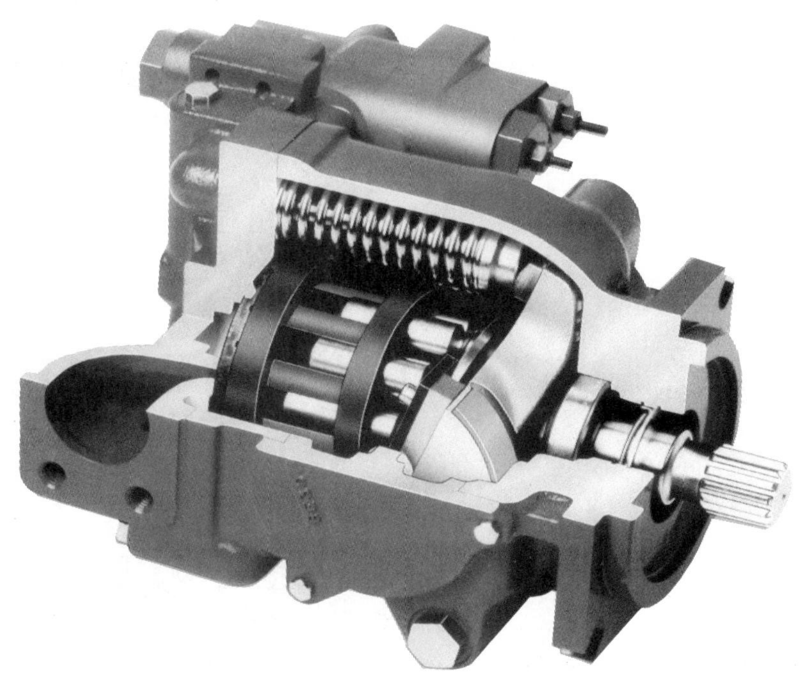

Aligned-axis axial piston pump. *(Courtesy Vickers, Inc. From Vockroth,* Industrial Hydraulics, *Albany, N.Y.: Delmar, 1994, p. 127.)*

of the cylinder chamber. The slot located above those pistons that are retracting in their bores is connected through the end cap to the external inlet port. Similarly, the slot located above those pistons that are advancing in their bores is connected to the external outlet port.

Sliding friction is evident in the aligned axial piston pump at the interface between the piston shoes and the swash plate. Therefore, a small hole is machined through each piston and shoe to allow lubrication at this interface. An external drain in the base chamber of the pump is used to relieve any possible back pressure that may be evident through an accumulation of fluid used for lubrication or air trapped under the pistons.

Another variation on the design is the **offset-axis axial piston pump** (see Figure 4-28). In this design, the basic arrangement of pistons located in a rotating cylinder chamber is maintained. However, the driveshaft, while still splined into the cylinder block on one end, is equipped with a segment having ball–socket joints (or universal joints) on both ends. With this arrangement, the pumping section is separated from the drive section. The offset necessary to cause reciprocation of pistons in their bores is attained through an angled relationship between the pumping and drive sections. If no relative angle is evident, then no reciprocation of pistons

FIGURE 4-28

Offset-axis axial piston pump. *(Photograph by the author.)*

(and no pumping) will occur. Increasing the relative angle between the two sections increases the reciprocating distance (throw) of each piston to a maximum pumping capacity, limited by the physical clearances necessary to house the ball–socket driveshaft segment (typically 30 degrees).

4.3.3.3 Piston pump displacement Displacement in the piston pump is a function of the cross-sectional area of one cylinder, the piston stroke, and the number of cylinders in the pump. In all piston pump designs, the cross-sectional area of the cylinder and the number of cylinders may be easily measured. However, the stroke of the piston may be more difficult to calculate.

In the crankshaft design, the stroke is determined by the **throw** of the crankshaft. The throw may be determined by measuring the distance between the center of the main bearing journal at one end of the crankshaft and the center of one crank journal. This measurement corresponds to one-half of the throw. To calculate the stroke of the cylinder, multiply the measured distance by two. In some cases, it may be possible to remove the face cap and simply measure the stroke from top-dead-center to bottom-dead-center by rotating the drive shaft.

In the radial piston pump, the piston stroke is often the distance between the center of the cylinder chamber and the center of the pump housing. More accurately, the stroke is the difference between the maximum and minimum clearances from the cylinder chamber to the inside of the pump housing. In some cases, these distances may be varied, resulting in a **variable-displacement radial piston pump**. When designing hydraulic circuits requiring accurate and variable actuator speed, it is often advisable to select a variable-displacement pump to compensate for decreased supply with age and pressure or increased supply rate as fluid temperature rises. Also, various applications may require varying delivery; and the variable-displacement pump will allow for adjustment that will match the requirement and potentially save power (horsepower) in that situation. In general, it is best to specify a variable-displacement pump that would supply the desirable output at mid-range.

Piston stroke in the aligned axial piston pump may be calculated by determining the difference between the maximum and minimum thicknesses of the swash plate bevel. Variability of displacement may be achieved by altering the angle of the bevel on the swash plate. This may be accomplished by placing the swash plate on a center pinion and varying its pitch through opposing adjustment screws (see Figure 4-29).

FIGURE 4-29

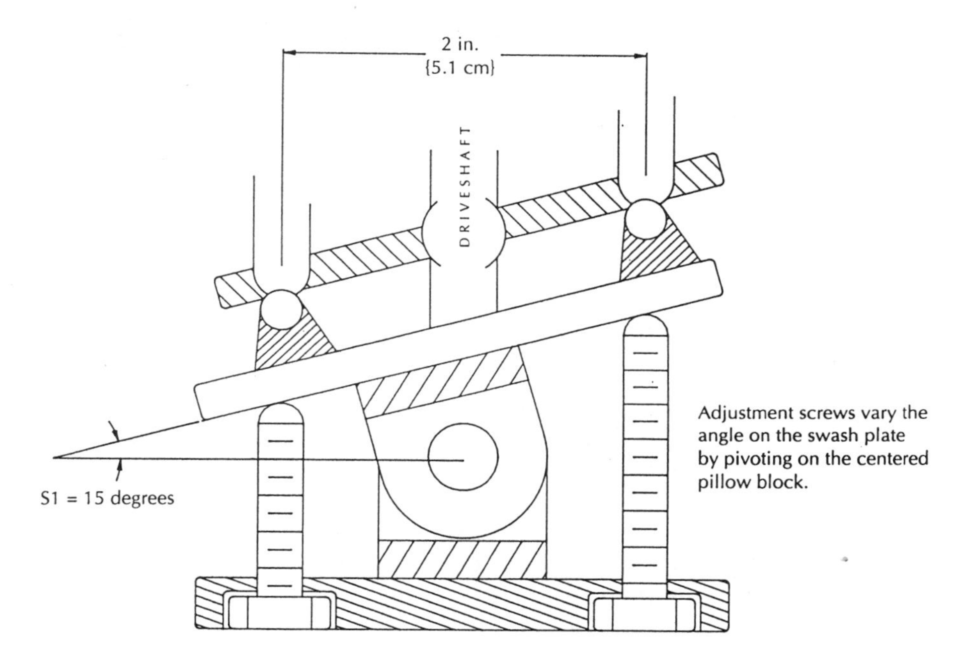

In most cases, the angle that may be achieved varies between 0 and 20 degrees. Within the range of pitch on the variable swash plate, the various piston strokes may be calculated using the diameter of the piston circle and the angle desired. For example, Figure 4-29 describes an axial piston pump having a 2-in. {*5.1-cm*} diameter piston circle. If the angle on the swash plate were set to 15 degrees, the piston stroke may be calculated as follows:

St1 (in.) {*cm*} = Tangent 15 degrees * 2 (in.) {*5.1 cm*}

Problem 4-11

If the piston circle in Figure 4-29 was 4 in. {*10.2 cm*} and the angle on the variable swash plate was set at 11 degrees, what would the resulting stroke be?

St1 (in.) {*cm*} = ____________

Piston pump displacement generally varies from low to medium range. However, the ability of piston pumps to operate at extremely high speeds (as high as 3,600 rpm) enables a low-displacement pump to deliver high volumes.

4.3.3.4 Piston pump characteristics Piston pumps are generally quiet in operation. They are also the most efficient pump design, having volumetric efficiencies as high as 95%.

The most attractive characteristics of piston pumps are their abilities to operate at high speed and under high pressure. Piston pumps typically operate in a pressure range of 1,500 to 3,000 psi {*10,342 to 20,684 kPa*}. Piston pumps have been recorded as operating at 5,000 rpm under pressures as high as 10,000 psi {*68,948 kPa*}. This combination allows for quick response and smaller hydraulic output devices (cylinders or motors) to manipulate high loads. This type of application is typical in aircraft and missile guidance controls.

The efficient operation of a piston pump is greatly affected by the cleanliness of the hydraulic fluid. Contaminated fluid will contain suspended solid particles that may clog the small lubrication passages in the piston pump. The most drastic result of the loss of lubrication would be the complete melting of the piston shoe through frictional heat generation. When this occurs, the shoe will often weld itself to the swash plate and the pump will lock up. Proper filtration and monitoring of the hydraulic fluid's cleanliness is critical in circuits using piston pumps. Sophisticated piston-type pumps are relatively expensive compared to gear and vane types; therefore, their usage in some equipment is limited.

4.4 HYDRAULIC POWER SUPPLY COMPONENTS

The simple hydraulic power supply includes not only the pump, driving device, and hydraulic fluid but also a well-designed reservoir, fluid conductors, filters, and a device to limit maximum pressure in the circuit. Although many prepackaged power supplies will include more sophisticated devices that monitor and adjust pressure, flow, and fluid condition, these components represent the basic hydraulic power supply.

4.4.1 The Fluid Reservoir

The hydraulic fluid **reservoir** is designed to contain and condition the fluid and to provide a mounting surface for other power supply components (see Figures 4-30 and 4-31). In a well-designed hydraulic circuit, the fluid should be circulating throughout the circuit only one-third to one-half of the time. The remaining quantity of fluid should be held in reserve in the reservoir to allow reconditioning (particulate separation and cooling) in a static environment. Therefore, as a container, the reservoir should hold a volume equal to two to three times the capacity that would be delivered by the pump in one minute. If the pump is delivering

FIGURE 4-30

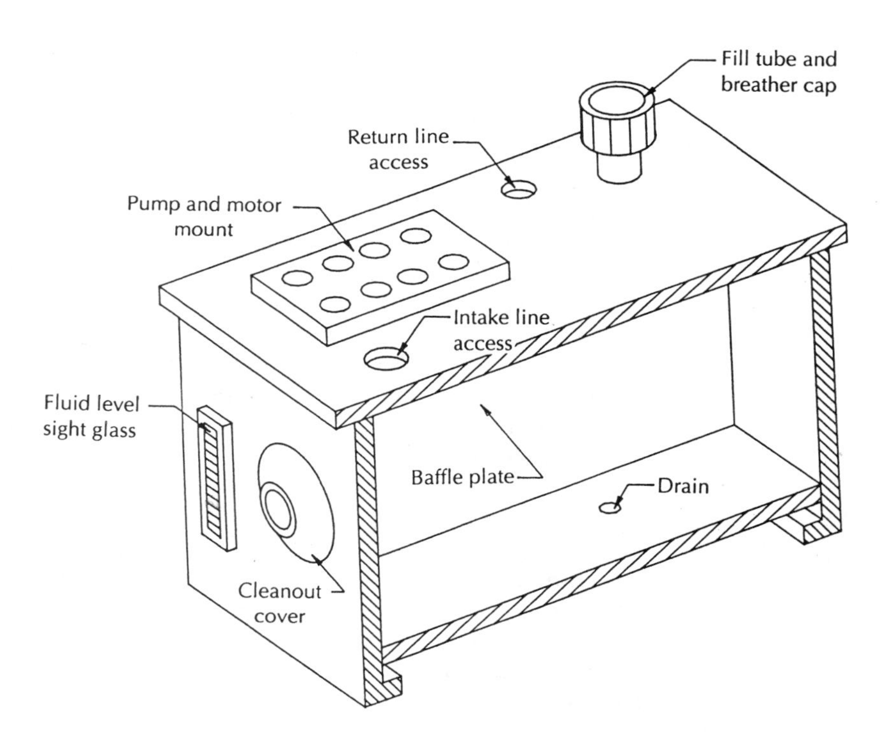

FIGURE 4-31

Hydraulic reservoir. *(Photograph by the author.)*

10 gal./min. {*37.9 l/min.*}, the reservoir should hold 20 to 30 gal. {*75.7 to 113.6 l*} of fluid. The extra fluid ensures that the supply held in the reservoir will have sufficient time to allow contaminants to settle from the fluid, and some of the heat generated during circuit operation will be dissipated through the steel reservoir casing to the atmosphere. Although hydraulic reservoirs come in all shapes and sizes, the more popular designs are basically rectangular cases. The major reason for rectangular designs is the necessity to mount a pump and an electric motor driving device in tandem on top of the reservoir.

Conditioning of the fluid actually begins before the fluid enters the reservoir at the **fill tube**. The fill tube contains a conical-shaped screen **strainer**, rated at 60 mesh or 60 openings per linear in. {*24 openings per linear cm*}, that separates coarse contaminants from supply fluid as it enters the system. The fill tube also acts as the passageway to allow air to be dispersed over the top of the fluid in the reservoir. It should, therefore, be terminated above the fluid level inside the reservoir and should be covered externally with a cap that contains air passageways and an **air filter** to remove fine airborne contaminants. Further conditioning of the fluid occurs within the reservoir through the use of a **baffle**. The baffle is a steel plate that separates the reservoir internally into two compartments and extends about two-thirds the height of the reservoir casing. In a correctly designed reservoir, the inlet to the pump is located in one of these chambers, while all return lines from the circuit and the fill tube are located in the other. In the return chamber, the turbulent return fluid is allowed to settle and release coarse contaminants to the bottom. When sufficient fluid accumulates in the return chamber, the fluid will flow over the baffle in a more streamlined manner and provide a relatively calm and clean supply to the pump inlet. To maintain proper fluid supply, the bottom of the reservoir is typically dished and a removable **cleanout cover** provided. A separate **drain plug** may also be provided, and a **sight level indicator** is included on one end cap of the reservoir to monitor supply quantity.

Another conditioning function of the reservoir is to act as a heat exchanger. The reservoir performs this function using two methods: **conductive** and **convective heat transferral**. The steel used in the casing of the reservoir has a higher specific heat than the hydraulic fluid. Therefore, fluid coming into contact with the casing will conduct the heat from the fluid to the steel material. The heat is then dissipated to the atmosphere. The major design consideration to be made, therefore, is to expose as much of the reservoir casing to the atmosphere as possible. For this reason, reservoirs are typically set on legs that raise the bottom surface for exposure. This may be critical because, oftentimes, the largest steel surface in the reservoir is the bottom. Other heat dissipation is evident in the reservoir through air exchange. The fill tube not only allows passage of air into the reservoir but also allows heated air inside the reservoir to be exhausted. The size (cross-sectional area) of the fill tube should increase as the pump delivery increases to accommodate this air exchange.

Finally, the reservoir houses and holds the power supply components. Filters and lines, baffles, and heat exchanger devices are often housed within the reservoir. However, the bulk of the power supply components are located and mounted on the outside of the top of the reservoir. A well-designed reservoir must allow for separation of inlet and return line porting according to baffle arrangement through the top as well as maximum isolation of heat-producing devices such as pumps, motors, and pressure controls from the fill tube and other vents. The components on top of the reservoir should be arranged to provide for easy, short, and efficient plumbing connection of components and easy mounting and removal of each component separately for maintenance purposes.

4.4.1.1 Filtration No matter how carefully hydraulic circuits are designed, they may be destroyed by particles so small that they may not be seen with the unaided eye. Because most industrial hydraulic circuits are not totally enclosed, contamination of the hydraulic fluid may begin with airborne particles taken in through the breather cap on the fill tube or through other vents located on the reservoir. Some contamination may be picked up by the fluid itself from the containers used for distribution. In any case, hydraulic fluid should be clean and free of particles before introduction into the system. According to the degree of accuracy and control needed within the hydraulic circuit, varying test and cleansing procedures will be incorporated before using a hydraulic fluid. Consider the cleanliness necessary for the hydraulic fluid in a missile guidance system versus that of the hydraulic fluid used in the automotive floor jack.

Fluid contamination will also occur within the hydraulic circuit through abrasion and wearing of mating parts in components, oxidation of these parts, and changes in the chemical and physical nature of the hydraulic fluid under heat and pressure. Two methods are used to remove these types of contamination from the fluid: **exclusion** and **absorption**.

In exclusion, a finely woven screen is used to stop the movement of coarse contaminants while allowing the liquid to continue to flow. Devices that incorporate the exclusion principle of filtration are known as **strainers** (see Figure 4-32) and are rated by **mesh size**. The mesh size relates to the number of openings in the screen within a 1-in. {*cm*} distance. Thus, a 60 mesh strainer would have 60 openings in 1 linear in. {*24 openings in 1 linear cm*} or 3,600 openings per sq. in. {*576 openings per cm*^{2}}.

FIGURE 4-32

Strainers and filters are currently available to remove contamination down to a size of 2 microns. *(Courtesy Parker-HanniFin, INC.)*

Those filtration devices using absorption principles are called **filters**. They are used to remove fine contaminants from the fluid stream. Filters are made from porous media such as cloth, paper, sintered metals and ceramics, and other materials such as diatomaceous earth. The fine contaminants are carried along within the fluid at nearly the same velocity. As the fluid stream flows through the filtration material, the solid contaminants are not able to change shape as the liquid fluid does. The result is that the contaminants impact upon the filter material and lose velocity or embed themselves in the filter media. After repeated impacts, even if they do not embed themselves, the contaminants lose sufficient velocity to be lost from the fluid stream and are deposited in pockets within the filter material. In either case, the fine contaminants are removed from the fluid stream. Filters are rated according to the **micron** or **micrometer size** of fine contaminants efficiently removed by the porous medium. *A micron is one-millionth of a meter,* or about 0.000039 in. Thus, a filter rated at 25 microns would efficiently remove contaminants down to a size of *0.000025 m* or 0.00098 in.

Filters should not be used in areas having low velocity of fluid flow. The impaction necessary to cause the absorption mechanism to work will be greatly reduced in such areas. Furthermore, the high density of the materials used in filters causes reduction of velocity in the fluid stream. This cannot be tolerated in low-velocity areas of the circuit. Strainers are, therefore, used in the fill tube and the inlet to the pump. Filters are most often found in exhaust lines that return fluid to the reservoir where high velocity and low pressure exists. Filters may also be incorporated into pressure lines. However, filters located in pressure lines may cause unnecessary and unwarranted back pressure in the circuit at the filter location. In general, 60-mesh {*24-mesh*} strainers and 25-micron filters are used in industrial hydraulic circuits, although finer filtration may be necessary in some instances. Modern-day **tribology** studies are typically performed to determine the type and size of contaminants existing in an operating hydraulic circuit. This information, in combination with known clearances among mating component parts, will provide a road map of the ratings and types of filtration that are appropriate using manufacturers' and ISO standards. Multiple filters and strainers are often incorporated into hydraulic circuits, especially when these circuits are used in dirty environments, such as earthmoving machinery, to protect components that require very precise mating parts or where high-velocity movement of the power output device is needed.

Simple linear circuits such as those used in log splitters or small hydraulic presses will generally incorporate one inlet strainer and one return line filter. More complicated circuitry, having multiple output devices, will generally add one return line filter for each loop or output device. Earthmoving equipment may also include in-line filters with easily removable elements. These filter packs are typically monitored by a combination pressure gauge that reads pressure drops from the filter inlet to the outlet. When the pressure drop becomes too high, the elements are removed and cleaned or replaced. In general, filters and strainers are also rated for delivery. As a general rule of thumb, a filter or strainer should be able to pass two to three times the delivery of the pump with no more than a 10% pressure drop. Figure 4-33 describes the setup of a test stand circuit used to evaluate the adequacy of the strainer to be used in a circuit having a 20 gal./min. delivery pump used to move an output load of 3,000 lbs. {*13.3 kN*} with a 3-in. {*7.6-cm*} diameter cylinder.

FIGURE 4-33

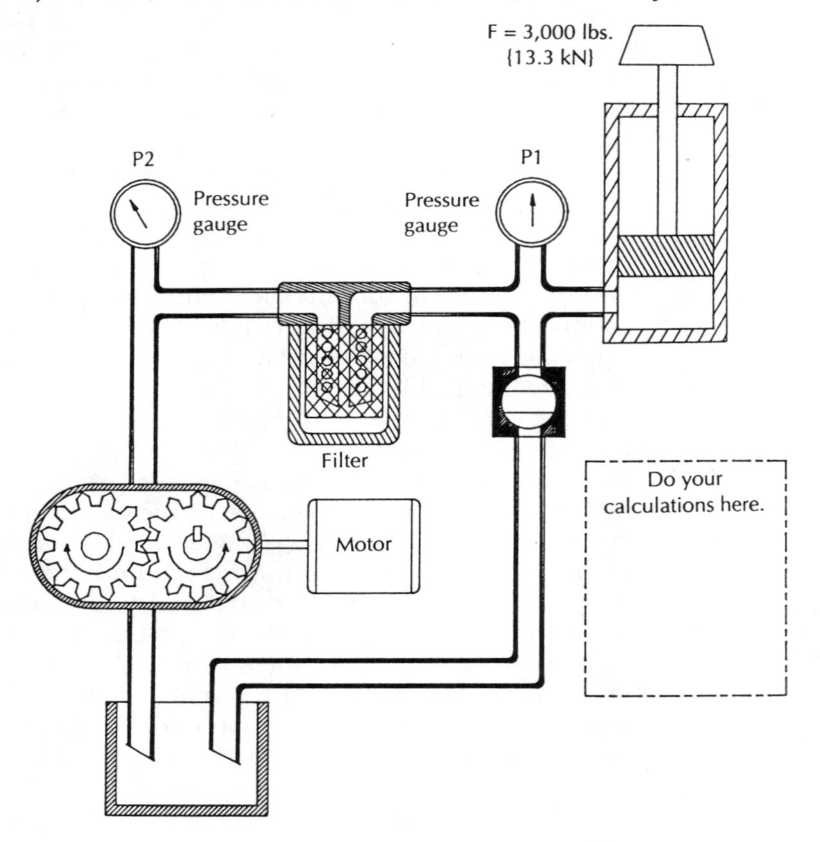

Problem 4-13

What will be the delivery range of the pump to evaluate the filter for a 20-gal./min. {75.7-l/min.} application?

Maximum delivery (gpm) {lpm} = ________

Minimum delivery (gpm) {lpm} = ________

Problem 4-14 What pressure will develop during movement of the 3,000-lb. {13.3-kN} load at P1?

P1 (psi) {kPa} = ________

Problem 4-15 What is the maximum allowable pressure reading at P2 for the filter to meet specifications?

P2 (psi) {kPa} = ________

An excessive pressure drop across a filter further reduces the efficiency of a hydraulic circuit by increasing the velocity of fluid flow. Inadequate or excessive fluid velocity within the circuit will cause turbulent (non-streamlined) flow to occur. Turbulence in a fluid stream is a function of the density and viscosity of the fluid as well as the size of the passageway through which the fluid is flowing and the rate at which the fluid is supplied (delivered). The specific conditions that relate to turbulence were extensively described by an English scientist named Osborne Reynolds. In the typical industrial hydraulic circuit, Reynolds's calculations set standard velocity limits for high- and low-pressure areas in a circuit. In the **inlet lines** and components located between the reservoir and the pump, velocities between 2 and 5 ft./sec. {*0.6 and 1.5 m/sec.*} should be maintained, while velocity in **pressure and exhaust lines** and their components should be maintained between 7 and 20 ft./sec. {*2.1 and 6.1 m/sec.*}.

4.4.1.2 Power supply plumbing Conductors that carry the fluid from the reservoir to the pump and in working lines throughout the circuit may include **pipes, tubes, hoses,** and accompanying **connectors** (see Figure 4-34). Design features and extensive descriptions of these conductors are presented in Chapter 11. However, some specifications of the fluid conductors greatly affect the operation and design of the power supply and are examined in this section.

Foremost of these considerations are the aforementioned smoothness of the inside surface and elimination of unnecessary changes of direction, along with the size of the conductor. If a conductor is too small or large in cross-sectional area, **turbulence** will occur in the fluid flow. Therefore, a conductor of the correct inside diameter must be matched with a specified velocity of fluid flow to assure streamlined flow. The following describes a primary step in devising a formula to determine the necessary cross-sectional area for the inside of a conductor:

$$\text{Area (cu. in.)} = \frac{\text{DELivery (gpm)} * 231 \text{ (cu. in./gal.)}}{\text{VELocity (ft./sec.)} * 12 \text{ (in./ft.)} * 60 \text{ (sec./min.)}}, \text{ or}$$

FIGURE 4-34

Hose, tube, and pipe are commonly used conductors for hydraulic systems. *(Photograph by the author.)*

$$Area\ \{cm^3\} = \frac{DELivery\ \{lpm\} * 1{,}000\ \{cm^3/l\}}{VELocity\ \{m/sec.\} * 100\ \{cm/m\} * 60\ \{sec./min.\}}$$

Simplified, this equation may be expressed as follows:

$$\text{Area (sq. in.)} = \frac{\text{DELivery (gpm)} * 0.3208}{\text{VELocity (ft./sec.)}}, \text{ or}$$

$$Area\ \{cm^2\} = \frac{DELivery\ \{lpm\} * 0.1667}{VELocity\ \{m/sec.\}}$$

The delivery in this equation relates to the pump supply. However, the apparent delivery in inlets is not identical to that which occurs in working lines. It has been established that all pumps are not 100% volumetrically efficient. Therefore, the delivery, when calculating inlet line sizes, should be adjusted through division by the pump's volumetric efficiency. The velocity will vary according to the specified permissible ranges for velocity in inlets (2 to 5 ft./sec.) {*0.6 to 1.5 m/sec.*} and working (7 to 20 ft./

sec.) {*2.1 to 6.1 m/sec.*} lines. *However, fluid conductors are normally sized by inside diameter*. Therefore, the findings from the cross-sectional area formula need to be applied to the formula for the area of a circle to accurately describe the size of the conductor. By combining the area equations, a composite formula may be achieved for use in determining the inside diameter of a fluid conductor:

$$\text{DIAmeter (in.)} = \sqrt{\frac{\text{DELivery (gal./min.)} * 0.3208}{\text{VELocity (ft./sec.)} * 0.7854}}$$

$$\text{DIAmeter } \{cm\} = \sqrt{\frac{\text{DELivery } \{l/min.\} * 0.1667}{\text{VELocity } \{m/sec.\} * 0.7854}}$$

Problem 4-16

A hydraulic power supply is being operated using a balanced vane pump rated at 80% volumetric efficiency, supplying 5-gpm {*18.93-lpm*} delivery. Describe the ranges of permissible inside diameters for inlet and working lines for this power supply.

A. Inlet

At lowest velocity, DIA (in.) {*cm*} = ________

At highest velocity, DIA (in.) {*cm*} = ________

B. Working

At lowest velocity, DIA (in.) {*cm*} = ________

At highest velocity, DIA (in.) {*cm*} = ________

A size rating for these conductors could be achieved by simply selecting any number within the permissible velocity range for inlet and working lines, then solving the equation using that value. However, the resulting inside diameter would rarely relate to a practical size specification for that conductor. For example, pipe is normally supplied in standard **nominal diameter** such as $^1/_4$ in. {*8 mm*}, $^3/_8$ in. {*11 mm*}, $^1/_2$ in. {*14 mm*}, $^3/_4$ in. {*20 mm*}, etc. The calculated inside diameter will rarely match any of these nominal sizes exactly. Therefore, it is more practical to determine the range of permissible size, then select a nominal pipe size having a decimal value that falls within that range. In general, Schedule 40 or standard pipe should be used in inlet lines and Schedule 80 should be used in working lines. Decimal equivalents for nominal pipe inside diameters appear in Appendix A. However, suitably sized tubing and hosing may also be used and, in many cases, will be the preferred type of hydraulic conductor. The nominal size for hydraulic hose, like pipe, relates to the inside diameter of the conductor; however, the nominal size for tube refers to outside diameter. The pressure environment existing within the circuit is a determining factor for specification of conductors. If you are interested in design and specification procedures based on pressure, see Chapter 11.

Selection of which nominal size to use when more than one size occurs within the range of permissible values is at the designer's discretion. However, the designer should consider increasing the diameter if factors that would result in corrosion are predicted in the conductor. Finally, selection of conductor size within the permissible range should be made in consideration of matching that size of porting available in valves and other components to be used in the circuit. For example, using the design procedure described, the inlet for a pump would necessarily always be larger than the outlet. However, manufacturers sometimes make inlet and outlet ports on pumps identically sized. This expedites the manufacturing process and allows the pump, in many cases, to be used as a bidirectional hydraulic motor if needed. In most circuitry, it would be more economical and efficient if the pump were selected according to necessary inlet size porting at both port locations. The outlet port may then be reduced in size by using a reducing bushing or sleeve and a correct working line diameter attained. This will reduce the cost for plumbing the circuit and ensure streamlined flow in working lines.

4.4.1.3 Power supply and pressure control A typical hydraulic power supply unit will also contain a device to limit the maximum pressure in circuitry to which it may be attached. Some pressure-limiting devices will be complicated; however, the basic and more commonly used device to limit maximum system pressure is a simple **pressure-relief valve**. Figure 4-35 describes the design and operation of a simple relief valve. One of the more attractive features of such a device is the fact that it is variable. Unlike the pop-off valve in a household water heater, the simple relief valve may be adjusted to allow an infinite degree of pressures within its range of operation. A typical operation range for a simple relief valve would be 250 to 1,200 psi {*1,724 to 8,275 kPa*}. With such a wide range of variable, controllable pressure, various adaptations of the hydraulic power supply may be made. A disassembled simple pressure-relief valve appears in Figure 4-36.

Simple relief valves are located in the working line leading from the pump outlet. They are typically set to allow system pressure to rise to 10% above the maximum pressure necessary to operate other valving and output devices within the circuit. Further applications of the simple relief valve and descriptions of other, more sophisticated pressure controls are presented in Chapter 7.

FIGURE 4-35

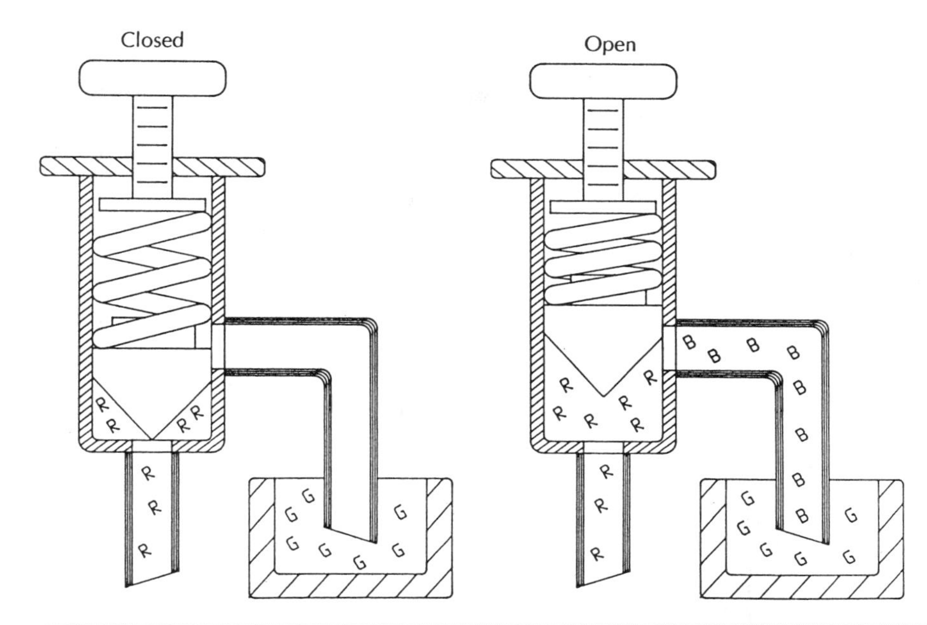

FIGURE 4-36

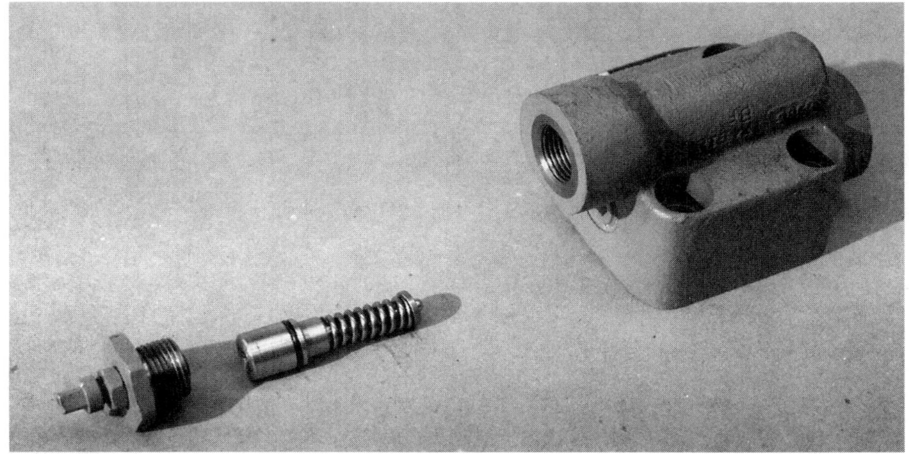

Simple pressure relief valve showing the poppet, spring, variable adjustment screw, and valve casing assembly. *(Photograph by the author.)*

4.4.2 The Packaged Power Supply

The total **power supply** package will include the **reservoir, filtration devices, pump, electric motor** (or other driving device)**, the pressure relief valve,** and **conductors**. The power supply may include a manifold attached to the working line leading from the pump outlet and a pressure gauge as illustrated in Figure 4-37 (see also Figure 4-38).

FIGURE 4-37

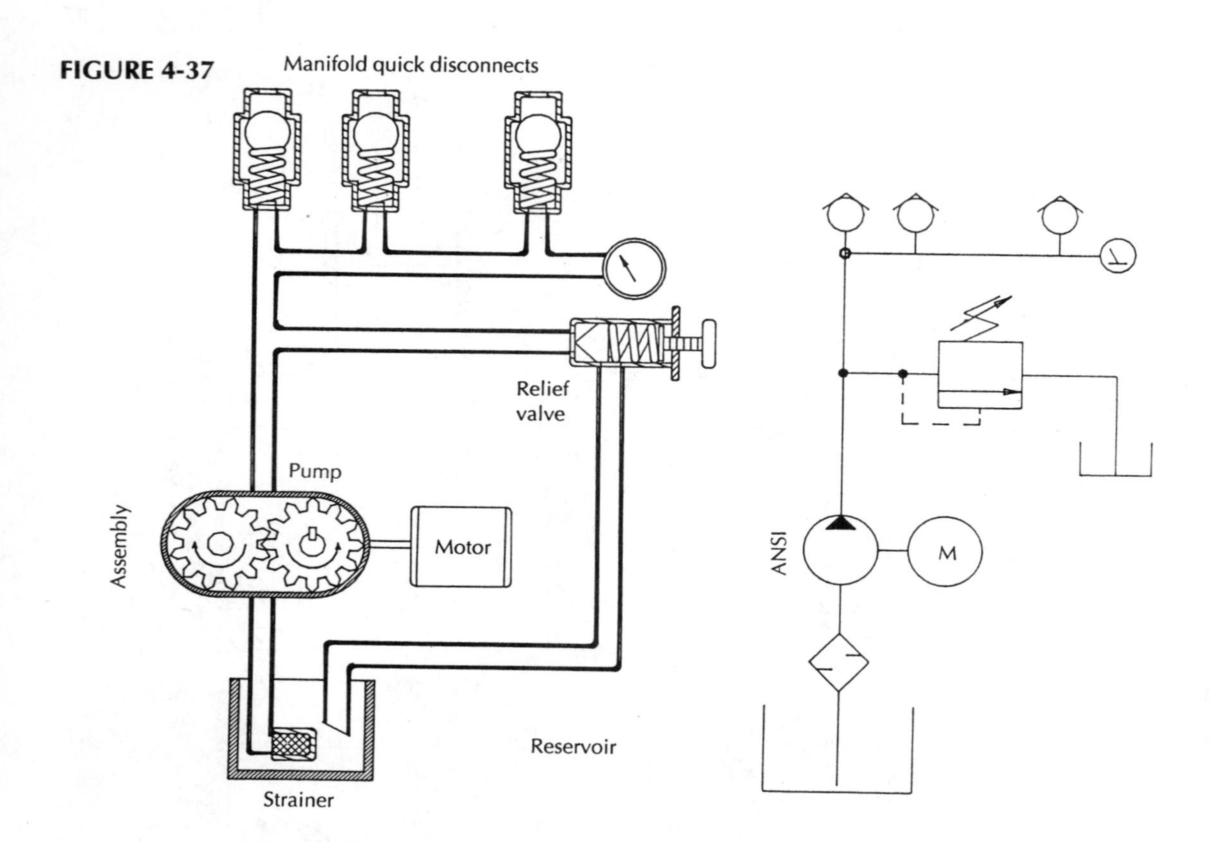

FIGURE 4-38

Hydraulic power supply package. *(Courtesy Polypac Systems Center, Continental Hydraulics. From Vockroth,* Industrial Hydraulics, *Albany, N.Y.: Delmar, 1994, p. 65.)*

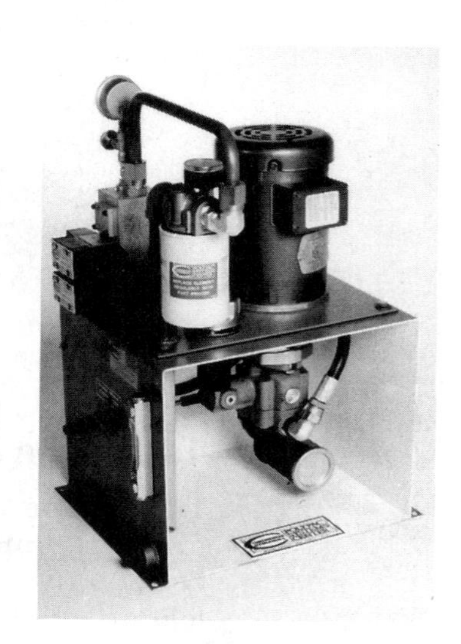

[] [] [] 4.5 PRACTICAL DESIGN PROBLEM

Representatives of a local municipal garage parking operation are considering installing hydraulic lifts to raise automobiles from one level of their building to another. They believe that this will reduce the vandalism that has occurred through the easy ramp access to each level. The racks and hydraulic cylinders necessary have been designed and specified.

Each rack will weigh 2,000 lbs. {*907 kg or 8.895 kN*} and will be raised by a 4-in. {*10.2-cm*} diameter cylinder having a 72-in. {182.9-cm} stroke. Building codes limit the maximum weight of automobiles allowed in this garage to 3,500 lbs. {*1,588 kg or 15.6 kN*}; thus, the total load possible will be 5,500 lbs. {*2,495 kg or 25 kN*} on extension of the cylinder.

The representatives believe that a great savings in cost can be achieved if two or three portable hydraulic power supplies could be used to operate these lifts, rather than having a built-in power supply dedicated to each lift. At this time, they are seeking your consultation through supplying design and specifications for a portable power unit that will operate these lifts to fully raise each automobile within 30 seconds.

Your report to them should include sizes, ratings, designs, types, and so on, of circuitry and components. Therefore, you will need to supply the information required in the following directions:

1. Calculate the maximum pressure that will be imposed upon the power supply during extension of the cylinder.

_____________________ (psi) {*kPa*}

2. Calculate the delivery required to fully extend the cylinder in 30 seconds.

_____________________ (gal./min.) {*l/min.*}

3. Determine the setting for the pressure relief valve in the power supply.

_____________________ (psi) {*kPa*}

4. If the motor in the power supply operates at 400 rpm, what will the necessary displacement for the pump be?

   ________________________ (cu. in./rev.) {*cm^3/rev.*}

5. What type of pump will be used in the power supply, and what volumetric efficiency will be designed for?

   ________________ (Pump) _________ (V.E. %)

6. Calculate the power rating for the electrical motor at the maximum pressure and delivery for the power supply.

   ______________________________ (hp) {*W*}

7. Determine the volume of fluid that should be contained in the reservoir.

   ______________________________ (gal.) {*l*}

8. Describe the rating for the strainer located in the pump inlet.

   __________________________________ (mesh)

   __________________________________ (gal./min.) {*l/min.*}

9. Describe the permissible range of inside diameter sizes for conductors in the power supply.

   a. Inlet Lines ________ (in.) {*cm*} to _______ (in.) {*cm*}

   b. Working Lines________ (in.) {*cm*} to ________ (in.) {*cm*}

10. Select nominal sizes of Schedule 40 {*STD*} piping to be used in the power supply.

    a. Inlet Lines ______________________ (in.) {*mm*}

    b. Working Lines______________________ (in.) {*mm*}

11. Complete the specifications for the portable power supply by providing the information in the following format.

A Pump

Type	Design	Delivery	Drive Speed	Volumetric Effic.
________	________	________ gpm{*lpm*}	400 rpm	________ %

Displacement at V.E.	*PORT SIZE* Inlet	Outlet
__________ cu. in./rev{*cm³/rev*}	_________ in.{*mm*}	________ in.{*mm*}

B Relief Valve

Type	Element	Setting	Mounting	Port Size
________	_________	________ psi{*kPa*}	_________	________ in. {*mm*}

C Motor (Drive Unit)

Type	Design	Arrangement	Drive Speed	Power at V.E. (%)
Int. Combustion	4 Cycle	Horizontal Shaft	400 rpm	_______hP{*W*}

D Reservoir

CAPACITY Minimum	Maximum
_________ gal.{*l*}	_________ gal.{*l*}

E Filtration

Type	Element	Element Rating	FLOW RATING Minimum	FLOW RATING Maximum
Strainer	Wire	________mesh	________ gpm{*lpm*}	________ gpm{*lpm*}

Port Size	Design
________ in.{*mm*}	Full Flow

F Conductors

Inlet Type	Inlet Description	INLETS (I.D. DECIMAL) Minimum	INLETS (I.D. DECIMAL) Maximum	Working Type	Working Description
______	________	______ in.{*mm*}	______ in.{*mm*}	________	________

WORKINGS (I.D. DECIMAL) Minimum	WORKINGS (I.D. DECIMAL) Maximum	LINES — NOMINAL Inlets	LINES — NOMINAL Workings
________ in.{*mm*}	________ in.{*mm*}	________ in.{*mm*}	________ in.{*mm*}

4.6 SUGGESTED ACTIVITIES

1. Secure a hydraulic equipment catalog. Find catalog inclusions for an external gear pump, an internal gear pump, an unbalanced vane pump, a balanced vane pump, a radial piston pump, and an axial piston pump. In a tabular form, list the manufacturer's model number, description, port sizes, displacement, drive speed, delivery, and rated pressure range for each pump.
2. Secure a hydraulic pump and disassemble the pump until you can identify the pumping element and chambers inside the pump. Sketch a diagram of the internal construction, identifying drive direction, and color-code all chambers according to location and fluid condition. Be sure to show internal passageways that lead to external porting and label each port in compliance with designated drive direction.

3. If you have a hydraulic power supply, examine the power supply to determine if the pump is directly driven from the motor (engine). If so, determine the drive speed of the pump and motor. If not, determine the ratio of drive speed reduction from the motor to the pump and determine the drive speed of the pump. Connect a flow gauge in series with the pump outlet and return the fluid from the gauge's outlet to the reservoir. Operate the power supply and determine the pump's delivery. Calculate the displacement of the pump from your measured delivery and determined drive speed. If your pump has a known rated displacement, calculate the volumetric efficiency of your pump under this "no-load" condition.
4. Assuming that you have a hydraulic power supply, connect a pressure gauge (with a plumbing "T") into the outlet side of the pump. Plug the open port on the plumbing "T." Assuming that you have a pressure-relief valve in your power supply, operate the power supply and record the pressure gauge reading as your relief valve setting. If possible, vary the setting on the relief valve and determine (or possibly calibrate) your relief valve to the readings on the pressure gauge. Leaving your relief valve set to one standard, determine the hydraulic horsepower {*or wattage*} using your pressure and flow gauge readings and your calculated volumetric efficiency for your pump. Compare this calculated value to the rating on the motor.
5. If you do not have a power supply, build one using a suitable container, strainer from an automotive engine, oil pump from an automotive engine, electric motor, and suitable plumbing. You will probably need to purchase a simple relief valve for your power supply. Then perform Activities 3 and 4.
6. If you do not have a power supply nor the ability, time, or the like to build yourself one, assume some values for drive speed, delivery, displacement, volumetric efficiency, and pressure. Then perform all calculations for Activities 3 and 4.

4.7 REVIEW QUESTIONS

1. Explain why a hydraulic circuit having an unpressurized reservoir must have a breather cap that is open to the atmosphere.
2. How does a negative head situation affect the pumping operation in a hydraulic circuit? What accommodations must be made to assure that optimal efficiency in the pumping operation is maintained even in a negative head situation?
3. Explain volumetric efficiency of a pump in terms of its void capacity versus the quantity of fluid that it will transfer from the inlet to the outlet in one revolution.
4. What happens to the delivery of a pump when either the displacement or the drive speed increases, assuming that volumetric efficiency remains constant?
5. Rationalize the necessary power input for a hydraulic circuit in terms of pressure and delivery when determining the minimum horsepower requirement for an electric motor or engine drive.
6. Describe the operation of the following pumps in terms of methods employed to create increasing and decreasing chamber sizes for charging and discharging fluid:

a. external gear
b. internal gear
c. unbalanced vane
d. radial piston
e. axial piston

7. What is side loading or thrust in an unbalanced vane pump, and how is it eliminated in the balanced vane design?
8. Describe design characteristics in unbalanced vane, radial piston, and axial piston pumps that may allow for variable displacement.
9. Describe components in the well-designed hydraulic reservoir.
10. Differentiate between strainers and filters.
11. What happens to the necessary inside diameter of a hydraulic conductor as the pump delivery for the circuit increases? What happens to the necessary inside diameter of a hydraulic conductor as the velocity of fluid flowing through that conductor increases?

5

Cylinders and Linear Circuitry

Although the hydraulic power supply may be described as the heart of a hydraulic circuit, there would be little practical application of hydraulics without the arms and legs of hydraulic output devices. Power output devices in hydraulics are known, collectively, as **actuators**. Hydraulic actuators are classified as linear, rotary, and combination. Many of the unique characteristics of hydraulic power that result in advantages over other power-transmission devices result from the unique design, operation, and application of hydraulic actuators.

5.1 LINEAR ACTUATORS

Linear actuation in hydraulics is produced by devices commonly known as **cylinders**. Cylinders produce movement in hydraulic circuitry similar to the linear movement of solenoids in electrical circuitry or beams in mechanical devices. Cylinder design varies according to size, features, and manufacturer; however, most cylinders contain a sleeve, caps on both ends of the sleeve, a piston fitted inside the sleeve, a rod or rods that attach to the piston and extend through the end cap or caps, a port or ports to allow fluid input or expulsion, and various seals and bearing surfaces (see Figures 5-1 and 5-2).

FIGURE 5-1

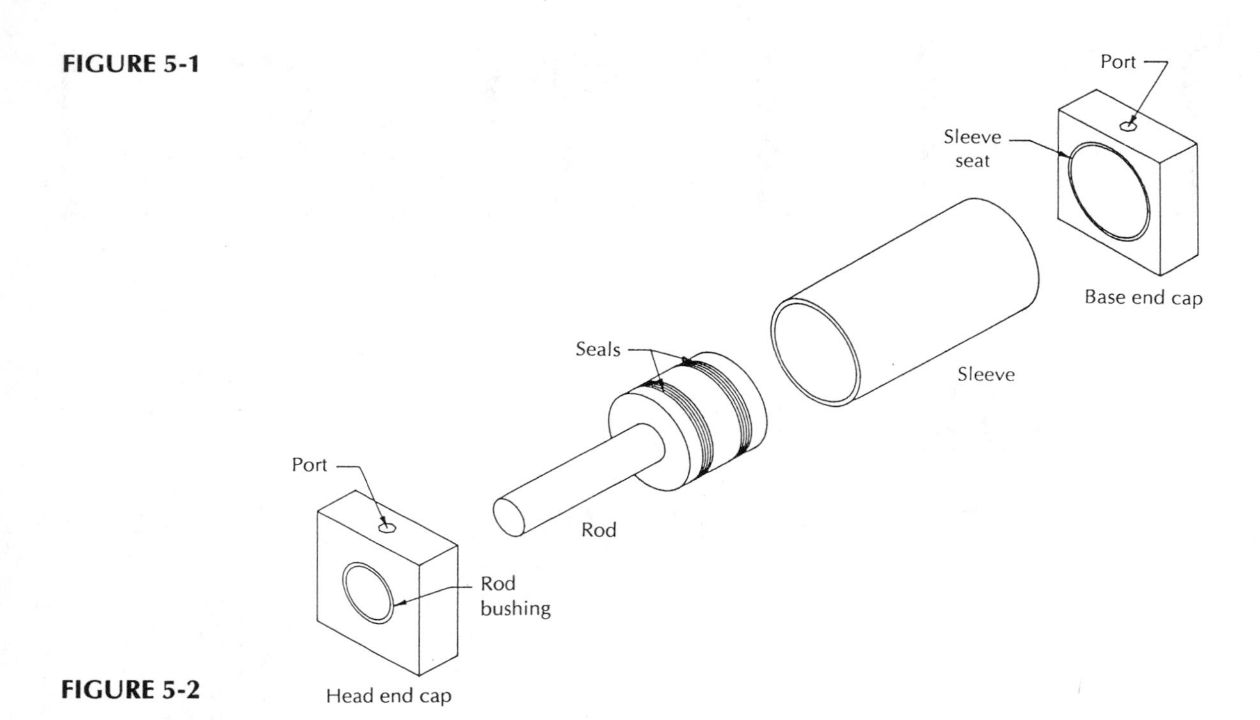

FIGURE 5-2

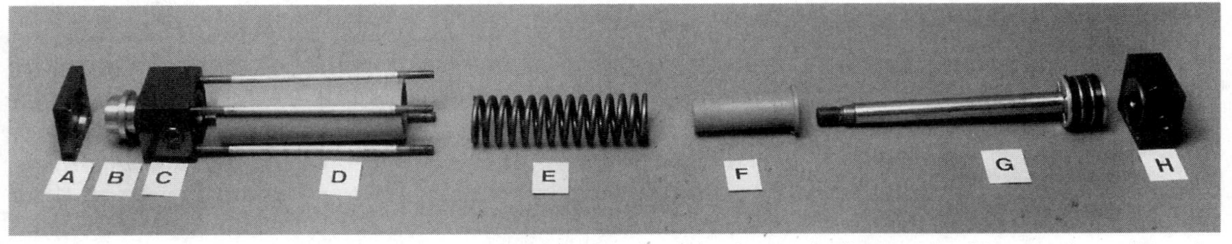

Typical construction of a hydraulic cylinder, including **(A)** bushing retainer, **(B)** rod busing, **(C)** head end cap, **(D)** cylinder sleeve, **(E)** return spring, **(F)** stop tube, **(G)** piston–rod assembly, and **(H)** base end cap. *(Photography by the author.)*

5.1.1 Single-Acting Cylinders

Single-acting cylinders, or **rams**, typically act under pressure only during extension. Gravitational forces acting on the load or other external forces may be used to cause retraction. Since hydraulic power is not necessary during retraction, rods in the ram-type cylinder are generally large in size relative to sleeve inside diameter, compared to other cylinder designs. This allows for uses of rams in circuitry where heavy loads require mechanical support from the cylinder rod (see Figure 5-3).

FIGURE 5-3

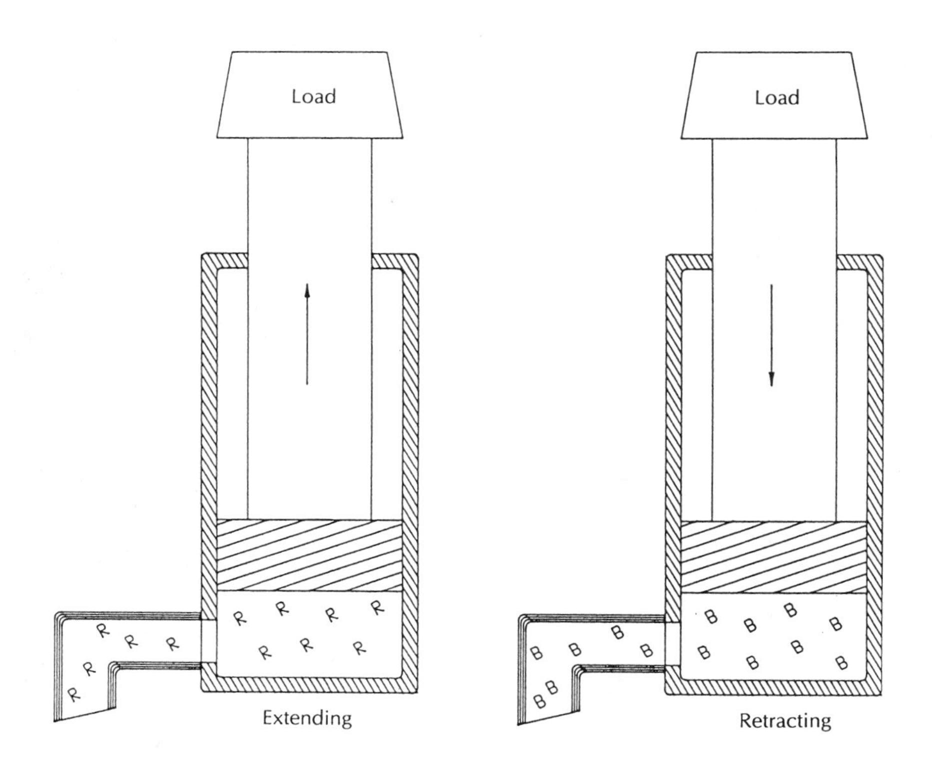

Force output in the ram cylinder during extension is a function of piston area and system pressure, while speed of actuation may be determined by piston area and system delivery. However, force and speed during retraction is difficult to determine accurately. Retraction force is usually of minor consequence in circuit design, while speed during retraction may be reduced through the use of flow-limiting devices (see Chapter 8). Most often, this control is incorporated to reduce damage resulting from impact or to reduce the rate of decompression for safety purposes. A mechanical spring may be inserted around the rod, between the piston and the rod end cap of the ram. The spring ensures retraction if the ram is used in any position other than vertical or if the load on the ram is only evident during extension. If spring return is used on the ram, the spring itself will impose a load on the cylinder during extension. Furthermore, ram cylinders with spring returns will extend through only one-half or less of the sleeve stroke distance because of the necessity for housing the spring mechanism (see Figure 5-4).

The **telescoping cylinder** is a special design ram that affords maximum extension stroke from a foreshortened sleeve. The telescoping cylinder is constructed of a series of rod-type sleeves fitted into the cylinder

FIGURE 5-4

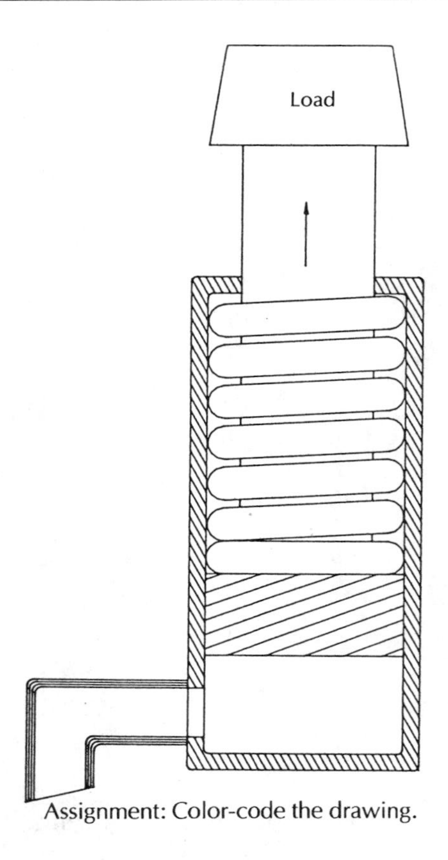

Assignment: Color-code the drawing.

sleeve. In operation, the rod sleeve having the largest effective cross-sectional area will extend first under pressure. The pressure will attempt to rise within the linear circuit to overcome loads encountered. As pressure rises, it will attain a value needed to overcome the loads using a larger area before a smaller area. Or, in formulation, if pressure equals force divided by area, then the larger area will result in a smaller value for pressure to overcome a given load or force. After this rod sleeve has fully extended, the other rod sleeves will extend in order of descending cross-sectional area. The net effect will be a total stroke nearly equal in length to the length of the cylinder sleeves multiplied by the number of rod sleeves. However, under a given load, *the telescoping cylinder will require incrementally increasing pressure* to cause displacement resulting from the reduction in effective area of each rod sleeve. Furthermore, when supplied with a constant delivery, *the telescoping cylinder will extend with increasing velocity* as each rod sleeve decreases in volumetric capacity. Since velocity may be expressed as supply rate (in cubic inches per minute or *cubic meters per minute*) divided by area (in square inches or *square centimeters*), the velocity will increase as area decreases (see Figure 5-5).

FIGURE 5-5

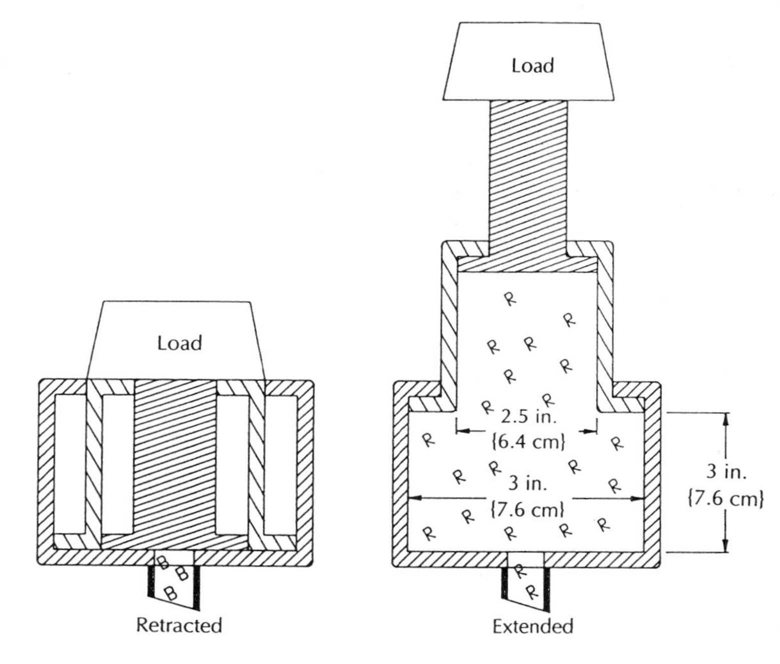

A telescoping cylinder has two rod sleeves of 3-inch and 2.5-inch {6.4-cm and 7.6 cm} diameters and a stroke of 3 inches {7.6 cm} per sleeve. The telescoping cylinder is being supplied at a rate of 2 gallons {7.6 liters} per minute, and its extension is being resisted by a 1700-pound {7.6 kN} load.

Problem 5-1

What pressure will develop during extension of each rod sleeve?

a. 3-in. {7.6-cm} sleeve: P1 (psi){kPa} = _________

b. 2.5-in. {6.4-cm} sleeve: P2 (psi){kPa} = _________

Problem 5-2

What will be the total time for one extension of the telescoping cylinder under the given supply?

Time (sec.) = _________

5.1.2 Double-Acting Cylinders

Often, hydraulic circuitry will be designed to overcome resistance in both extension and retraction of the actuator. The ram-type cylinder discussed in the previous section is referred to as a single-acting cylinder. A cylinder that has the capability to allow hydraulic pressure to act upon both sides of the cylinder's piston is referred to as a double-acting cylinder. In the **single end rod, double-acting cylinder**, the effective area for power transmission is reduced during retraction by the reduction of piston area due to the rod being attached. The **annulus area**, as it is known, may be calculated by subtracting the cross-sectional area of the rod from the cross-sectional area of the piston. The net effect is that under a specified load and at a constant delivery, *the single end rod, double-acting cylinder will require more pressure to operate in retraction than extension.* However, increased velocity is evident in retraction as compared to extension because of the reduced effective area (see Figure 5-6).

FIGURE 5-6

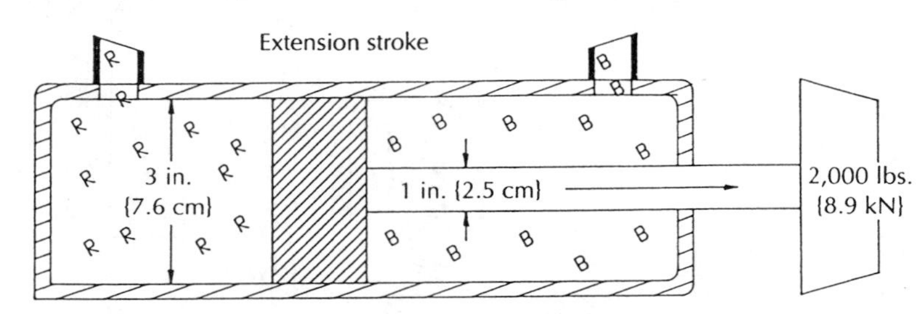

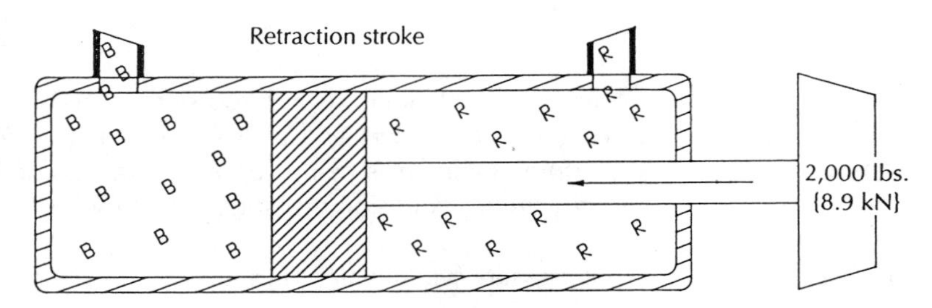

The single end rod, double-acting cylinder has a piston diameter of 3 inches {7.6 cm}, a rod diameter of 1 inch {2.5 cm}, and a stroke of 10 inches {25.4 cm}. The cylinder is being supplied at a rate of 5 gallons {18.9 liters} per minute, and movement is being resisted by a 2,000-pound {8.9-kN} load in both extension and retraction.

Problem 5-3

What pressures will develop during extension and retraction?

Extension pressure (psi){kPa} = _________

Retraction pressure (psi){kPa} = _________

Problem 5-4

At what velocities will the cylinder extend and retract?

Extension velocity (ft./sec.){m/sec.} = _________

Retraction velocity (ft./sec.){m/sec.} = _________

The **double end rod, double-acting cylinder** has rods extending through both end caps attached to both faces of the cylinder piston. Standard design calls for both cylinder rods to be of the same diameter (see Figure 5-7). Therefore, under a specific load and a constant delivery, *the double end rod, double-acting cylinder will operate with equal speed and pressure on both extension and retraction. (Note: The terms* extension *and* retraction *are relative when discussing a double end rod, double-acting cylinder; either stroke may be considered to be extension or retraction.)* This type of cylinder may be used to move and position a feed table for a machining operation. In this application, the end rods serve not only to transmit power but also to activate other controls that automatically change direction at the end of the stroke. In another application, the end rods may be used as a sensing mechanism in combination with other hydraulic, pneumatic, or electronic devices to accurately locate and position a feed table.

FIGURE 5-7

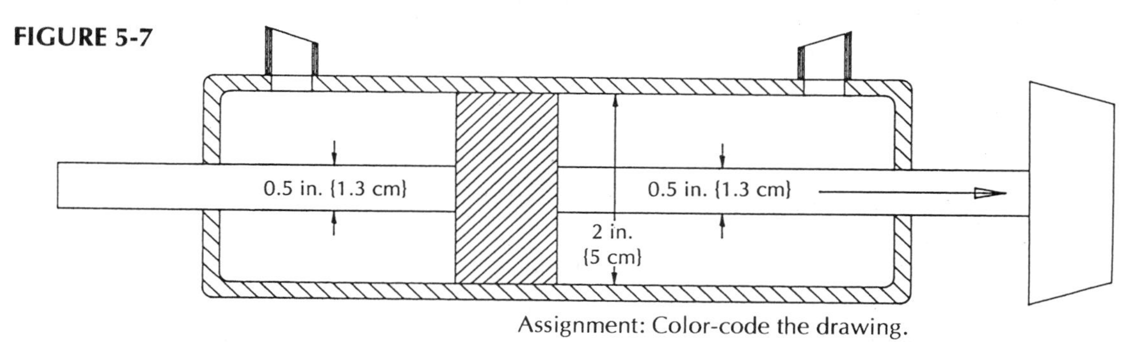

Assignment: Color-code the drawing.

Not only is it possible to attach more than one rod to the cylinder piston, but it is also possible to stack pistons on the rods. In the **multiple-piston cylinder**, pistons are attached to the rod and are separated from one another by internal barriers. Independent ports are located on either side of the piston within the pocket created by the cap ends and internal barriers (see Figure 5-8). The net effect is that *the multiple-piston cylinder operates much like series cylinders acting to assist one another in both extension and retraction.* If the multiple-piston cylinder is a double-end rod design, the effective area to transmit power may be calculated using the annulus area multiplied by the number of integral pistons. The supply needed for a cylinder stroke is a function of this annulus area times the clearance between the piston and the barrier times the number of integral pistons. In calculations for the single end rod, multiple-piston cylinder, allowances must be made for the full piston area and cylinder volume in one pocket in combination with the other annulus areas and volumes. The only practical use of the multiple-piston cylinder is in applications where space limitations require a small-diameter cylinder requiring high-force output with low pressure. In other low-pressure applications, increased force output can be more effectively achieved by increasing the cross-sectional area of the piston. For example, *a 4-in. diameter piston has twice as much cross-sectional area as two pistons having 2-in. diameters*. Furthermore, the potential stroke length of the multiple-piston cylinder is greatly foreshortened, considering the overall length of the cylinder package.

FIGURE 5-8

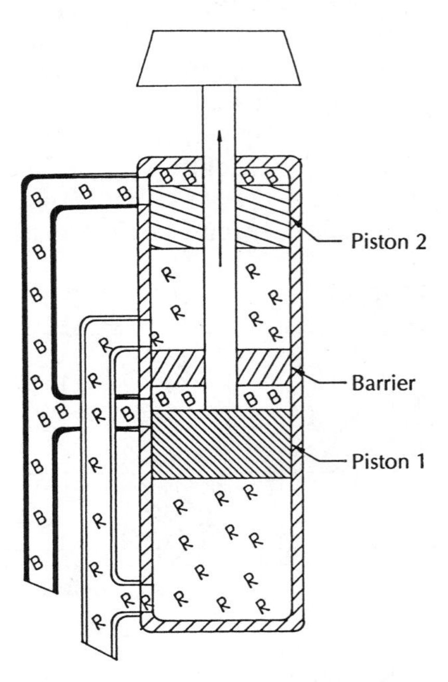

The multiple-piston cylinder has 2-in. {5-cm} diameter pistons, a 3/8 in. {0.9 cm} diameter rod, and a stroke of 5 in. {12.5 cm}. The cylinder is operating in a circuit having a maximum pressure of 750 psi {5,171.1 kPa} in both extension and retraction.

Problem 5-5 What is the maximum load that may be moved in extension and retraction?

Extension force (lbs){kN} = ________

Extension force (lbs){kN} = ________

Problem 5-6 What supply must be made to cause one full cycle of the cylinder?

Extension volume (cu. in.){cm³} = ________

Retraction volume (cu. in.){cm³} = ________

Total cycle volume (cu. in.){cm³} = ________

5.1.3 Double-Acting Cylinder Design Features

The sealing mechanisms on the piston and at the interface between the head and the base end cap are critical to efficient operation of the double-acting cylinder. Cup seals and compression seals made from materials compatible with the hydraulic fluid are typically included on the piston, while lip seals are used in the rod bushing. Hydraulic seals (or packing) are commonly of the O-ring, U-ring, V-ring, or cup designs.

5.1.3.1 O-ring seals The common O-ring seal has the appearance of a torus or doughnut and is typically round in cross-section in the unloaded condition. When exposed to squeezing forces (usually about 5% to 10% of the size of the O-ring cross-sectional diameter), it will deform to an oval-shaped cross-section. During inital assembly, the interference caused by the geometry of the groove that the O-ring fits into and its mating surface will result in this static loading condition. Under moderate hydraulic pressure the O-ring will distort to a D-shaped cross-section and, in extreme conditions, may even extrude into the clearance between mating surfaces (see Figure 5-9).

It is best to avoid the extrusion condition, for most O-ring damage occurs during rapid shifting of force direction when the O-ring is extruded or when insufficient lubrication is afforded for the O-ring. O-rings may be purchased in a variety of graded resistances to deformation. These grades are referred to as durometer hardnesses. Needless to say, higher-durometer O-rings will deform less under higher pressures than lower-durometer hardness O-rings. As a general rule, O-rings having a rating of 70 durometer should be used in circuitry operating up to 1,500 psi {*10,343 kPa*}, increasing to 80 durometer for circuitry operating between 1,500 and 2,500 psi {*10,343 and 17,238 kPa*} and 90 durometer for circuitry operating at more than 2,500 psi {*17,238 kPa*}. In most cases, antiextrusion or backup rings of metal should be used to eliminate excess extrusion of the O-ring.

5.1.3.2 U-ring seals Another type of sealing device that provides only one level of protection is the U-ring. The U-ring has thin flexible lips as contact points for sealing. Although the U-ring will typically reduce frictional contact as compared to the O-ring, it provides little (if any) support to the bearing interface. U-rings are typically used as a lip seal between the end caps of the cylinder and the rods extending through them. A typical application of the U-ring is illustrated in Figure 5-10.

5.1.3.3 V-ring seals Unlike O-ring or U-ring seals, the V-ring seal provides multiple levels of defense against leakage. Perhaps the greatest advantage of V-ring seals is their ability to automatically adjust to changes in pressure. In simple terms, the higher the pressure, the harder the V-ring will force against the sealing surface. Inversely, when pressure is

FIGURE 5-9

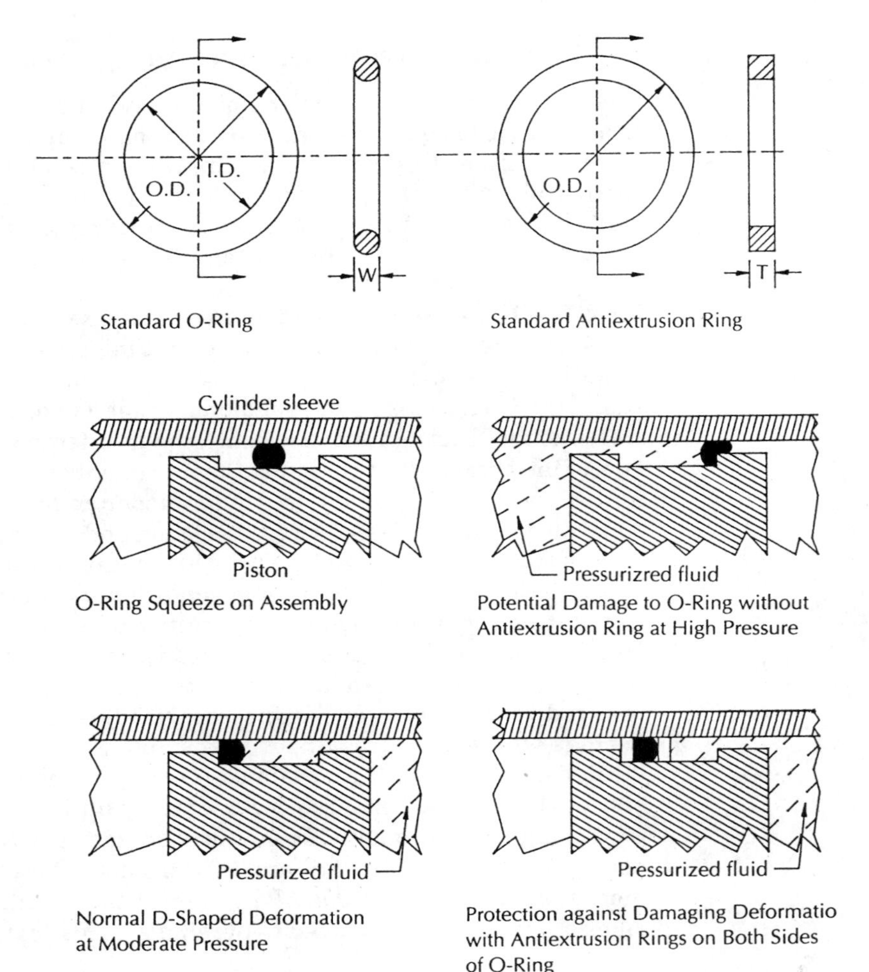

reduced, the V-ring will relax, greatly reducing wear during cyclical operation. V-ring designs may be used at the piston–sleeve interface or at the rod–bushing interface in the end cap or caps. In either application, it is important to reduce the interface between the mating surfaces to a minimum to ensure that the V-ring lips do not extrude into the clearance space. As a general rule, clearances of about 0.0015 in. {*0.0038 cm*} per inch {*2.54 cm*} should be allowed for diameters less than 2 in. (*5.08 cm*}; and 0.001 in. {*0.0025 cm*} clearance per inch {*2.54 cm*} for diameters of more than 2 in. {*5.08 cm*}. The V-ring seal design is illustrated in Figure 5-11.

FIGURE 5-10

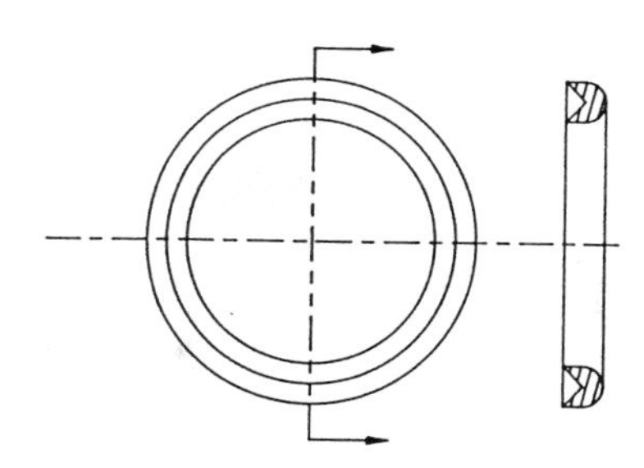

Standard U-Ring

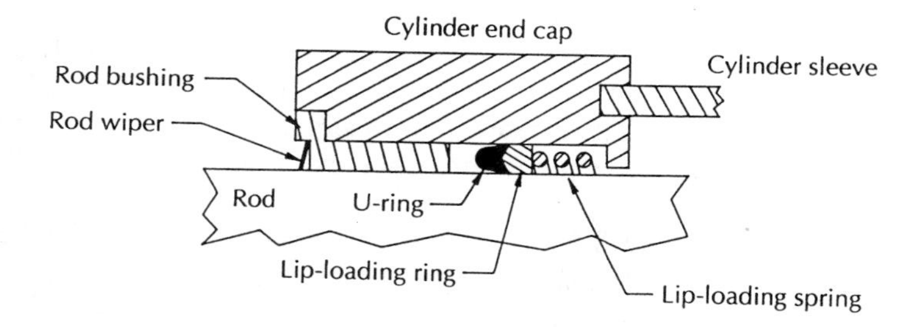

U-Ring Seal Assembly Used
at Rod–Bearing Interface

FIGURE 5-11

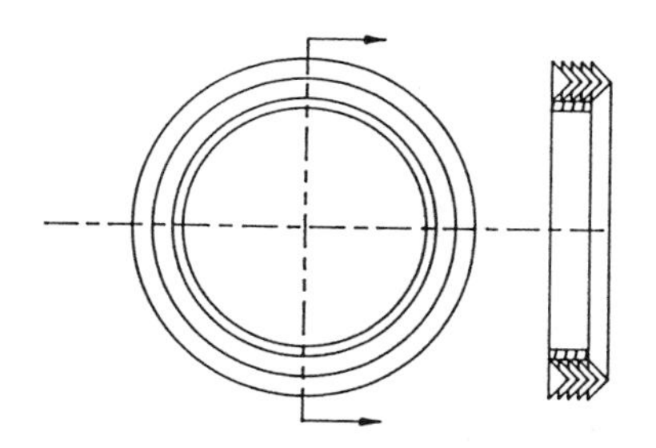

Standard V-Ring

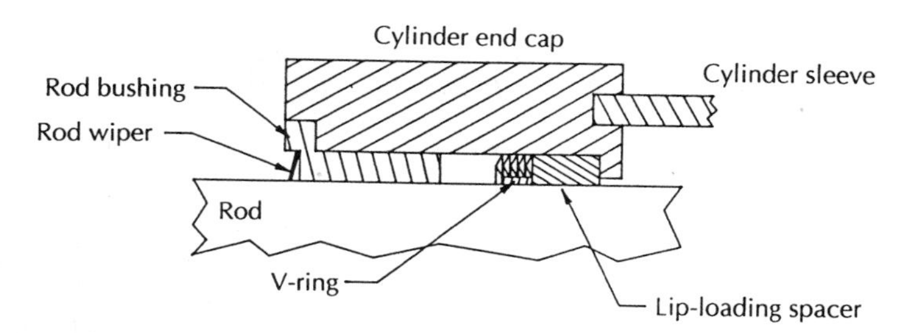

V-Ring Seal Assembly Used
at Rod–Bearing Interface

5.1.3.4 Cup seals Most sealing mechanisms require a highly accurate degree of roundness to exist between mating surfaces to operate properly. The cup seal having long lips will self-adjust to compensate for some degree of out-of-roundness. In all cases, the cup seal's extreme flexibility demands that backup mechanisms be included because the cup seal is not self-supporting. If properly supported, the cup seal will only contact the sealing surface at the ends of the lips. Otherwise, the cup will collapse and will not work properly as a sealing device. All sealing devices are dependent on good bearing support from other parts of the cylinder package. The seal itself should not provide this support, and it is poor practice to simply replace seals in a cylinder if the problem is being caused by wear on the bearing surfaces. The cup seal is illustrated in Figure 5-12.

The other major design feature of the double-acting cylinder is the inclusion of an additional port on the rod end cap of the cylinder (see Figure 5-13).

5.1.3.5 Stop tubes Double-acting cylinders may also include a mechanism that limits the stroke of the cylinder (see Figure 5-13). Although stop tubes decrease the overall stroke capacity, they do provide needed

FIGURE 5-12

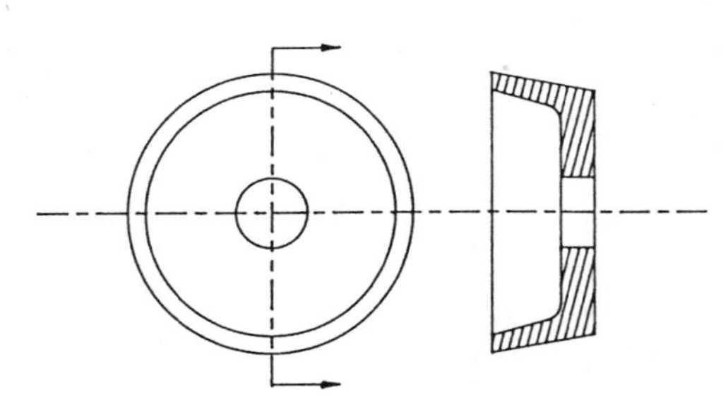

Standard Cup Seal

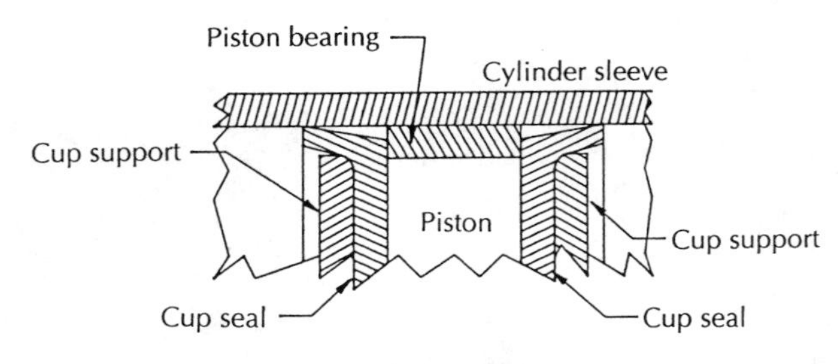

Cup Seal/Support Assembly
at Piston–Sleeve Interface

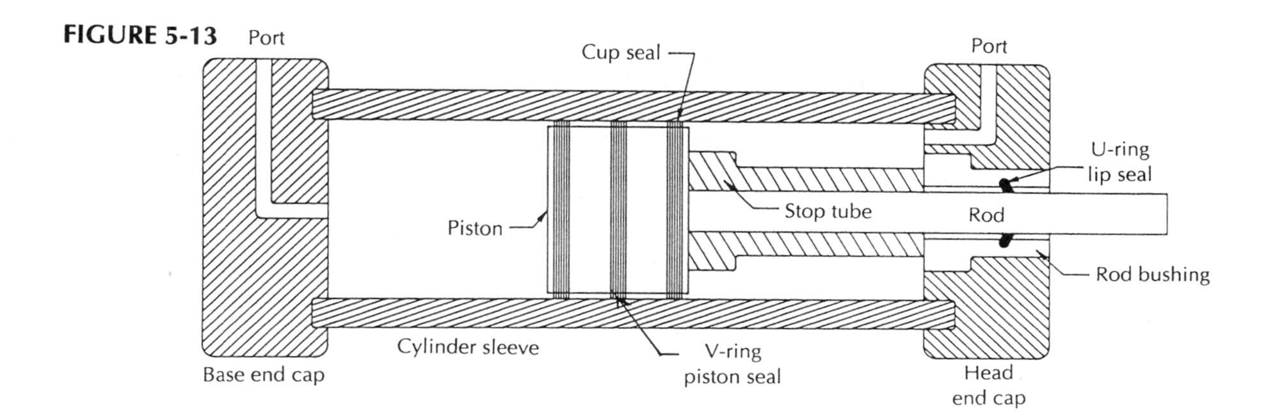

FIGURE 5-13

support for horizontally mounted cylinders under heavy loads. The design of the double-acting cylinder provides little mechanical support of such loads when the piston is allowed to come into close contact with the rod end cap. Binding and kinking of the piston against the cylinder sleeve will occur in many cases, and increased wear and loss of efficiency will be evident. By limiting the cylinder stroke through the use of a stop tube, mechanical lever advantage is gained and the potential for binding and kinking is greatly reduced. In general, stop tubes will decrease the stroke capacity of cylinders in which they are used by one-half.

5.1.3.6 Cylinder construction Another important design consideration for the double-acting cylinder is its body style. Basically, the construction of the cylinder is determined by the method used to attach the end caps to the cylinder sleeve. Probably the most popular body style of cylinder is the **tie-rod** type. In the tie-rod cylinder, elongated bolts or threaded rods are passed through clearance holes drilled through one end cap. These rods pass parallel to the outer perimeter of the cylinder sleeve and thread into tapped holes in the other end cap. Tightening of the tie-rod threading mechanism draws the cylinder sleeve into circular slots in each cap end. O-rings, specified to be compatible with the hydraulic fluid used, may be fitted into the slots to seal the interfaces between the cylinder sleeve and the end caps. Small tie-rod cylinders may only incorporate two tie-rods, while larger cylinders may contain as many as eight or more equally spaced tie-rods, with the most common arrangement being for four. Care should be taken to equally and sequentially tighten the tie-rods to the specified torque to ensure proper alignment of the cylinder sleeve and end caps and to eliminate possible damage to the cylinder sleeve.

Another popular body style, especially for large-diameter cylinders, is the **flange** type. The flange cylinder incorporates the use of integral bolt rings on the ends of the cylinder sleeve. The end caps are designed to mate with these flanges. Attachment is made with bolts that extend through the end cap and thread into aligned holes on the flange or with mating nuts on the rear of the flange. Flat gaskets are used to seal the interfaces between the flanges and the end caps. Alignment of flange cylinder components is made through the matched machining of the bolt holes. However, proper torque and tightening sequence should be observed to assure proper sealing and to eliminate potential warping or cracking of the flange or the end cap.

Recent developments in the machine tool industry have resulted in body styles that use threaded end caps and cylinder sleeves. Dry-seal threads that require little or no extra sealing are typically incorporated into the **screw**-type cylinder. The major advantage of this type of cylinder is the ease of assembly and disassembly for manufacturing and maintenance purposes. Tie-rod and flange-type cylinders may also be disassembled and field serviced. However, this feature normally requires the use of sealing devices that may leak or allow external contaminants to enter the fluid stream.

The **one-piece cylinder** provides nearly perfect seals and reduces fluid contamination. It is also the least expensive construction type of cylinder. The one-piece cylinder uses cast cylinder sleeves that include an integrally cast end cap on the base end. The head end cap is typically machined from a sleeve that tightly fits over the cylinder sleeve and is sealed with a continuous weld after assembly of the piston, rod, and rod bushing. Although the internal construction and piston seals are similar to other types of cylinders, the one-piece cylinder cannot easily be disassembled for repair. One-piece cylinders are typically used where extremely high pressures are necessary or in dirty environments such as in earthmoving machinery. Figure 5-14 graphically describes four body styles for cylinders: the tie-rod, flange, screw, and one-piece types.

5.1.3.7 Cylinder mounts Another important consideration when specifying cylinders is to describe the method by which they will be attached to supporting members. Figure 5-15 illustrates various mounts used for cylinders and other components as well.

Threaded end cap, lug, foot, and subplate mounts are referred to as flush mounts. Such cylinder mounts are normally used for down-hand work like clamping or for board mounting. In most cases, loading mechanisms applied to flush-mount cylinders have a low profile, are partially supported by sources other than the cylinder rod, or both. **Lug mounts** provide the most effective support for the cylinder when located on the

FIGURE 5-14

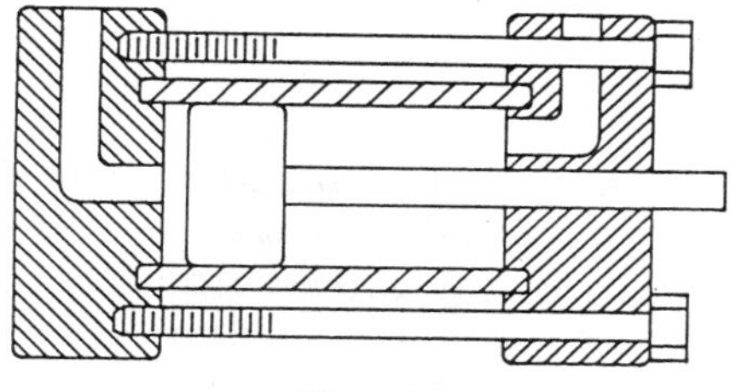

Tie-rod

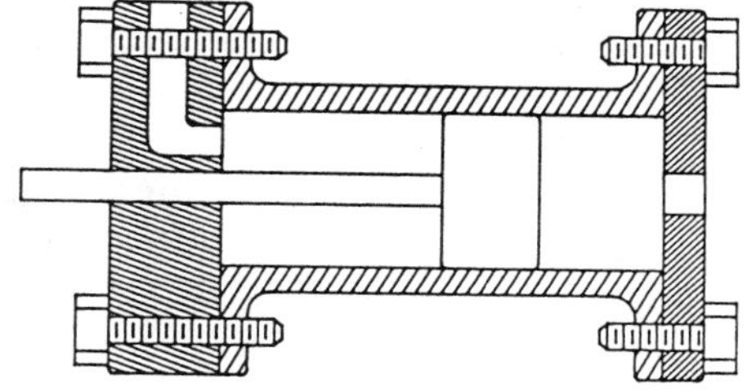

Flange

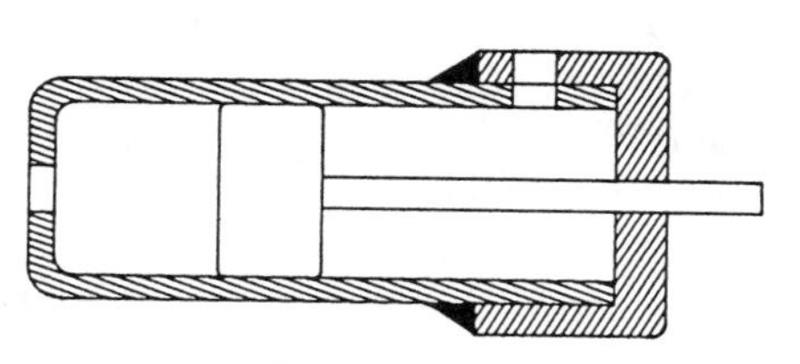

One-piece

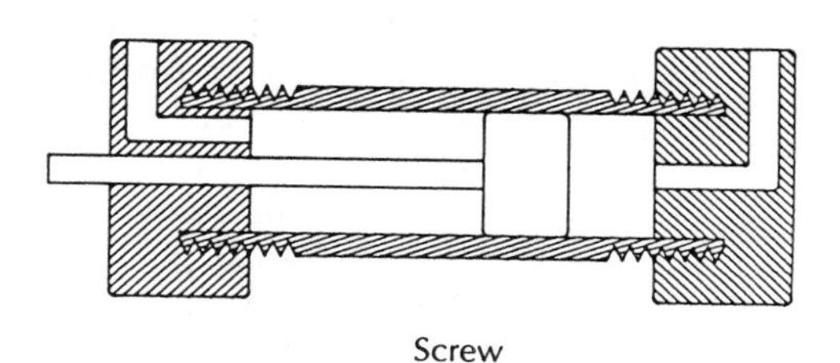

Screw

Assignment: Establish a stroke direction for each cylinder by drawing an arrow on its rod; then color-code each drawing.

FIGURE 5-15

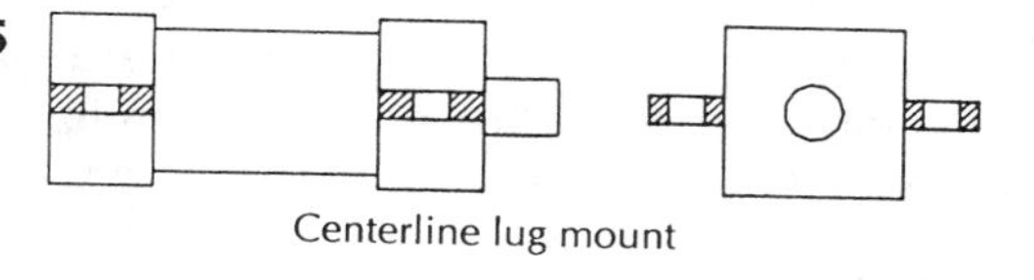

Centerline lug mount

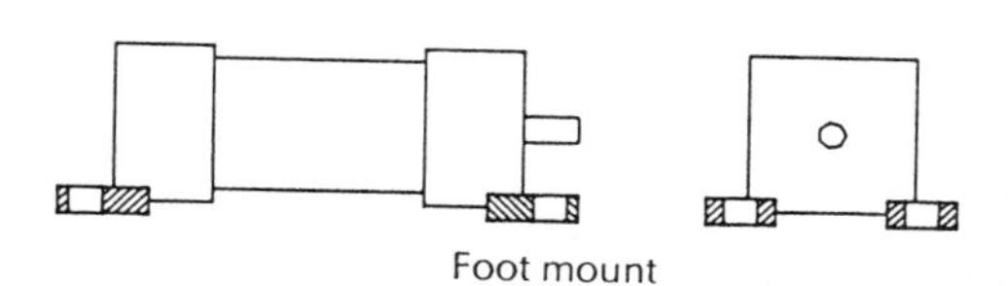

Foot mount

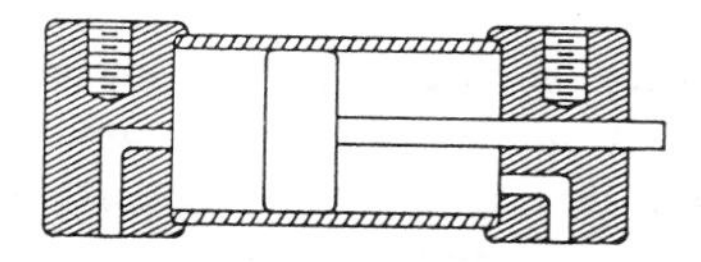

Threaded end cap mount

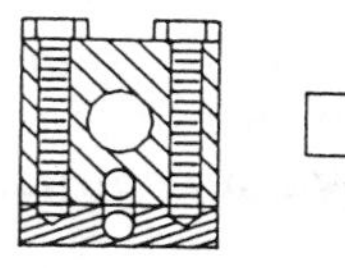

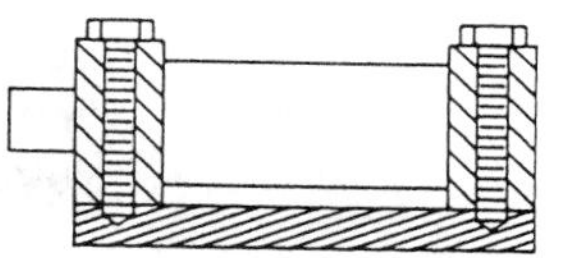

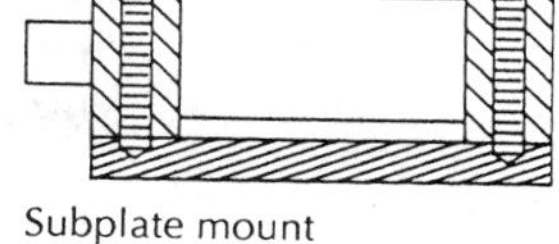

Subplate mount

Flange mount

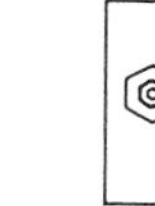

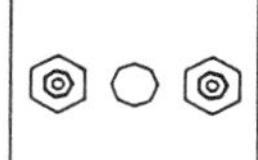

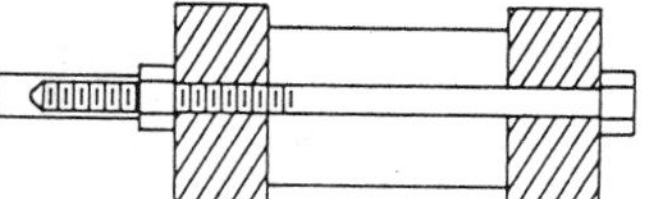

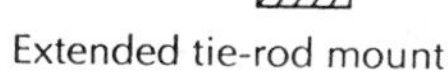

Extended tie-rod mount

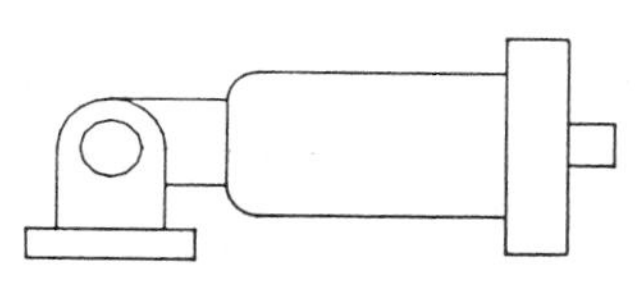

Clevis mount

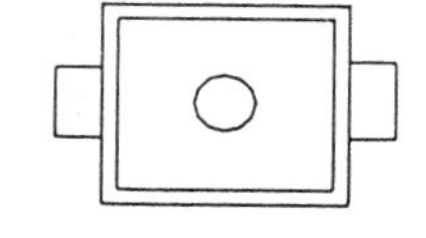

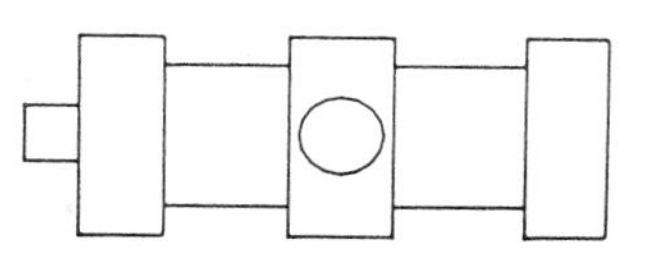

Intermediate trunion mount

centerline of the sleeve compared to the **foot mount**, which may allow more bucking of the cylinder under load. However, the foot mount is probably the least expensive type of mount. **Threaded end cap mountings** must be carefully aligned with holes in the mounting base to secure them with bolts. Threaded end cap mountings typically allow easy access for mounting and removal, while taking up little space beyond that needed for the cylinder. The **subplate mount** is a special type of threaded end cap that has the inlet and outlet ports located on the same face of the end caps as the threaded mounting holes. With this arrangement, the cylinder can be bolted to a plate beneath the cylinder that contains passageways for plumbing. The cylinder–subplate interface is typically separated by a gasket that acts as a seal when the cylinder is mounted. The advantage of the subplate mount is that it accommodates easy removal of the cylinder from the circuit for repair purposes.

Extended tie-rod and flange mounts are face mounting mechanisms. Face mounts are typically used when tension or compression loads are encountered on the rod. In face mounts, alignment between the cylinder mount and the supporting mechanism is critical. Any slight misalignment will result in stresses other than compression or tension to occur at the mounting and accelerate failure. Of course, the **extended tie-rod mount** only occurs with tie-rod-type cylinders, while **flange mounts** are the face mounting mechanisms for other construction-type cylinders.

Trunion and clevis mounts are lever-action mounting mechanisms. They are used to locate a cylinder and allow varying degrees of cylinder pitch to occur during extension and retraction. Lever mounts are typically used with other mechanical lever arms that result in alteration of direction and force from rod output, such as toggle action or yoke rotation. The **trunion mount** may be located on either end cap or on an intermediate yoke fitted around the cylinder sleeve. Trunion mounts may be used with double-acting, double end rod cylinders, while clevis mounts may not. However, **clevis mounts** afford longer lever action because the clevis is typically located on the rear of the cap end. In applications such as that for a dump bed on a truck, this is an important consideration. Various body styles and mounts are illustrated in Figure 5-16.

5.1.3.8 Other features Other design features of double-acting cylinders may include flow-limiting devices and integral directional controls. Electronic sensing packages may also be attached to either end cap. These devices serve as proximity switches for accurate electronic positioning control or sequential switching. The operation of these features is discussed in Chapters 6 and 8 of this book, where more detailed descriptions of flow controls and directional controls are presented.

FIGURE 5-16

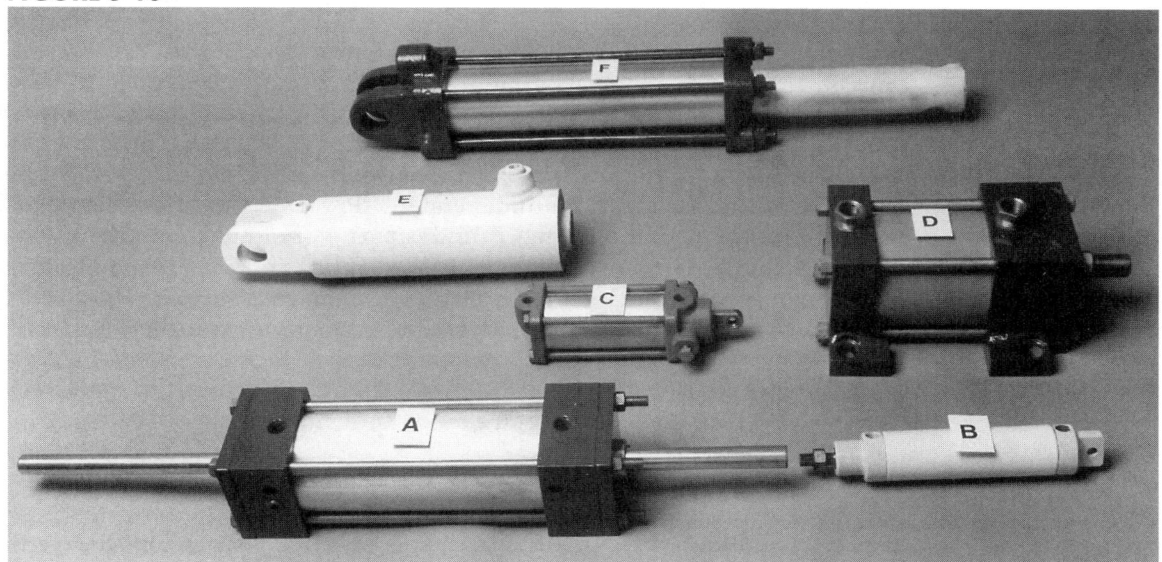

A variety of body styles, mountings, and types of hydraulic cylinders: **(A)** double end, double-acting, tie-rod body style, extended tie-rod mount; **(B)** single end rod, double-acting, screw body style, clevis mount; **(C)** single end rod, double-acting, tie-rod body style, trunion mount; **(D)** single end rod, double-acting, tie-rod body style, foot lug mount; **(E)** single end rod, double-acting, one-piece body style, clevis mount; and **(F)** single end rod, single-acting, tie-rod body style, clevis mount. *(Photograph by the author.)*

5.2 DOUBLE-ACTING CYLINDERS AND LINEAR CIRCUITRY

Unlike hydraulic jacks that may use ram cylinders, circuits that incorporate double-acting cylinders must supply fluid to both ends of the cylinder, selectively. The gate valve used in the hydraulic jack directs hydraulic fluid selectively either to the piston or to the tank. When the flow is directed to the piston, displacement will occur if circuit pressure acting upon the piston area is great enough to overcome the load imposed. When the flow is directed to the tank, the load acting under gravitational forces exhausts fluid from the ram and retracts the cylinder. The directional control for the double-acting cylinder circuit must be more complicated. While the cylinder is extending, the fluid in the head end of the cylinder will need to be exhausted to the tank. Inversely, when the cylinder retracts, the exhaust will be passed through the base end of the cylinder to the tank.

A directional control is required that can be shifted into two positions to allow the selective flow paths. Such directional control is possible with a sliding cylindrical spool seated in a bore. The spool is undercut in specific locations to open and close passageways inside the valve selectively. These directional controls typically have four external ports: a pressure port, a tank port, and two cylinder ports. In one position, the pressure port and the first cylinder port are connected while the second cylinder port is connected independently to the tank port. Fluid is, therefore, allowed to flow two different ways in this position. Shifting the spool to the second position reverses the independent flow paths to allow flow from the pressure port to the second cylinder port while independently allowing flow from the first cylinder port to the tank port. Thus, two more flow ways exist in this position. A description of the **two-position, four-way, lever-activated directional control** appears in Figure 5-17. *The simplest double-acting linear circuit is composed of a power supply; a two-position, four-way directional control valve; and a single end rod, double-acting cylinder* (see Figure 5-18).

5.2.1 Linear Circuit Design Parameters

Since the double-acting linear circuit serves as a power output device in both stroke directions, selection and specification for circuit components may need to accommodate various or combination parameters. In general, *the purposes of a linear circuit are to convert hydraulic energy into linear motion, to overcome any load encountered or exert a force during or at the end of the linear motion, and to return to the previous condition or position to allow repetition of the function*. Therefore, the parameters under which a linear circuit is designed normally include a force to be exerted (or a load to be overcome) and a rate at which cylinder movement will occur **(feed-rate parameter)** or a number of cycles to be completed by the cylinder in a given period of time **(piece-rate parameter)**. In most instances, the selection of the size of the cylinder under the assumed or given parameters will result in all other specifications for components in the circuit.

Primary attention should be made as to whether the parameters will be evident in both stroke directions or only extension or only retraction of the cylinder. For example, if a load of 1,000 lbs. {*4.5 kN*} is resisting the extension only of a single end rod, double-acting cylinder having a 2-sq. in. {*12.9-cm^2*} cross-sectional area piston and a 1-sq. in. {*6.5-cm^2*} cross-sectional area rod, a pressure of 500 psi {*3,488.4 kPa*} will develop in the circuit. Normally, the relief valve for this linear circuit would be set at 10% above this pressure, or at 550 psi {*3837.2 kPa*}. However, if the load on the linear circuit resists both extension and retraction of the cylinder, a pressure of 1,000 psi {*6,923.1 kPa*} would be required to retract the cylinder. With the pressure-relief valve set at 550 psi {*3,837.2 kPa*}, re-

FIGURE 5-17

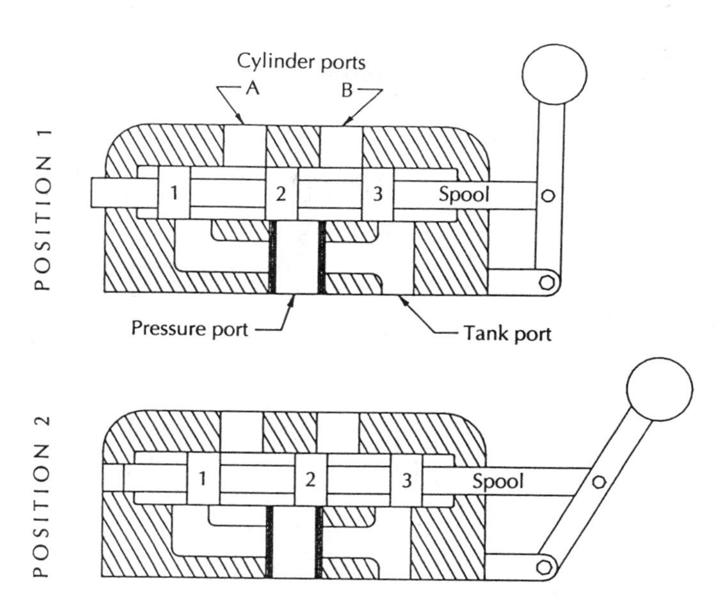

In Position 1, the pressurized flow from the pump is directed to the pressure port, through the valve to the cylinder B port, while landing 2 excludes this flow from the cylinder A port and landing 3 blocks the passage back to the tank port. At the same time, return flow is directed from the circuit to the cylinder A port and returned to the tank port through the valve. Landing 2 excludes this flow from the pressure chamber.,

In Position 2, the pressurized flow from the pump is directed to the pressure port, through the valve to the cylinder A port, while landing 2 excludes this flow from the cylinder B port and from one tank passageway. Landing 1 excludes pressurized flow from returning to tank through the other passageway. At the same time, return flow is directed from the circuit to the cylinder B port and returned to the tank port through the valve. Landing 2 excludes this flow from the pressure chamber.

Assignment: Color-code the drawings.

traction would not occur. The hydraulic fluid would take the path of least resistance and return to the tank over the relief valve rather than traveling to the head end of the cylinder to cause retraction. In this case, the pressure-relief valve should be set at 1,100 psi {*7,615.4 kPa*} (10% more than the maximum system pressure required in the circuit). Of course, system pressure will rise to 1,100 psi {*7,615.4 kPa*} at the end of the extension stroke also. If this pressure could cause damage or malfunction during the extension stroke, other pressure-controlling devices may be necessary to limit extension pressure. Such may be the case if the linear circuit was being used as a clamp. The excessive force, in this case 2,200 lbs. {*9.8 kN*} total, might cause damage to the clamped piece of material.

FIGURE 5-18

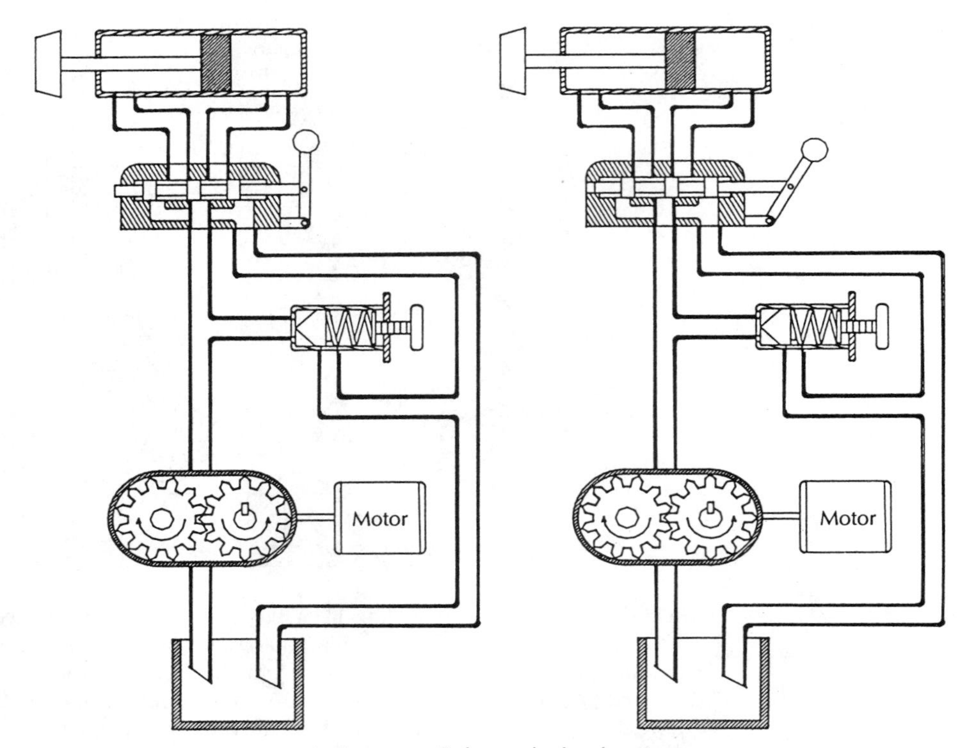

Assignment: Color-code the drawings.

Feed- or piece-rate parameters will determine the supply needed from the pump in the linear circuit. **Consider the cylinder described in the previous example having a 10-in. {*25.4-cm*} stroke. In this example, 20 cu. in. {*327.7 cm³*} of fluid would be needed to extend the cylinder and 10 cu. in. {*163.8 cm³*} of fluid would be needed to retract the cylinder, for a total of 30 cu. in. {*491.5 cm³*} of supply fluid during one cylinder cycle. If this cylinder needed to be cycled five times during a one-minute period, it would need to be supplied with 150 cu. in. {*2,457.5 cm³*} of fluid during that minute. The pump would need to deliver 0.65 gpm {*2.5 lpm*} (see Figure 5-19).**

Problem 5-7

What pump delivery will be required if the 3-inch {3.8-cm} diameter cylinder shown in Figure 5-19, having a 1-inch {1.3-cm} rod and a 6-inch {7.5-cm} stroke,is to be cycled ten times during a one-minute period?

DEL (gpm){lpm} = ________

FIGURE 5-19

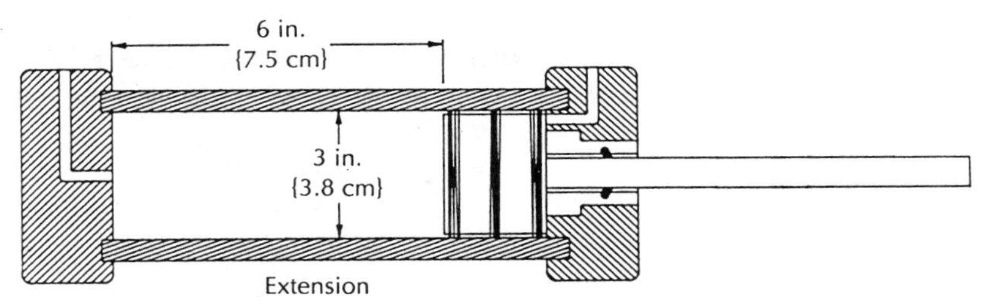

Extension

The volume necessary for extension is determined by the cross-sectional area of the piston and the stroke.

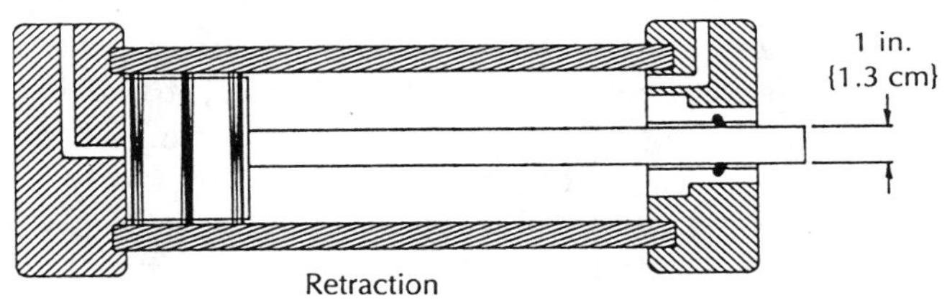

Retraction

The volume necessary for retraction is determined by the annulus area and stroke.

Other linear circuits are designed to cause the cylinder to extend or retract at a specified rate (feed-rate parameter). This rate is determined by the effective area of the cylinder and the delivery of the pump. A desired feed rate may be achieved by selecting a cylinder and specifying a pump supply to achieve that velocity. If a cylinder having a 2-sq. in. {*12.9-cm²*} cross-sectional area piston and a 1-sq. in. {*6.5-cm²*} cross-sectional area rod is required to operate at a rate of 30 in./min. {*76.2 cm / min.*}, the following formula could be used to determine the necessary pump supply:

Supply Rate (cu. in./min.) = VEL (in./min.) * A (sq. in.), or

Supply Rate {$cm^3/min.$} = VEL {$cm/min.$} * A {cm^2}

If the desired velocity is necessary during extension of the single rod, double-acting cylinder, the piston area should be used in this formula (resulting in a determined supply rate of 60 cu. in./min. {*983* $cm^3/min.$}). If the desired velocity is necessary during retraction, the annulus area should be used (resulting in a determined supply rate of 30 cu. in./min. {*492* $cm^3/min.$}). In either case, the determined pump supply rate will result in a variation of velocity in the opposite stroke. The resultant velocity may be determined by solving the aforementioned equation for velocity using the determined supply rate and the opposite appropriate area:

$$\text{VEL (in./min.)} = \frac{\text{Supply Rate (cu. in./min.)}}{\text{A (sq. in.)}}, \text{ or}$$

$$\text{VEL \{cm/min.\}} = \frac{\text{Supply Rate \{cm}^3\text{/min.\}}}{\text{A \{cm}^2\text{\}}}$$

For example, if the cylinder described previously required 60 cu. in./min. {*983.2 cm^3/min.*} to extend at a rate of 30 in./min. {*76.2 cm/min.*}, the resultant velocity during retraction would be 60 in./min. {*152.4 cm/min.*}. If a specified velocity is required in both stroke directions, a double end rod, double-acting cylinder may be used or some other control incorporated.

The determined pump supply rate should be expressed in gal./min. {*l/min.*} as a **pump delivery** specification. Rarely will a pump be available that has a rated delivery exactly matching a determined delivery. If an absolutely accurate cylinder velocity is necessary, a variable-displacement pump having the determined delivery near mid-range may be used. The pump may then be "tuned" to adjust to the desired cylinder velocity (see Figure 5-20).

Problem 5-8

The ANSI schematic in figure 5-20 represents a linear circuit using a 4-inch {10.6-cm} diameter cylinder having a 1-inch {2.5-cm} rod.

What delivery will be required from the variable-displacement pump if the cylinder needs to extend at a velocity of 15 in./min. {37.5 cm/min.}?

DEL (gpm){lpm} = ________

[] [] [] 5.3 LINEAR CIRCUIT DESIGN PROBLEMS

5.3.1 Piece-Rate Parameter

In a large metal fabrication factory, I-beams are welded together to form sections for bridge supports. Each weld is examined by X ray in an inspection room. Currently, the I-beams are pushed into position at the inspection station by four operators using a cart that reduces the effective load to 2,000 lbs. {*8.9 kN*}. After the X ray, the operators remove the sections from the station. Currently, this operation requires 80 seconds to perform.

A new contract has been signed that will necessitate the production and inspection of four of these I-beams in the same period of time. Your

FIGURE 5-20

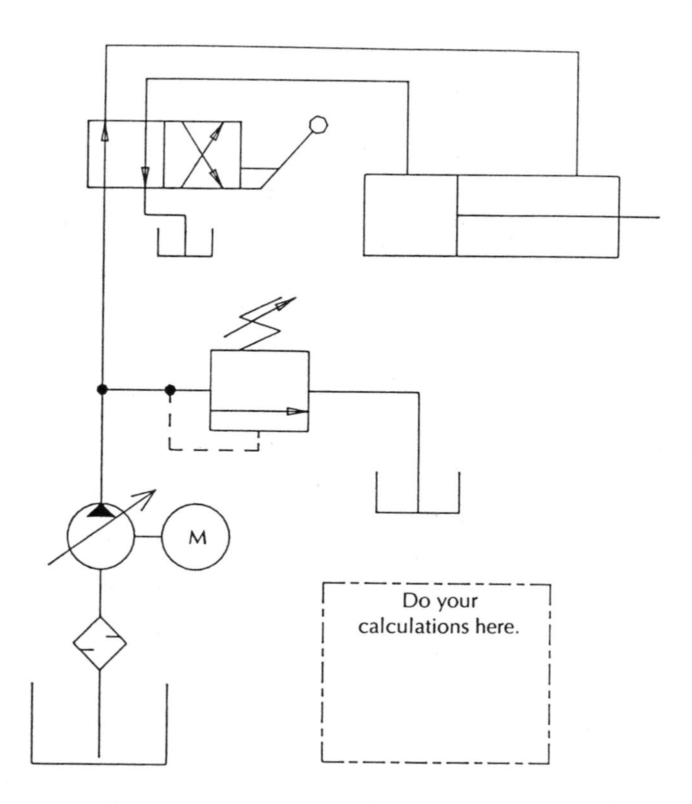

assignment is to design a hydraulic circuit that will meet this production requirement if it takes 5 seconds total time to connect and disconnect a cart to the circuit's cylinder. Thus, the actual circuit must perform 4 cycles per minute. The carts will move through a distance of 5 ft. {*1.3 m*} into and from the inspection station. The operator will, therefore, be removed from the X-ray station at all times during the operation for safety reasons. Consider the actual time to take the X ray to be negligible. Complete your assignment by specifying all components for the circuit in the following format:

A Single End Rod, Double-Acting Cylinder

PISTON		*ROD*			
Diameter	Area	Diameter	Area	Annulus Area	Stroke
3 in. {*6 cm*}	_____ in.^{2} {*cm*^{2}}	1 in. {*2.5 cm*}	_______ in.^{2} {*cm*^{2}}	________ in.^{2} {*cm*^{2}}	60in. {*132 cm*}

A Single End Rod, Double-Acting Cylinder *(continued)*

FORCE		*PRESSURE*		*SUPPLY*	
Extension	Retraction	Extension	Retraction	Extension	Retraction
2,000 lbs. {*8.9 kN*}	2,000 lbs. {*8.9 kN*}	_______ psi {*kPa*}	_______ psi {*kPa*}	________ in.^{3} {*cm*^{3}}	________ in.^{3} {*cm*^{3}}

			VELOCITY	
Total Supply	Rate	Supply Rate	Extension	Retraction
_______ in.^{3} {*cm*^{3}}	4 cyc/min.	_______ in.^{3}/min. {*cm*^{3}/*min.*}	_______ in./min{*cm/min*}	_______ in./min {*cm/min*}

Port Size	Body Style	Mounting
_______ in.{*mm*}	__________	__________

B Pump

Type	Design	Delivery	Drive Speed	Volumetric Effic.
__________	__________	________ gpm{*lpm*	___________ rpm	___________

	PORT SIZE	
Displacement at V.E.	Inlet	Outlet
______ in.^{3}/rev. {*cm*^{3}/*rev.*}	________ in. {*mm*}	________ in. {*mm*}

C Directional Control

Positions	Ways	Element	Activation	Port Size
_______	________	Sliding Spool	__________	________ in. {*mm*}

D Pressure Relief Valve

Type	Element	Setting	Mounting	Port Size
Simple	Poppet	_______ psi {kPa}	__________	_______ in. {*mm*}

E Filtration

Type	Element	RATING: Element	RATING: Flow	Port Size
Strainer	Wire	________ mesh	________ gpm {*lpm*}	________ in. {*mm*}

Description	Mounting
Full Flow	________

F Reservoir

CAPACITY	
Minimum	Maximum
______ gal. {*l*}	______ gal. {*l*}

G Electric Motor

Speed	Voltage	Phase	Power at V.E.
________ rpm	________	________	________hp {*W*}

H Conductors

Inlet Type	Inlet Description	*INLETS (I.D. DECIMAL)* Minimum	Maximum	Working Type	Working Description
________	________	______ in. {*cm*}	______ in. {*cm*}	________	________

WORKINGS (I.D. DECIMAL) Minimum	Maximum	*LINES—NOMINAL* Inlets	Workings
______ in. {*cm*}	______ in. {*cm*}	______ in. {*mm*}	______ in. {*mm*}

I Draw an accurate schematic diagram of the circuit described in the previous specification.

J The following are variations for the Piece-Rate Parameter—Linear Circuit Design Problem (5.3.1):

Component: Single End Rod, Double-Acting Cylinder

	A	B	C	D	E
Piston Diameter	4 in.	3 in.	4 in.	10 *cm*	8 *cm*
Rod diameter	2 in.	2 in.	1 in.	5 *cm*	2.5 cm
Stroke	55 in.	48 in.	36 in.	75 *cm*	1.5 *m*
Extension Force	6,000 lbs.	2,500 lbs.	8,500 lbs.	14 *kN*	10 *kN*
Retraction Force	6,000 lbs.	2,500 lbs.	8,500 lbs.	14 *kN*	10 *kN*
Cycle Rate	3 cyc/min.	4 cyc/min.	2 cyc/min.	2 *cyc/min.*	4 *cyc/min.*

5.3.2 Feed-Rate Parameter

A single-operator, two-station slitting saw operation is being considered at a local foundry for removal of castings from the runner system. A shuttle system that allows the operator to position one set of castings into a machining jig while another set is being sawed is required. Total load including weight of parts and jig, as well as the resistance of the sawing operation, has been calculated to be 1,500 lbs. {*6.7 kN*}. The operation requires a feed rate of 20 in./min. {*51 cm/min.*} through a 25-in. {*72-cm*} stroke. Design a linear hydraulic circuit that will act as the shuttle-feed mechanism. This will require the selection of an appropriately sized **double-acting, double end rod cylinder** and specifications for all other components in the circuit according to the following format:

A Double End Rod, Double-Acting Cylinder

PISTON		*ROD*			
Diameter	Area	Diameter	Area	Annulus Area	Stroke
_____ in. _____ {*cm*}	_____ in.^{2} {*cm*^{2}}	_____ in. _____ {*cm*}	_____ in.^{2} {*cm*^{2}}	_____ in.^{2} {*cm*^{2}}	25in. {*72 cm*}

FORCE		*PRESSURE*	
Extension	Retraction	Extension	Retraction
1,500 lbs. {*6.7 kN*}	1,500 lbs. {*6.7 kN*}	_________ psi {*kPa*}	_________ psi {*kPa*}

FEED RATE			*TIME*	
Extension	Retraction	Supply Rate	Extension	Retraction
20 in./min. {*51 cm/min.*}	20 in./min. {*51 cm/min.*}	_______ in.^{3}/min. {*cm*^{3}*/min.*}	_______ sec.	_______ sec.

Port Size	Body Style	Mounting
_______ in.{*mm*}	__________	__________

B Pump

Type	Design	Delivery	Drive Speed	Volumetric Efficiency
__________	__________	________ gpm {*lpm*}	__________ rpm	_______ %

Displacement at V.E.	*PORT SIZE*	
	Inlet	Outlet
______ in.^{3}/rev. {*cm*^{3}*/rev.*}	________ in. {*mm*}	________ in. {*mm*}

C Directional Control

Positions	Ways	Element	Activation	Port Size
________	________	Sliding Spool	___________	________ in. {*mm*}

D Pressure Relief Valve

Type	Element	Setting	Mounting	Port Size
Simple	Poppet	_______ psi {kPa}	__________	_______ in. {*mm*}

E Filtration

		RATING		
Type	Element	Element	Flow	Port Size
Strainer	Wire	_________ mesh	________ gpm {*lpm*}	________ in. {*mm*}

Description	Mounting
Full Flow	________

F Reservoir

CAPACITY	
Minimum	Maximum
______ gal. {*l*}	______ gal. {*l*}

G Electric Motor

Speed	Voltage	Phase	Power at V.E.
________ rpm	__________	___________	_________hp {*W*}

H Conductors

		INLETS (I.D. DECIMAL)			
Inlet Type	Inlet Description	Minimum	Maximum	Working Type	Working Description
_______	_________	______ in. {*cm*}	______ in. {*cm*}	__________	_________

H Conductors *(continued)*

WORKINGS (I.D. DECIMAL)		*LINES—NOMINAL*	
Minimum	Maximum	Inlets	Workings
______ in. {*cm*}	______ in. {*cm*}	______ in. {*mm*}	______ in. {*mm*}

I Draw an accurate schematic diagram of the circuit described in the previous specification.

J The following are variations for the Feed-Rate Parameter—Linear Circuit Design Problem (5.3.2):

Component: Double End Rod, Double Acting Cylinder

	A	B	C	D	E
Stroke	12 in.	18 in.	20 in.	40 *cm*	50 *cm*
Extension Force	4,000 lbs.	3,800 lbs.	3,200 lbs.	18 *kN*	12 *kN*
Retraction Force	2,000 lbs.	1,500 lbs.	1,300 lbs.	8 *kN*	5 *kN*
Feed Rate	25 in./min.	15 in./min.	12 in./min.	6 *cm/min.*	5 *cm/min.*

5.4 REGENERATION

The previous design problem illustrates one method of achieving equal speed in both extension and retraction of a cylinder. It is not always feasible to include a double end rod cylinder in a linear circuit design. This may be true when space is limited or the expense of such a cylinder cannot be justified. A single end rod cylinder can achieve equal speed in both stroke directions if a flow control, which reduces the supply rate to the rod end of the cylinder, is included. The added expense of the flow-control device may not be justified, especially if constant cylinder feed rate is more important than variability of feed rate. However, equal velocity of the single end rod cylinder is possible with a simple design alteration to the linear circuit. This circuit is called a **regenerative circuit** (see Figure 5-21).

FIGURE 5-21

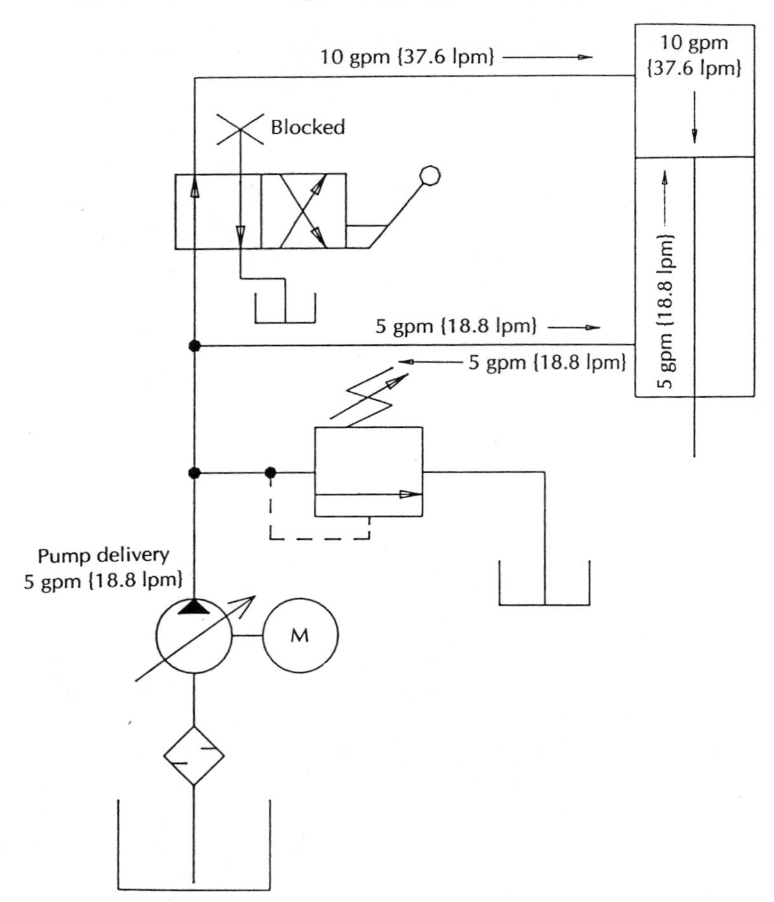

In regeneration, during extension, the differential cylinder's 5-gpm {18.8 lpm} pump delivery is supplemented by an additional 5-gpm {18.8 lpm} delivery forced from the head end of the cylinder (for a total effective delivery of 10 gpm {37.6 lpm} going to the base end of the cylinder). During retraction, only the 5-gpm {18.8=lpm} delivery is effective to the head end of the cylinder, since the base end is exhausting its fluid to the reservoir through the directional control. Since extension effective delivery is twice that of effective retraction delivery and the piston has twice the effective area of the annulus, equal velocity on both strokes is attained.

5.4.1 Regenerative Operation

In the regenerative circuit, the exhausting fluid from the head end of the cylinder is directed back into the supply side of the extending cylinder. Since the exhaust is leaving the head end of the cylinder at the same rate that pump supply is entering the base end, the net effect is a doubling of the supply rate during extension of the cylinder. The cylinder velocity will, therefore, be doubled. During retraction, the exhaust from the base end of the cylinder is directed back to the tank through the directional control. System pressure acting on the head end will retract the cylinder at a rate determined by the pump supply and the annulus area.

5.4.2 Differential Cylinders

Most regenerative circuits operate using a specially designed cylinder that has a piston to annulus area ratio of two to one. This type of cylinder is known as a **differential cylinder**. In most cases a cylinder having a standard (nominal) size piston is fitted with a rod machined to displace one-half of the piston area. Special rod end caps, bushings, and seals must be made for the differential cylinder. Thus, a cylinder with a 2-in. diameter piston (3.14-sq. in. area) would require a rod that would displace 1.57 sq. in. {*A cylinder with a 5.1-cm diameter piston (20.3 cm²) would require a rod that would displace 10.15 cm².*} Using the following formula, the diameter that the rod will need to be machined to can be calculated:

$$\text{DIA (in.)}\ \{cm\} = \sqrt{\frac{\text{A (in.}^2\text{)}\ \{cm^2\}}{0.7854}}$$

Problem 5-9

Calculate the rod diameter for a differential cylinder having a 2-in. {*5.1-cm*} diameter piston.

DIA (in.) {*cm*} = ________

When the regenerative circuit is fitted with a differential cylinder, cylinder velocity will be equal in both stroke directions. Furthermore, *if the load is constant on both extension and retraction, system pressure will be equal during both strokes* (see Figure 5-22).

5.4.3 Specifications in Regeneration

Differential cylinders simplify calculations and specifications for regenerative circuitry. Since the circuit acts with equal pressure and velocity in both stroke directions, retraction components can be used to calculate

FIGURE 5-22

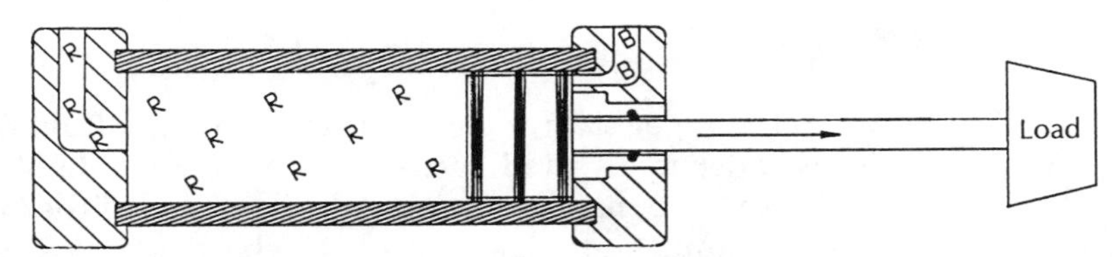

During extension, the differential cylinder in a regenerative circuit will have system pressure acting on both the piston and the annulus areas. If the circuit is operating at 500 psi {3447.4 kPa} with a 2-in.^{2} {12.9-cm^2} piston area and a 1-in.^{2} {6.5-cm^2} rod area, the cylinder will have an extension force of 1,000 lbs. {4.4 kN} [F = 500 (lbs./in.^{2}) * 2 (in.^{2})] minus 500 psi acting on the annulus area [F = 500 (lbs./in.^{2}) * 1 (in.^{2})], for a net extension force of 500 lbs.

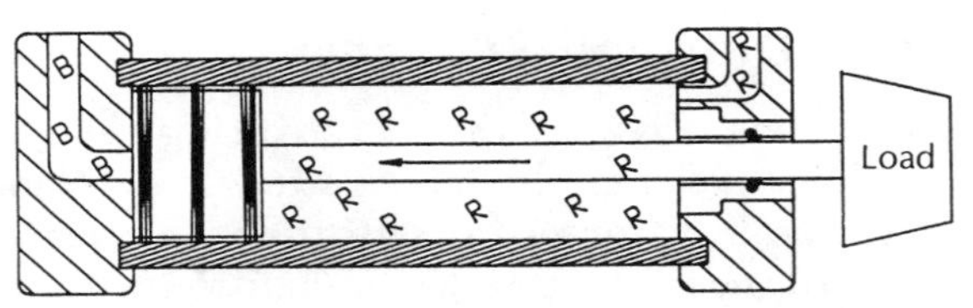

During retraction, the differential cylinder will have system pressure acting on the annulus area only. The piston area is exhausted to the tank through the directional control valve. If the circuit is operating under 500 psi {3,447.4 kPa}, the net retraction force will be determined by the annulus area and system pressure [F = 500 (lbs./in.^{2}) * 1 (in.^{2})].

other circuit specifications. During retraction, the regenerative circuit is operating like a linear circuit. Thus, *supply rates and operating pressure can be calculated as would any linear circuit during retraction using annulus areas and volumes*. During extension, other specifications are simply set equal to these values.

5.4.4 Regenerative Circuit Design Problem

A manufacturing plant produces sinks made from type 308 stainless steel. In the manufacturing process, it is necessary to punch two large holes in the bottom of each basin plus four other holes in the sink skirt to allow for attachment of the faucet and spray attachment. The punch currently being used to perform this operation is a unique horizontal press. The press must overcome a resistance of 2,000 lbs. to punch each part. Also, the press die itself weighs 1,500 lbs. Thus, 3,500 lbs. {*15.6 kN*} of force are encountered during extension of the cylinder and 1,500 lbs. {*6.7 kN*} of force are encountered during retraction.

Lately, there have been complaints from press operators that the linear circuit being used to operate the press is dangerous because it extends and retracts at different rates. They claim that this upsets their work rhythm. Your assignment is to redesign the hydraulic circuitry to allow for an equal speed of 15 in./sec. {*38.1 cm*/sec.} in both extension and retraction of a cylinder having a 12-in. {*30.5-cm*} stroke. Complete this assignment by selecting an appropriately sized **differential cylinder** and specifying all components necessary for the regenerative circuit according to the following format:

A Double End Rod, Double-Acting Cylinder

PISTON		*ROD*			
Diameter	Area	Diameter	Area	Annulus Area	Stroke
____ in. ____ {*cm*}	____ in.^{2} {*cm*^{2}}	____ in. ____ {*cm*}	____ in.^{2} {*cm*^{2}}	____ in.^{2} {*cm*^{2}}	12 in. {*30.5 cm*}

FORCE		*PRESSURE*	
Extension	Retraction	Extension	Retraction
3,500 lbs. {*15.6 kN*}	1,500 lbs. {*6.7 kN*}	________ psi {*kPa*}	________ psi {*kPa*}

FEED RATE			*TIME*	
Extension	Retractiion	Supply Rate	Extension	Retraction
15 in./sec. {*38.1 cm/sec.*}	15 in./sec. {*38.1 cm/sec.*}	______ in.^{3}/min. {*cm*^{3}*/min.*}	______ sec.	______ sec.

Port Size	Body Style	Mounting
______ in.{*mm*}	________	________

B Pump

Type	Design	Delivery	Drive Speed	Volumetric Efficiency
________	________	________ gpm {*lpm*}	________ rpm	________ %

	PORT SIZE	
Displacement at V.E.	Inlet	Outlet
______ in.^{3}/rev. {*cm*^{3}*/rev.*}	________ in. {*mm*}	________ in. {*mm*}

C Directional Control

Positions	Ways	Element	Activation	Port Size
________	________	Sliding Spool	________	________ in. {*mm*}

D Pressure Relief Valve

Type	Element	Setting	Mounting	Port Size
Simple	Poppet	________ psi {kPa}	________	________ in. {*mm*}

E Filtration

		RATING		
Type	Element	Element	Flow	Port Size
Strainer	Wire	________ mesh	________ gpm {*lpm*}	________ in. {*mm*}

Description	Mounting
Full Flow	________

F Reservoir

CAPACITY	
Minimum	Maximum
________ gal. {*l*}	________ gal. {*l*}

G Electric Motor

Speed	Voltage	Phase	Power at V.E.
________ rpm	________	________	________hp {*W*}

H Conductors

Inlet Type	Inlet Description	INLETS (I.D. DECIMAL) Minimum	INLETS (I.D. DECIMAL) Maximum	Working Type	Working Description
______	__________	________ in. {*cm*}	________ in. {*cm*}	__________	__________

WORKINGS (I.D. DECIMAL) Minimum	WORKINGS (I.D. DECIMAL) Maximum	LINES—NOMINAL Inlets	LINES—NOMINAL Workings
________ in. {*cm*}	________ in. {*cm*}	________ in. {*mm*}	________ in. {*mm*}

I Draw an accurate schematic diagram of the circuit described in the previous specification.

J The following are variations for the Regenerative Circuit Design Problem (5.4.4):

Component: Differential Cylinder

	A	B	C	D	E
Stroke	35 in.	20 in.	36 in.	80 *cm*	50 *cm*
Extension Force	2,600 lbs.	3,500 lbs.	4,000 lbs.	8 *kN*	10 *kN*
Retraction Force	2,600 lbs.	3,500 lbs.	4,000 lbs.	8 *kN*	10 *kN*
Feed Rate	25 in./sec.	15 in./sec.	10 in./sec.	33 cm/sec.	25 cm/sec.

5.5 SUGGESTED ACTIVITIES

1. Perform an inspection by disassembling, measuring, and reassembling a cylinder noting the following:
   a. type of cylinder (i.e., ram; single end rod, double-acting; double end rod, double-acting; telescoping; etc.)
   b. body style (i.e., tie-rod, flange, screw, one-piece, etc.)
   c. mounting (i.e., extended tie-rod, threaded end cap, trunion, clevis, flange, lug, etc.)
   d. size of port
   e. size of piston
   f. size of cylinder sleeve
   g. size of rod(s)
   h. size of bushing(s)
   i. stroke length
   j. other features
2. Calculate the capacity (supply) necessary to extend and retract the cylinder. Explain any assumptions that you may have made concerning these calculations.
3. Set up and operate a linear circuit on a training panel. Even if unloaded, note the following:
   a. pump delivery
   b. time for extension
   c. time for retraction
   d. pressure during extension
   e. pressure during retraction
   f. pressure relief valve setting
4. Determine the piston and annulus areas for the cylinder in Activity 3. Determine the power necessary to drive the circuit in Activity 3. Check the drive motor for the determined power. Does it match the calculation. If not, why not?

5.6 REVIEW QUESTIONS

1. Describe the general construction of a single end rod, double-acting cylinder.
2. Discriminate among the operations of a single end rod, single-acting cylinder; a single end rod, double-acting cylinder; and a double end rod, double-acting cylinder.
3. Describe the operation of a telescoping cylinder.
4. Describe the operation of a multiple-piston cylinder.
5. What is the major purpose of a stop tube used in cylinder construction?
6. List names for four types of cylinder construction.
7. Describe four types of mounts used for cylinders.
8. Are there any combinations of mounts, constructions, and types of cylinders that should not or could not be combined? If so, what are they?

6

Directional Control

As mentioned in Chapters 4 and 5, one of the primary considerations in any fluid power circuit is to control the direction of fluid flow. In early fluid devices, the direction of flow was the only control parameter within the circuitry. The movable masts on sailing ships and watergates in irrigation systems adjust fluid movement and divert it to use the fluid in a desired location. *The functions of a directional control are to divert flow from one location to another or to stop it.*

6.1 TWO-POSITION VALVES

Many descriptors are typically used to designate a given directional control. Of primary concern are the **number of different flow paths** and the **number of separate positions** into which the valve may be shifted. Together these descriptions identify the **fluid logic** in a valve. For example, a valve that may be shifted into two positions and diverts all flow from one path to another when shifted is called an **"or"** gate. The two-position valve that both diverts flow to a second path and maintains flow in the primary path is called an **"and"** gate. Other valve logic may allow multiple flow paths or totally exclude flow when shifted.

6.1.1 Check Valves

The simplest type of exclusionary device in a hydraulic circuit is a **check valve**. The composite ANSI symbol for the check valve is rather complicated and is nearly always replaced by the simplified symbol (see Figure 6-1).

The composite symbol is useful to explain the operation of a check valve. The two-position valve is normally held in the closed position by a light

FIGURE 6-1

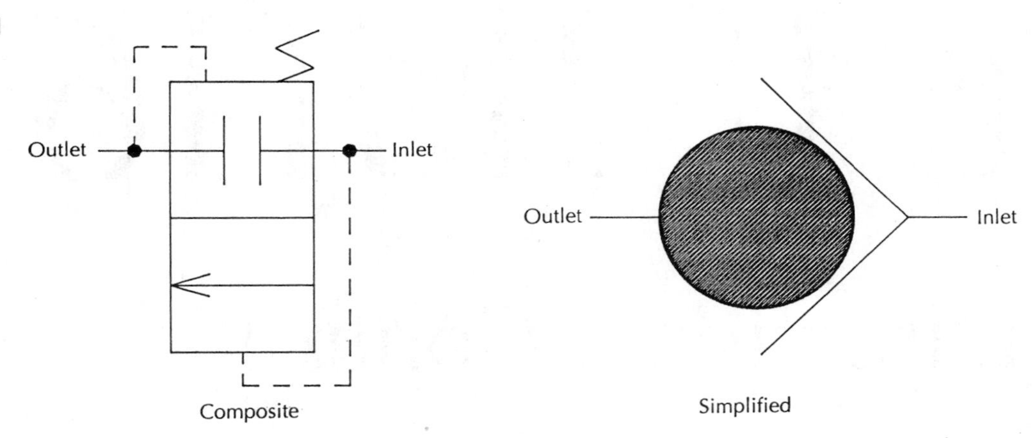

spring, and any pressure resulting from a resistance to flow from the outlet to the inlet will only assist the spring and exclude flow in this direction. In the other position, flow directed to the valve inlet will develop enough pressure to overcome the light spring; and flow will occur from the inlet, through the valve, to the outlet. The valve will allow flow to occur only one way. Thus, *the check valve is a two-position, one-way, normally closed directional-control valve.*

6.1.1.1 Types of check valves The two most common types of check valves include either a **ball** or **poppet element**. The **in-line check valve** described in Chapter 2 uses a ball element. The **right-angle check valve** described in Figure 6-2 uses a poppet element. The angle on the face of

FIGURE 6-2

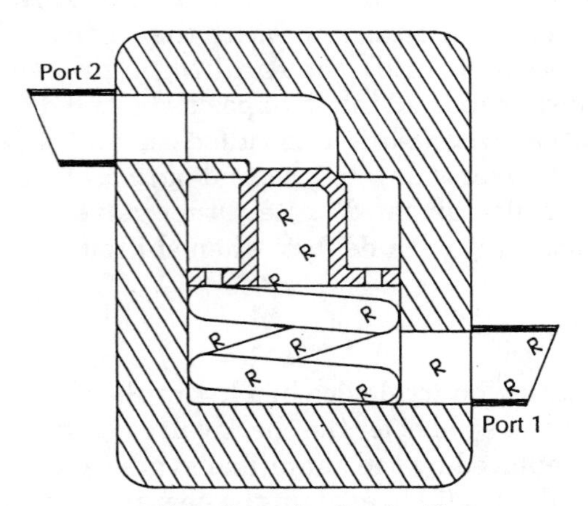

Flow directed to Port 1 will assist the spring and exclude flow through the valve.

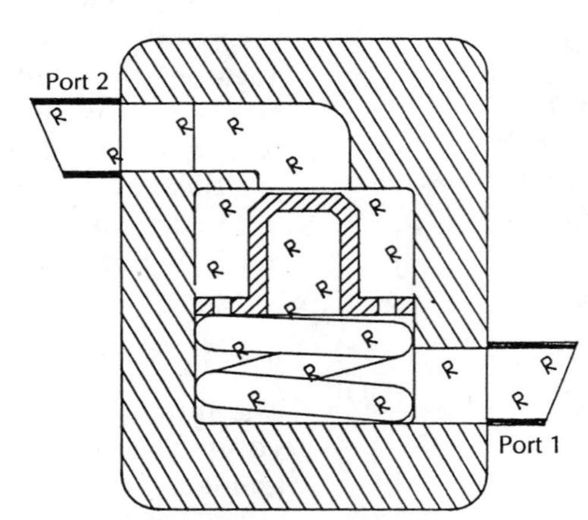

Flow directed to Port 2 will overcome the spring and allow flow through the valve.

the poppet element should be ground to a different angle than the corresponding seat ground into the valve casing. Typically, the interference angle is between 2 and 5 degrees. This allows a fine line interface to develop, and a superior seal occurs. In many cases, check valves are used in combination with directional controls having other types of elements because the check valve has this superior sealing capacity. The right-angle check valve may also be referred to as an **L-design check valve.** This design of check valve may be used to eliminate an elbow in circuit plumbing.

The right-angle design also allows some other design features to be incorporated into the check valves. For example, a separate piston may be included into the inlet port that will allow the check valve to be overridden and opened when flow is directed from the circuit to the valve outlet. A pilot pressure taken from a remote pressure location acting on the piston area will result in a greater force than is evident through the combination forces of system pressure operating on the area of the poppet and the light duty spring. The piston will advance in its bore and force the poppet from its seat, allowing reverse free flow in the **pilot override check valve** (see Figure 6-3).

FIGURE 6-3

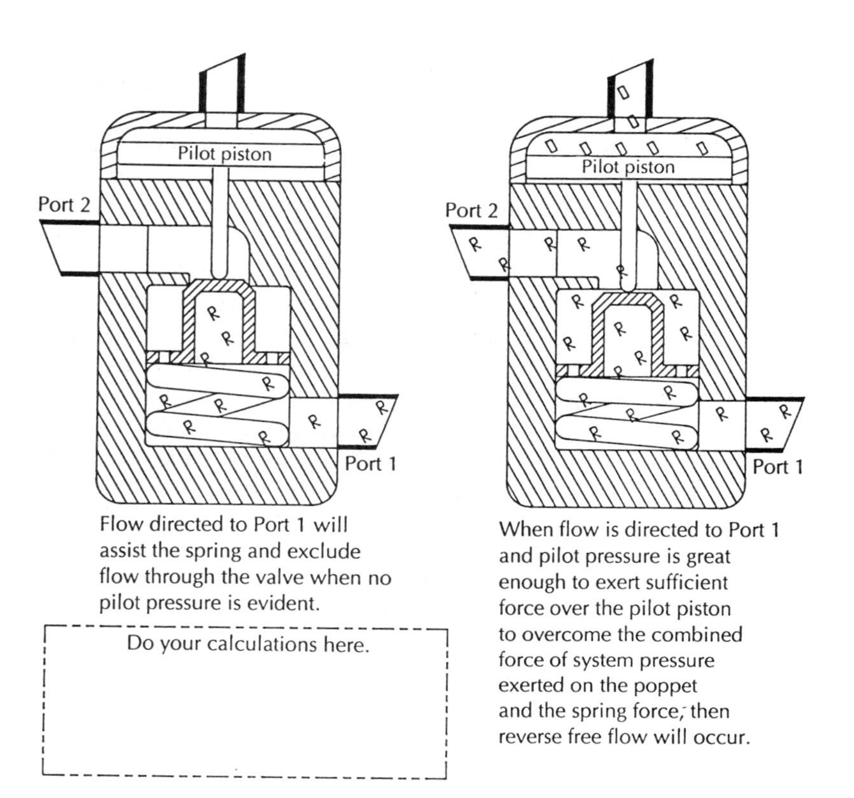

Flow directed to Port 1 will assist the spring and exclude flow through the valve when no pilot pressure is evident.

When flow is directed to Port 1 and pilot pressure is great enough to exert sufficient force over the pilot piston to overcome the combined force of system pressure exerted on the poppet and the spring force, then reverse free flow will occur.

Problem 6-1

The pilot-operated check valve described in Figure 6-3 has a spring that exerts 25 lbs. {110 N} of force over the poppet that has a cross-sectional area of 1 in.^{2} {3.2 cm^2}, and system pressure is 500 psi {3447.4 kPa}. What pressure must be evident in the pilot section to allow reverse free flow if the pilot piston has a cross-sectional area of 2 in.^{2} {12.9 cm^2}?

Pressure (psi){kPa} = ________

6.1.1.2 Uses of check valves The use of check valves in the hand pump has been previously described. Check valves are also common in more sophisticated industrial circuits. Often, they are included to exemplify or specify the actions of other circuit components. They may be separate components or integral to those other components as illustrated in a cylinder cushion device.

The **cylinder cushion** actually contains two integral components: a check valve and a flow-control valve. Figure 6-4 illustrates a cushion device used to decelerate the retraction of a cylinder during the final portion of the stroke. During the final phase of retraction, the rod extending through the rear of the piston closes the main passageway for fluid exhaust. Two other passageways exist for fluid exhaust. However, the exhausting fluid assists the spring in the integral check valve and excludes flow though this passageway. The other passageway is partially blocked by the needle valve flow control. This restriction of the exhaust flow reduces the speed of cylinder retraction only during the final phase.

The importance of the integral check valve becomes apparent during extension of the cylinder. Without the check valve, the flow would occur either through the flow control or through the major port currently sealed by the rod extension. In this case, the major purpose of the cylinder cushion is to decrease cylinder impact at the end of a rapid retraction stroke. If the fluid is directed through the flow control during extension, the speed will be greatly reduced during the beginning of the stroke until the rod projection clears its seat. In most instances, this action would have no favorable results and serve no useful purpose. Furthermore, if the fluid attempts to extend the cylinder by passing through the main port, it will act on the small area of the rod projection to overcome any load imposed during extension. This would cause an initial pressure surge to occur until the projection clears its seat. However, with the check valve, pressurized flow will not pass through the flow control nor the main passageway but will overcome the light duty spring on the check valve and act on both the rod projection area and the remainder of the piston area to apply maximum force under pressure during the full-extension stroke. After the rod projection clears its seat, the pressurized fluid will act on the full piston area through the main port.

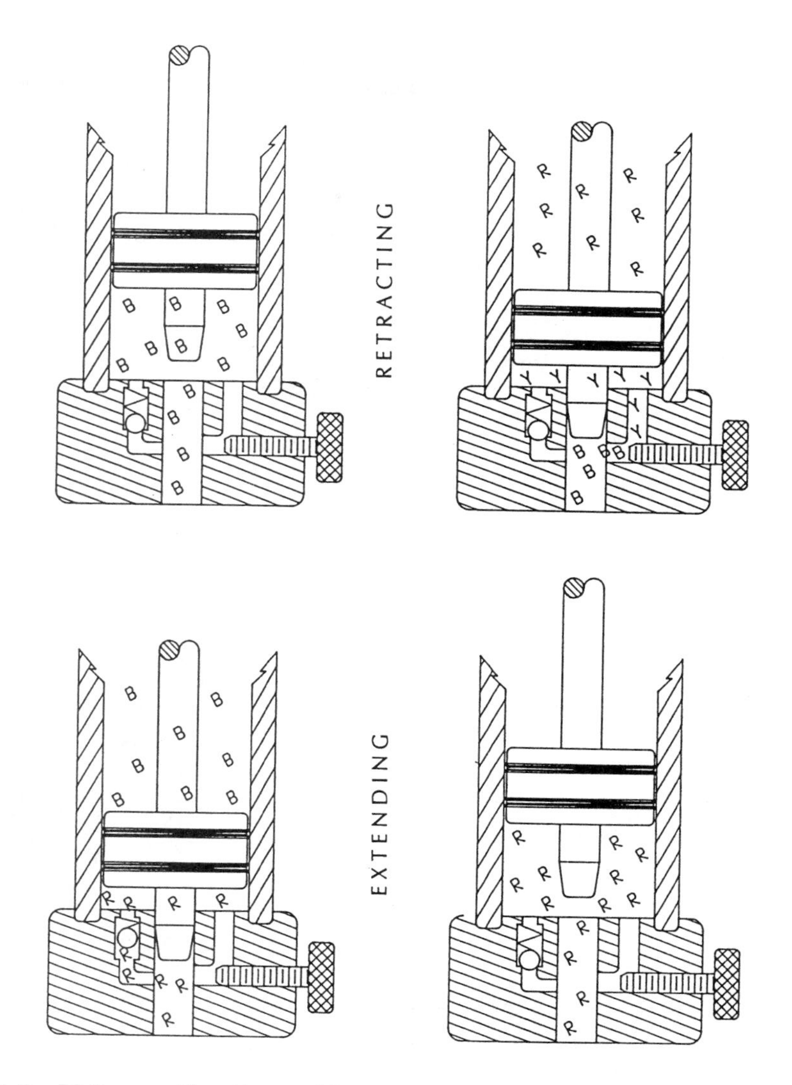

FIGURE 6-4

6.1.1.3 Valve activation Check valves use internal pilots (available hydraulic pressure) and spring activation to shift the valving element from one position to another. However, external activation methods are available for two-position valves. Although normally considered to be a flow control, the gate valve described in Bramah's jack may be described as a two-position, one-way, **lever-activated** directional control. Other methods of shifting valves include *electrical solenoids, foot treadles, palm buttons, and heat and mechanical types*. Combinations of these shifting methods are also used.

A two-position directional control that uses mechanical activation typically relies on physical contact with another moving component to cause shifting. One method of accomplishing this is to attach an enlarged rod extension, called a *cam*, to a cylinder (see Figures 6-5 and 6-6). The two-position, one-way, plunger-activated directional control described in Figure 6-5 uses a **sliding spool element**. The relationship between the cam and the physical mounting location of the directional control can be altered to cause activation during an intermediate portion of the extension or retraction stroke. The output from the directional control after shifting may be used to alter the action of the cylinder during the end of the extension stroke or the beginning of the retraction stroke. It may also be used to actuate another cylinder.

FIGURE 6-5

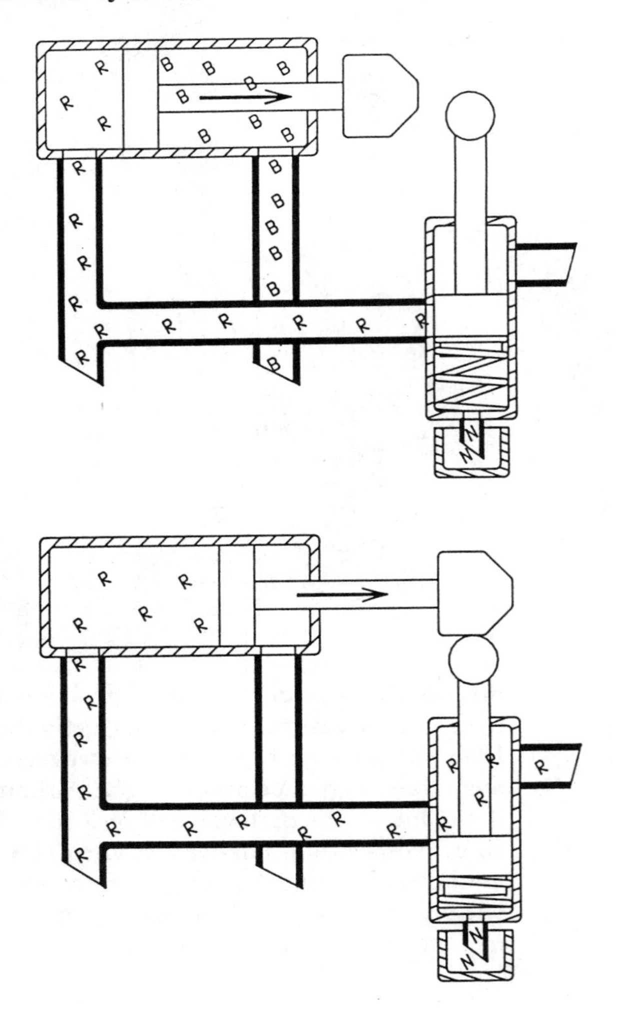

FIGURE 6-6

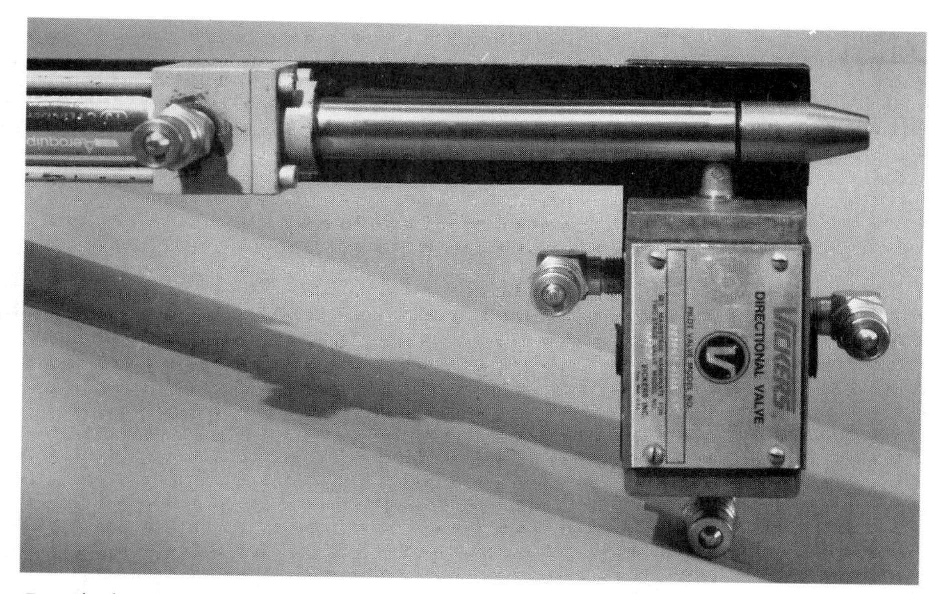

Detail of a cylinder cam activating the plunger shifting mechanism on a two-position, one-way, directional-control valve. (*Photograph by the author.*)

6.2 DIRECTIONAL-CONTROL APPLICATIONS OF CARTRIDGE VALVES

Traditionally, hydraulic components—including valves—have been rather heavy and bulky mechanisms, compared to electrical and electronic or even pneumatic control devices. This characteristic has been rationalized by citing the need for materials to withstand high-pressure environments and addition of stability through mass to high-powered, high-speed machinery. Recently, there has been a change in this philosophy in the hydraulic component industry. Borrowing from the electronic industry and even aerospace applications of hydraulics, component manufacturers have begun to downsize components. Using more exotic materials and more precise design tolerances, a new series of control components have emerged that are still capable of withstanding high pressure. Exaggerating this condition is a trend toward the use of smaller actuators (cylinders and motors). This results in the need for higher operating pressure to overcome heavy load or apply high forces, and machine stability is becoming a function of mounting rather than mass. At the forefront of this movement is the emergence of small, multifunctional control components called **cartridge valves**. Although cartridge valves may be used as pressure controls and flow controls, most cartridge valves are employed as directional controls.

Also breaking with tradition, cartridge valves have been named according to design rather than function. The small cylindrically-shaped casings for these valves somewhat resemble a rifle cartridge and are designed to be fitted into a central manifold block housing many such valves in what might best be described as a control module. Typically, they either slip into position and are held by a cover plate or may be screwed into place. Like electronic control modules operating on low-voltage circuitry, the cartridge valve control module typically operates at a reduced pressure from the remainder of the hydraulic power package. The net result is a centralized, safer, and more easily maintained control package.

6.2.1 Cartridge Valve Design

The cartridge valve is similar in design to the pilot override check valve in that it employs a control element (poppet or spool) held into a normal position by a lightly biased spring and opposing areas capable of operating under hydraulic pressure to cause shifting (see Figure 6-7).

Three functional areas exist within the design of the cartridge valve. The control area (A_{con}) is ported through the spring chamber and operates in conjunction with the spring to hold the valve normally closed. In potential opposition to this area are the signal port (A_{sig}) and the pilot port (A_{pil}) areas. The area combination of the signal port and the pilot port is equal to the area of the control port:

$$A_{con} = A_{sig} + A_{pil}$$

Cartridge valves are further designated by the ratio of signal area to control area; that is, cartridge valves having equal control and signal areas are designated as 1:1. In this design, there is little or no appreciable pilot area. Other popular designs are the 1:1.1 and 1:1.2 arrangements, having an increasing order of potential pilot area.

6.2.2 Cartridge Valve Operation

Being two-position, normally closed directional controls, cartridge valves are capable of performing as check valves. In these operations, control port pressure (P_{con}) operating over the control port area (A_{con}) in combination with the spring force operating over the control area serves to hold the valve closed. If no control port pressure is evident and signal port pressure (P_{sig}) and/or pilot port pressure (P_{pil}) operating over the appropriate area(s) create(s) forces greater than the spring force, then the poppet will move and expose the pilot port, allowing flow through the valve

FIGURE 6-7

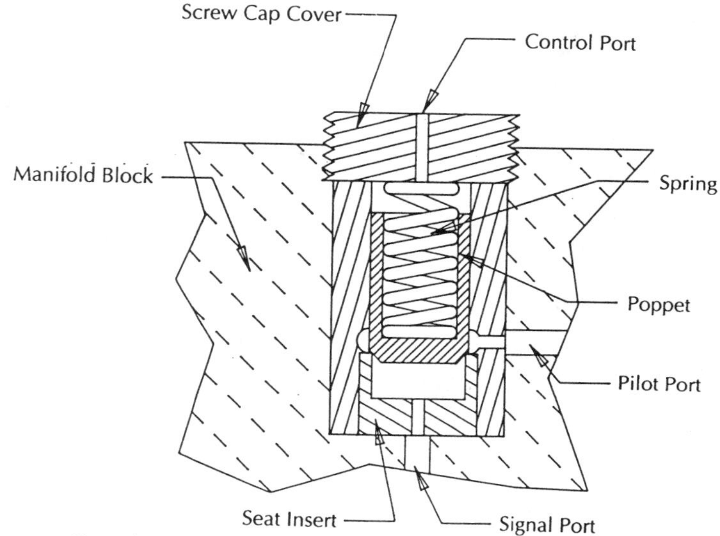

Control
Pilot
1 : 1
Signal

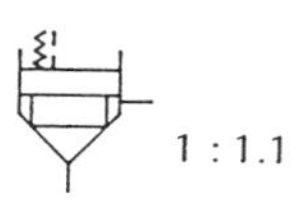

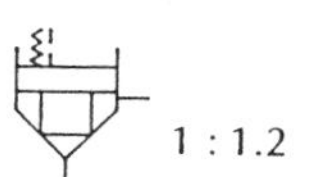

The multifunctional cartridge valve is designed to take advantage of multiple ports operating in conjunction and opposition to one another at the control, signal, and pilot locations. Further variability is attained by adjusting the inside diameter of the seat insert pocket to increase or decrease the effective sizes of the poppet at the signal and pilot ports. Increasing its inside diameter to maximum will make the signal port area equal to the control port area (1:1). Decreasing its diameter will expose less of the poppet face to the signal port, resulting in other designs (1:1.1 and 1:1.2), and increase the poppet area exposure at the pilot port.

from signal port to pilot port. Inversely, if pilot port pressure operating over its area is less than spring force (and/or control pressure operating over its area) and no pressure is evident at the control port, then flow is excluded from pilot port to signal port. Although relatively simplistic in this application, combinations of cartridge valves operating in a coordinated fashion can perform sophisticated logic procedures similar to four-way directional control (see Figure 6-8).

FIGURE 6-8

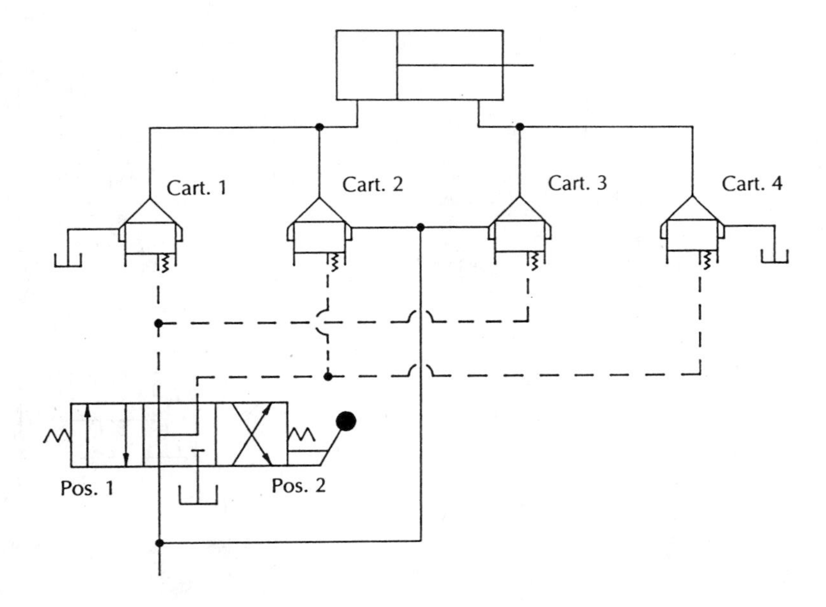

In this linear circuit, when the lever on the three-position, four-way directional control is unattended, the valve will shift into its normally centered position. At that time, supply fluid is directed to the control ports on all cartridge valves and supply fluid is unconditionally directed to the pilot ports on cartridge valves 2 and 3. In this condition, no fluid will be directed to the cylinder, since the resulting force holds these valves closed owing to greater control area than pilot area. Since cartridge valves 1 and 4 are also held closed, no fluid may escape from either end of the cylinder and it remains held rigidly in position.

When the three-position, four-way directional control is shifted into position 1, supply is directed to the control ports of cartridge valves 1 and 3 and unconditionally to pilot ports on cartridge valves 2 and 3. The control ports on cartridge valves 2 and 4 are open to the tank through the three-position, four-way directional control. This allows cartridge valve 2 to open and direct supply to the base end of the cylinder and extend it because cartridge valve 4 will open, allowing return to tank through its signal port to its pilot port.

When the three-position, four-way directional control is shifted into position 2, supply is directed to the control ports of cartridge valves 2 and 4 and unconditionally to pilot ports on cartridge valves 2 and 3. The control ports on cartridge valves 1 and 3 are open to the tank through the three-position, four-way directional control. This allows cartridge valve 3 to open and direct supply to the head end of the cylinder and retract it because cartridge valve 1 will open, allowing return to tank through its signal port to its pilot port.

6.3 SEQUENCING

6.3.1 Sequencing with Two-Position Directional Controls

If the output from the directional control is used to actuate another cylinder after the first cylinder finishes its stroke, sequencing occurs. Many industrial circuits rely on the timed actuation of multiple cylinders (see Figures 6-9 and 6-10).

FIGURE 6-9

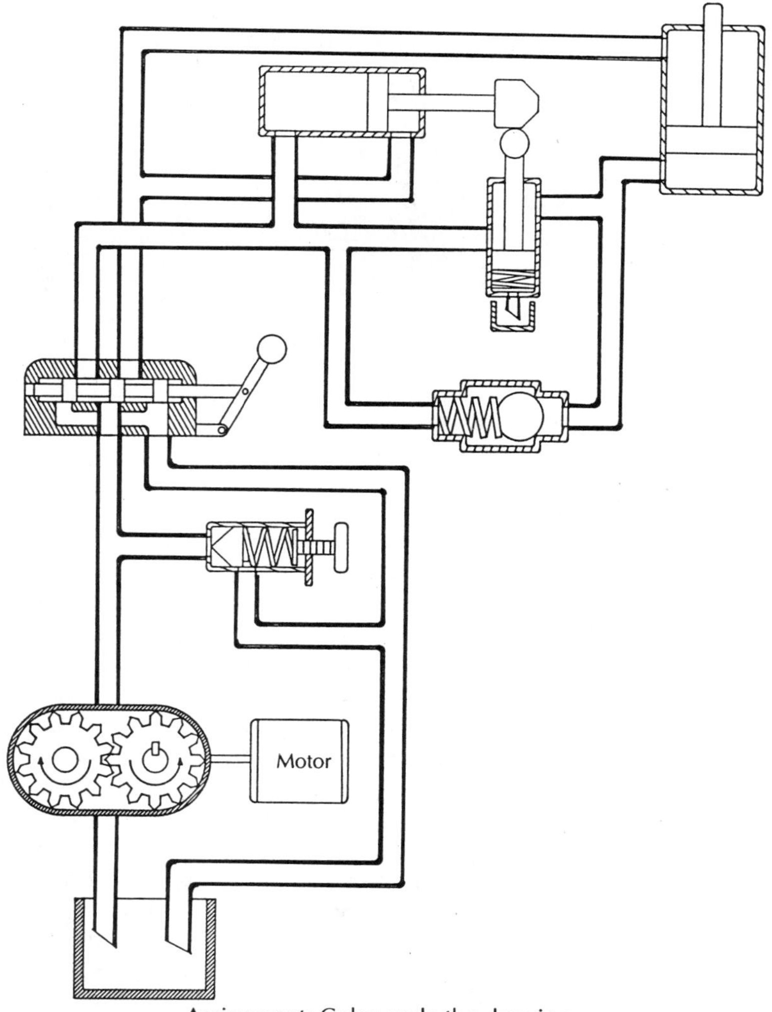

Assignment: Color-code the drawing.

FIGURE 6-10

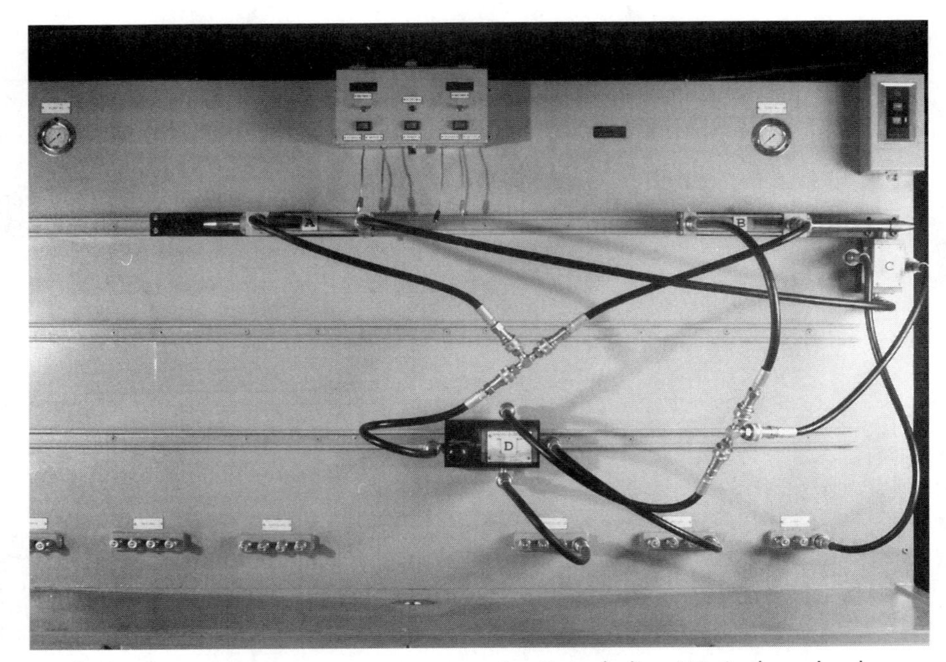

Mock-up of a position parameter sequence circuit, including **(A)** single end rod, double-acting cylinder (secondary extension); **(B)** single end rod, double-acting cylinder (primary extension); **(C)** two-position, one-way, normally closed, plunger-activated, directional-control valve; and **(D)** three-position, four-way, lever-activated, directional-control valve. (*Photograph by the author.*)

The check valve included in the circuit illustrated in Figure 6-9 serves no practical purpose during extension of the cylinders. However, the retraction of the cylinders is not positively sequenced. Either cylinder may retract first, depending on the loads encountered and other factors that determine the path of least resistance for fluid flow. Hydraulic fluid will travel in the path of least resistance. If that path is to retract Cylinder 1 first, then the plumbing connection to Cylinder 2 will be closed when the cam retracts from the mechanical actuator. Cylinder 2 could not retract because trapped fluid in the base end of the cylinder could not exhaust to the tank through the circuit's major directional control. The check valve allows free exhaust flow to occur from the base end of Cylinder 2 regardless of the position of Cylinder 1. A check valve may be integrated into the casing of the two-position, one-way, plunger-activated directional control instead of being a separate component. Figure 6-11 shows a schematic representation of the total sequence circuit using a two-position, one-way, plunger-activated directional control having an integral check.

FIGURE 6-11

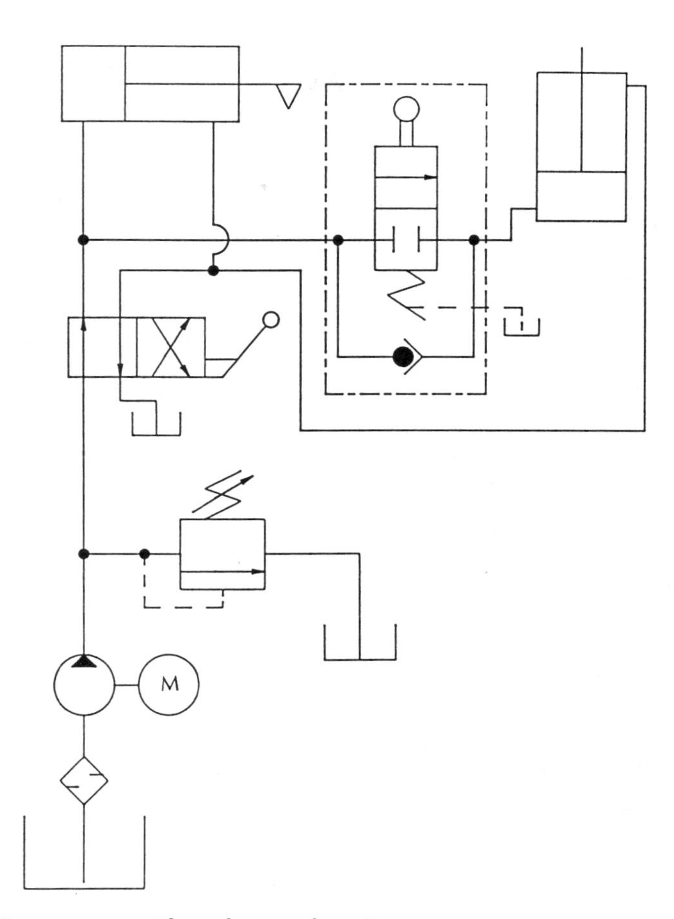

6.3.2 Sequence Circuit Design Parameters

Sequence circuits that use plunger-activated directional control are designed from a position determinant. Typically, a primary cylinder is allowed to extend to a definite position; then, a secondary cylinder is allowed to extend. Therefore, the stroke of the primary cylinder becomes an important determinant of other specifications for this type of sequence circuit. Other cylinders may be added to the simple two-cylinder sequence circuit. The stroke of each subsequent cylinder becomes an added design factor.

6.3.2.1 Piece-rate parameter Sequence circuits may be designed to perform a number of cycles in a given period of time. Typically, machinery using this parameter is set up to process a number of operations or produce a number of pieces of work in a period of time. These sequence circuits are designed from a **piece-rate parameter**.

The two-cylinder piece-rate sequence circuit will need to supply sufficient fluid to extend and retract both cylinders in a given period of time. Since both cylinders must perform a full cycle, a simple summation of the extension and retraction volumes of each cylinder should be used to describe the supply necessary to achieve one cycle of the circuit. The theoretical delivery of the pump may be calculated by multiplying this supply by the number of cycles desired in the specified time period (typically one minute). Of course, the delivery should be expressed in gallons per minute {*liters per minute*}.

Problem 6-2

A dual-cylinder sequence circuit is operating under a piece-rate parameter of 5 cycles/min. The primary cylinder has a 3-in. {*7.6-cm*} diameter piston with a 1-in. {*2.5-cm*} diameter rod and a 20-in. {*50.8-cm*} stroke. The secondary cylinder has a 2-in. {*5.1-cm*} diameter piston with a 1-in. {*2.5-cm*} diameter rod and a 12-in. {*30.5-cm*} stroke. Describe the pump delivery necessary for this circuit.

A. Primary Cylinder Extension Volume (cu. in.) {*cm³*} = ________

Retraction Volume (cu. in.) {*cm³*} = ________

B. Secondary Cylinder Extension Volume (cu. in.) {*cm³*} = ________

Retraction Volume (cu. in.) {*cm³*} = ________

C. Cycle Volume (cu. in.) {*cm³*} = ________

D. Supply Rate (cu. in./min.) {*cm³/min.*} = ________

E. Delivery (gpm) {*lpm*} = ________

6.3.2.2 Feed-rate parameter Other position-determinant sequence circuits require one cylinder to operate at a specified rate of movement. Necessary pump delivery for these circuits is determined by the rate of movement and the cross-sectional area of the feed cylinder's piston. If the desired **feed rate** is expressed in inches per minute {*centimeters per minute*}, pump supply is calculated simply by multiplying the feed rate by the piston area. The rate at which this cylinder retracts may be calculated by dividing the pump supply rate by its annulus area. Cycle times and feed rates of other cylinders in the circuit are also functions of the pump supply rate necessary to cause the feed cylinder to extend as specified.

Problem 6-3

Another dual-cylinder sequence circuit operates using a primary cylinder having a 2-in. {*5.1-cm*} diameter piston, a 0.5-in. {*1.3-cm*} diameter rod, and a 6-in. {*15.2-cm*} stroke. The secondary cylinder has a 4-in. {*10.2-cm*} diameter piston, a 1-in. {*2.5-cm*} diameter rod, and a 30-in. {*76.2-cm*} stroke. The circuit is designed so that the secondary cylinder must extend at a rate of 5 in. {*13 cm*} per second. What pump delivery is necessary for this circuit to operate correctly?

DEL (gpm) {*lpm*} = __________

6.4 MULTIPLE-PATH DIRECTIONAL CONTROLS

The **two-position, two-way directional** control may be described as an **"or"** type logic valve. It typically includes a cylinder port, a pressure port, and a tank port. When used in a ram or single-acting cylinder circuit, this valve allows supply fluid to be directed to the ram through the cylinder port in one position while the tank port is internally blocked. In the other position, flow is allowed to occur from the ram through the valve's cylinder port to the tank while the supply fluid is internally blocked at the pressure port (see Figure 6-12).

Some two-position valves have internal passageways that allow multiple flow paths to exist. Combining the logic of single-path directional

FIGURE 6-12

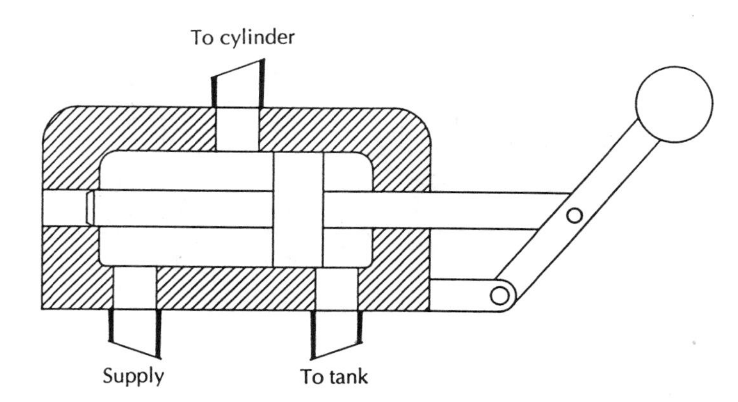

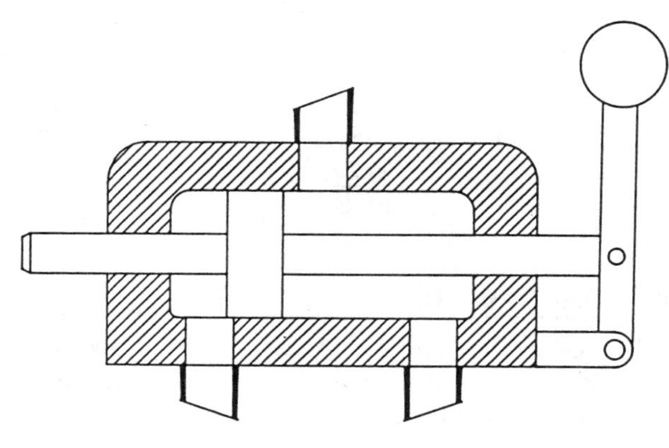

Assignment: Color-code the drawing.

control, **multiple-path directional controls** may be described as having **"and/or"** logic. The two-position, four-way valve, first described in Chapter 5, is an example of such valving. Designed with two cylinder ports, a pressure port, and a tank port, the two-position, four-way valve is typically used in double-acting cylinder circuits. In one position, flow is allowed from the pressure port to one cylinder port *and* from the other cylinder port to the tank port; *or* in the other position, the pressure port and tank port connections to the cylinder ports are reversed.

Other two-position valves include three-way logic (not typically used in hydraulics) and **distribution valves** that may allow as many as six different flow paths to exist through the valve. The internal flow paths may also be consolidated to produce combination effects in circuitry. The regenerative circuit described in Chapter 5 could be equipped with such a valve to cause combination of the supply and exhaust fluid in one position while operating as a normal four-way valve in the other position to cause cylinder retraction (see Figure 6-13).

FIGURE 6-13

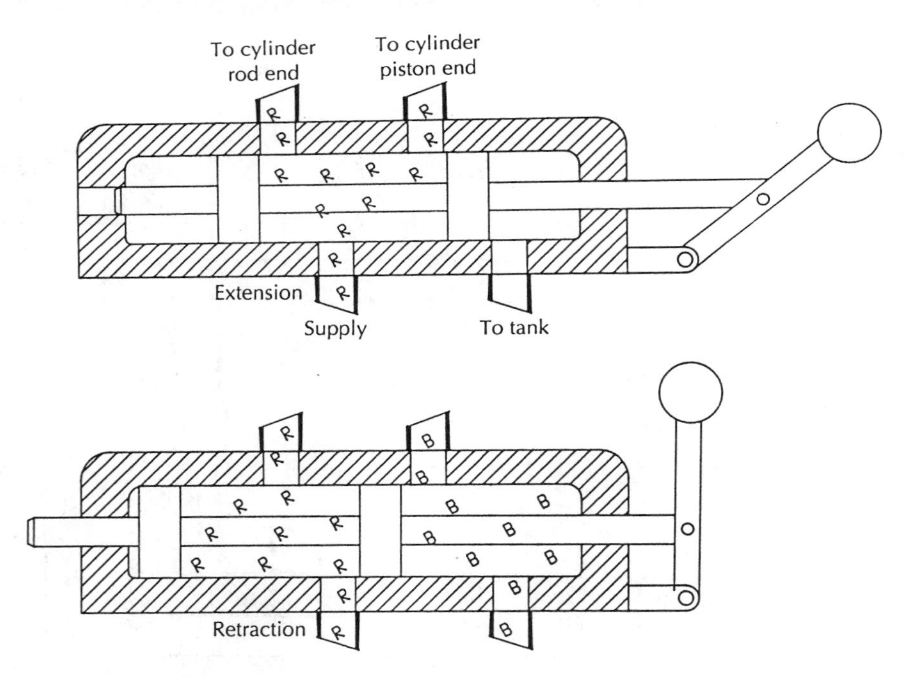

6.5 MULTIPLE-POSITION DIRECTIONAL CONTROLS

Directional controls may also be designed to have more than two distinct positions. The most commonly used multiple-position directional control is the **three-position, four-way valve** (see Figure 6-14).

FIGURE 6-14

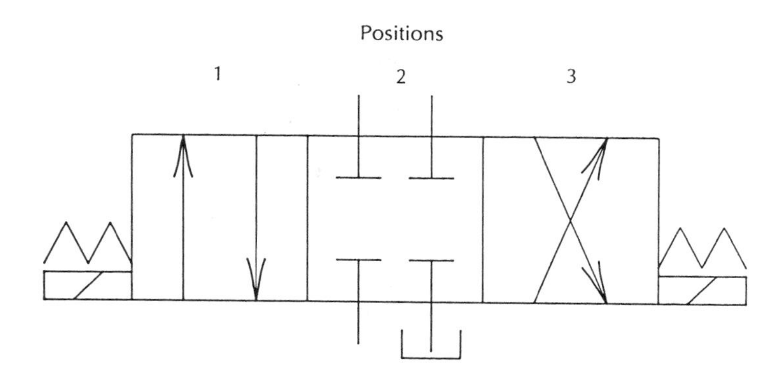

The ANSI schematic in Figure 6-14 describes a directional control that has outside positions identical to two-position, four-way valves. In the center position, all ports are blocked from external flow and to each other. The activation symbols outside the extreme envelopes describe the **activation** mechanism used to place the valve into the position adjacent to those symbols. Thus, when the **solenoid** on the right side is activated, the valve is shifted into Position 3; and when the solenoid on the left side is activated, the valve is shifted into Position 1. When neither solenoid is energized, the **springs** operating against one another shift the valve into Position 2. *Verbally, the ANSI symbol in Figure 6-14 describes a three-position, four-way, normally-centered, closed-centered, solenoid-activated directional control.*

6.5.1 The Rotary Spool Valve

One method of achieving flow logic and multiple positions in a directional control is by using a **rotary spool–type element**. The rotary spool valve has multiple ports located radially around the periphery of the valve casing. A freely rotating spool is fitted concentrically into the casing. The spool has passageways machined into it that allow flow to occur between some ports and exclude flow from others. This valve is typically shifted by a lever that is splined into the spool and rotated around the outer casing. Positions within the rotary spool directional control are normally defined by ball bearings (called detents) that ride in a race machined into the spool and the inner valve casing (see Figures 6-15 and 6-16).

The directional control illustrated in Figure 6-15 shows a **closed-center position**, which is typical of rotary spool–type valves. Rotary spool directional controls may incorporate more than four flow paths and are easily adapted to distribution purposes. This ability in a small-size component accounts for the widespread use of rotary spool valves in portable hydraulic equipment that uses many cylinders independently.

FIGURE 6-15

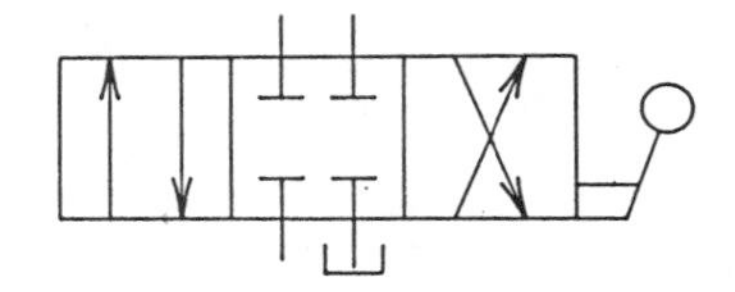

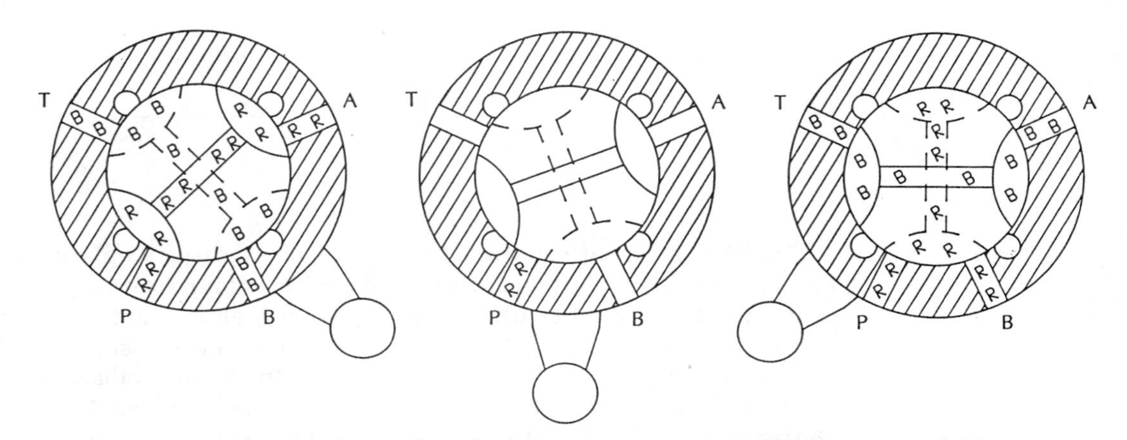

FIGURE 6-16

Rotary spool element—three-position, four-way, directional-control valve. (*Photograph by the author.*)

6.5.2 Sliding Spool Valves

The sliding spool directional control, described previously, is easily adapted to include multiple positions by altering landing locations on the spool. However, sliding spool directional controls seldom achieve more than four separate positions, and most commonly only three. Two positions are used to direct fluid flow to and from actuators and other valving. The third position, typically the center one, is referred to as a *neutral* position. Although fluid is flowing in this position, it is not typically used to operate actuators or valving. However, when included, it constitutes a necessary function or enhancement of circuit operation. A detailed description of a **three-position, four-way, open-center directional control** appears in Figure 6-17 (see also Figure 6-18).

FIGURE 6-17

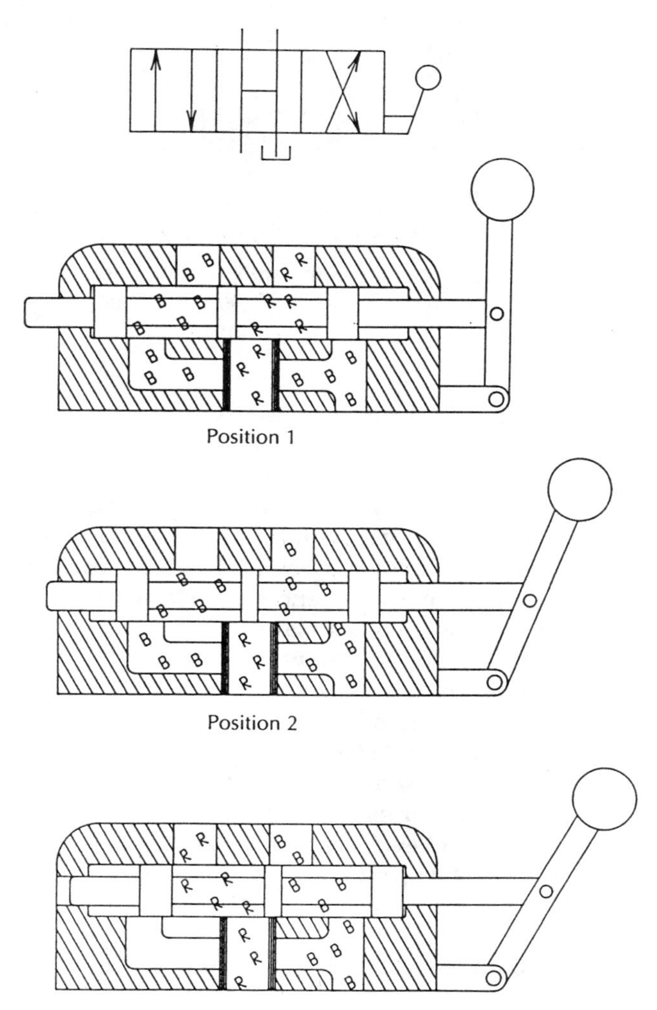

FIGURE 6-18

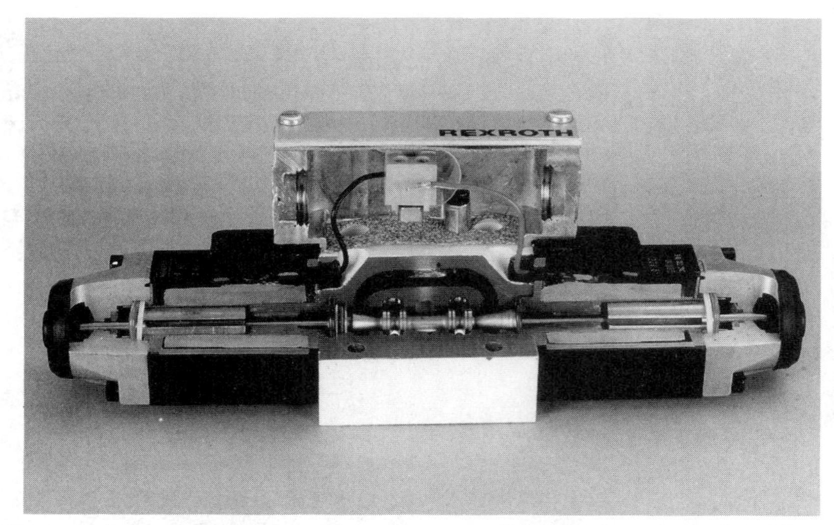

Sliding spool element—three-position, four-way, directional-control valve. (*Courtesy Rexroth Corporation. From Vockroth,* Industrial Hydraulics, *Albany, N.Y.: Delmar, 1994, p. 152.*)

6.5.2.1 Center positions in three-position, four-way valves The **open center** allows all ports to be open to the tank and to one another. With the low pressure evident, little horsepower is consumed by the circuit in the open-center position. If the circuit using an open-centered valve is operating a cylinder, the cylinder is free floating in the center position. The cylinder may be positioned manually because the exhaust in either retraction or extension mode is always being directed to the tank (the path of least resistance).

In some circuitry the advantage of the open path between the pressure port and the tank port is desirable; however, the cylinder needs to be held rigidly when in the center position. The type of logic used to produce this effect is called a **tandem center**. The tandem center has landings similar to the closed-center valve, with a channel cored through the spool to connect the pressure and tank ports (see Figure 6-19).

The closed, open, and tandem designs represent the most popular types of centers for three-position, four-way directional controls. Other common center types are represented in Figure 6-20.

6.5.2.2 Center position selection The choice of which type of center to use in any circuit is determined by the application and, to a lesser extent, by costs. Open passageways from other ports to the tank port allow unrestricted movement of fluid in those portions of the circuit. This

FIGURE 6-19

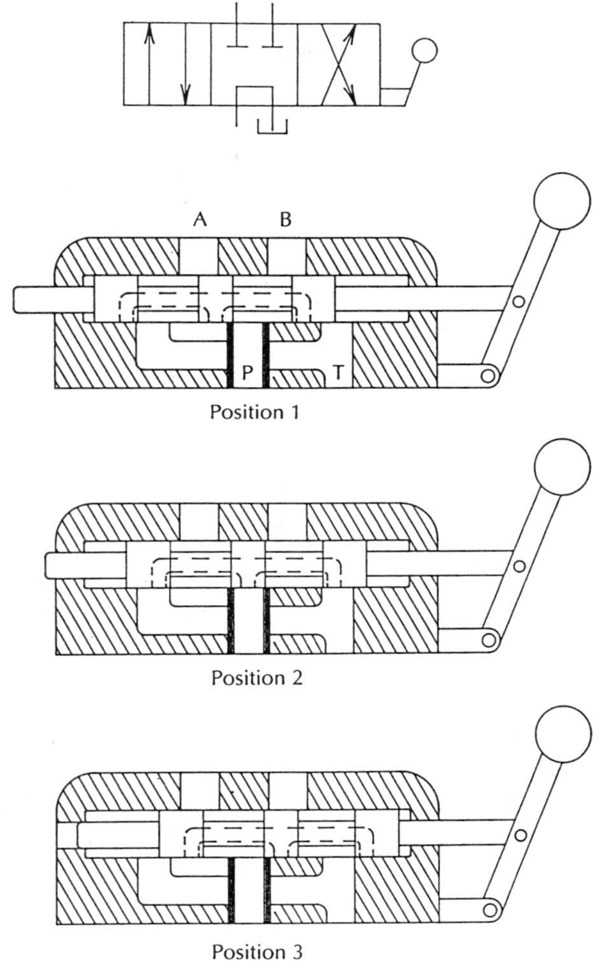

Assignment: Color-code the drawings.

results in both low pressure and resistance at those points. When a load needs to be held rigidly at any intermediate position during cylinder stroke, the exhaust during that stroke should be closed from the tank port at the cylinder port. Inversely, if free movement of the cylinder is required, that cylinder port should be opened to the tank port.

If the pressure port in the directional control's center position is closed to the tank port, circuit pressure will rise to a level set by the pressure-relief valve. Since the pump will be constantly supplying fluid at a given rate and pressure will be maintained, horsepower will be consumed even

FIGURE 6-20

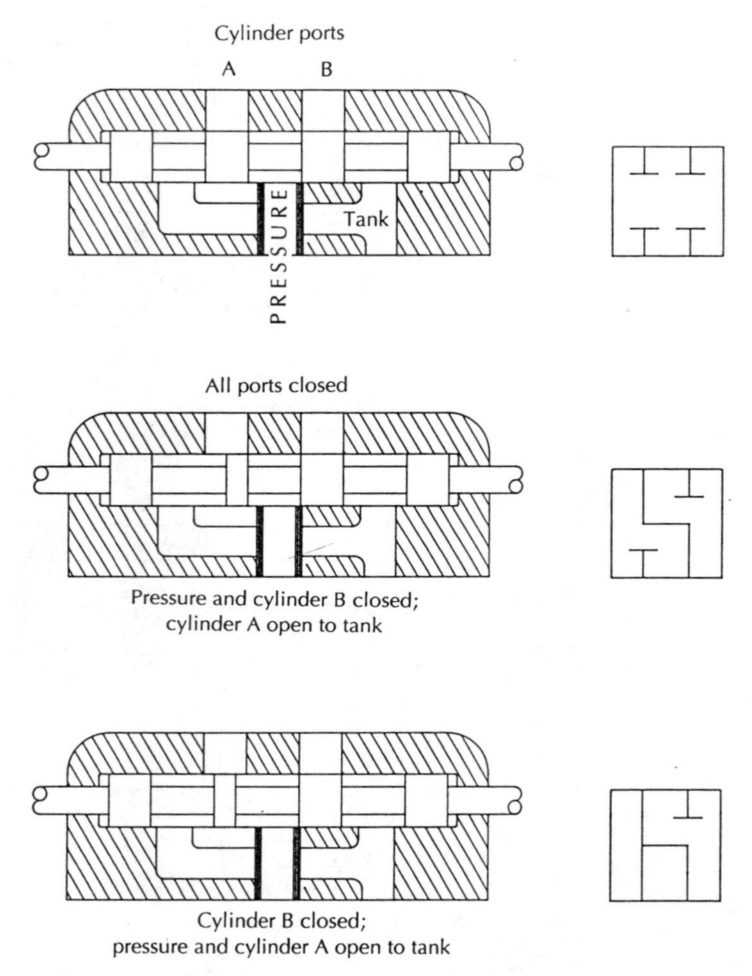

Assignment: Color-code the drawings.

in what would normally be a neutral center position. If the pressure port is opened to the tank port, minimal pressure is evident in the power supply and horsepower consumption drastically reduced. However, branch circuits taken from locations before the directional control will not operate with this center design, since the flow will return to the tank through this center condition rather than traveling to the branch circuit. Through combinations of pressure and cylinder port conditions, relative to passage to the tank port, a great variety of circuitry operation can be achieved (see Figures 6-21 to 6-23).

FIGURE 6-21

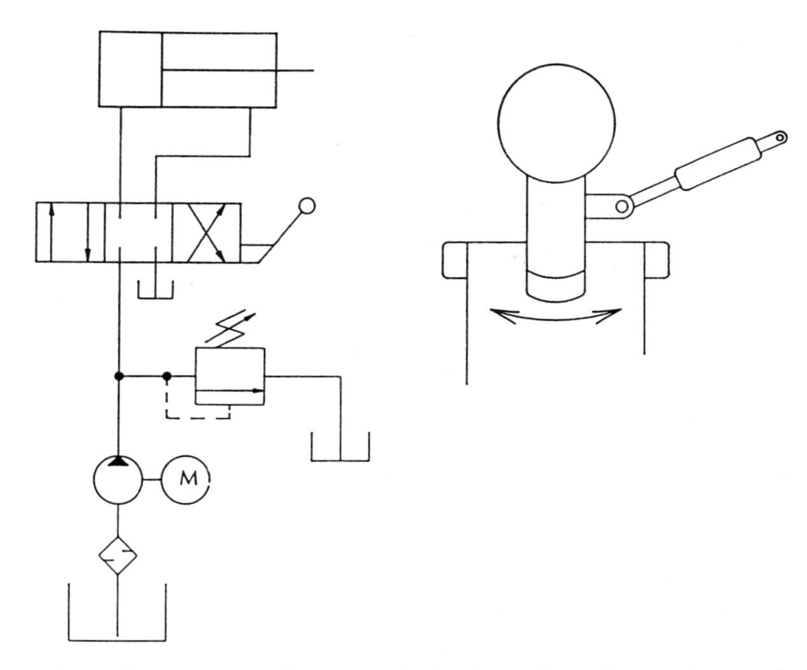

This linear circuit is used to swing a loading chute from side to side to distribute parts onto a drying belt after cleaning. It is designed so that the operator may shift the directional control into the center position and manually push the chute in both directions if desired.

Problem 6-4

Describe the type of center that should be used in this circuit by completing the ANSI symbol's center position for the directional control.

Problem 6-5

Verbally describe the fluid flow in the circuit when the directional control is shifted into the center position.

____

____

____

____

____

____

FIGURE 6-22

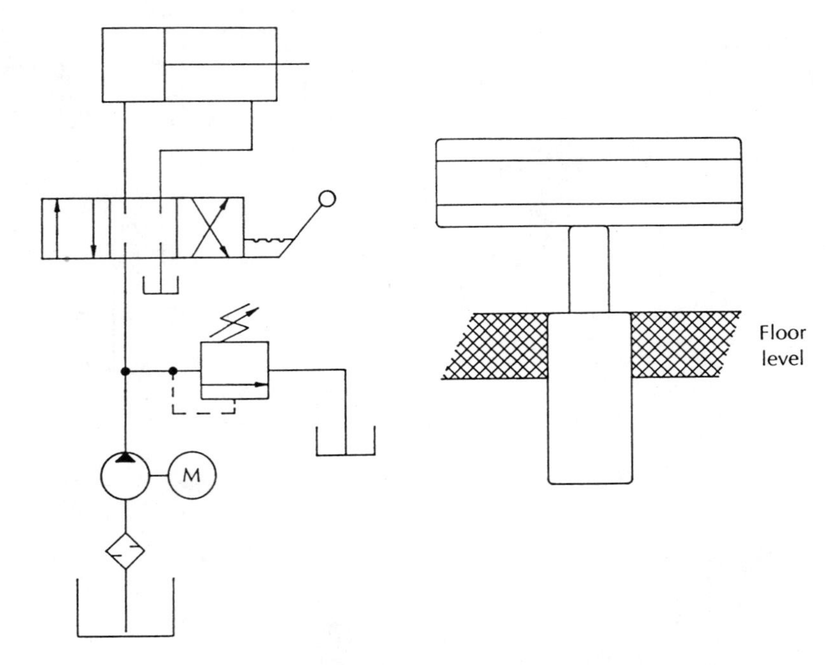

This linear circuit is used to raise heavy sections of steel I-beams into various positions to weld them into framing networks. It is designed so that it may be stopped at any point during extension or retraction and should hold the load rigidly at that point. Since some welding operations require long periods of time, the power supply postion of the circuit is designed to use a minimum amount of horsepower during welding.

Problem 6-6 Describe the type of center that should be used in this circuit by completing the ANSI symbol's center position for the directional control.

Problem 6-7 Verbally describe the fluid flow in the circuit when the directional control is shifted into the center position.

____

____

____

____

____

FIGURE 6-23

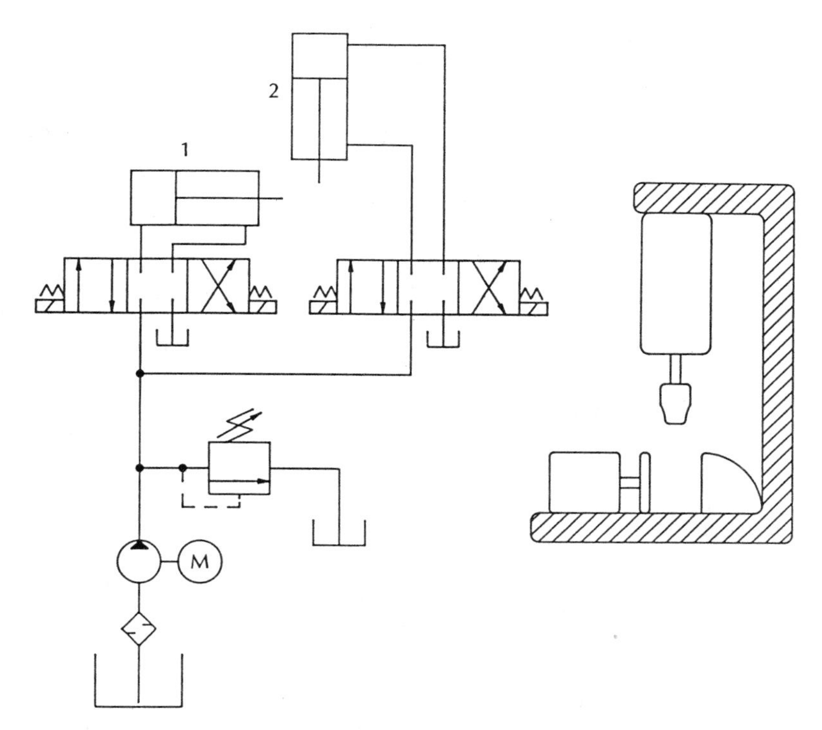

This combination circuit is used to position a piece part with cylinder 1 during extension, then punch holes in that piece part with cylinder 2 during extension. Notice that each directional control is normally centered and solenoid activated.

Problem 6-8 Describe the type of centers that should be used in this circuit by completing the ANSI symbols' center positions for the directional controls.

Problem 6-9 Verbally describe the fluid flow in the circuit when the directional controls are shifted into the center position.

____

____

____

____

____

6.5.3 Combination Activation

With the advent of control panels for large industrial machinery and the desire to remove the operator from direct contact with the machine for safety purposes, solenoid activation of directional controls has become increasingly important. The quick response and positive action of these electromechanical switches are necessary in much modern machinery.

Although efficient and fast acting, solenoids have relatively low thrust capability. Therefore, by themselves, they are typically used only in directional controls that have relatively small valving elements. However, **combination valves** using a small **piloting valve** that is solenoid activated and a large slave valve that is pilot activated from the pilot valve section overcome this deficiency. These combination valves are commonly known as **piggy-back directional controls** because of their physical arrangements (see Figures 6-24 and 6-25).

FIGURE 6-24

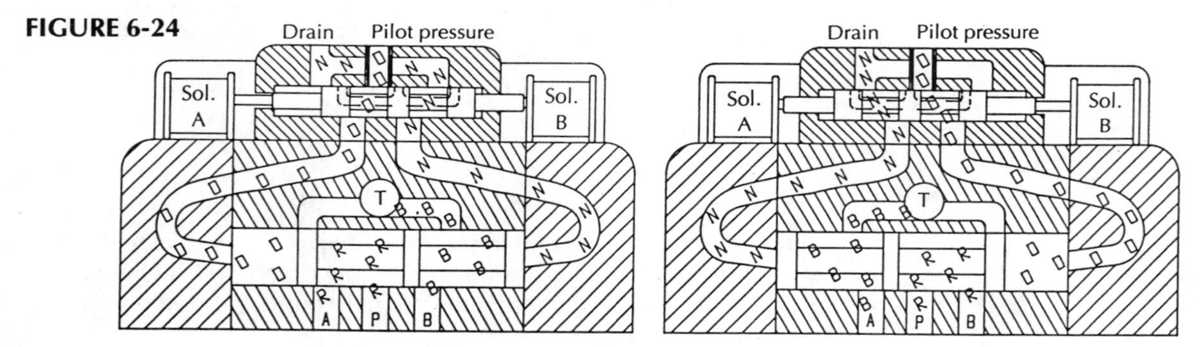

When solenoid A is energized (left drawing), the piloting spool is shifted to the right. Pilot pressure is directed to the left side of the large slave valve's spool, shifting it so that pressure is directed to cylinder port A while exhause flow from cylinder port B is directed to the tank port.

When solenoid B is energized (right drawing), the piloting spool is shifted to the left. Pilot pressure is directed to the right side of the large slave valve's spool, shifting it so that pressure is directed to cylinder port B while exhaust flow from cylinder port A is directed to the tank port.

Other complex arrangements of directional controls are readily apparent through the combination of independently and dependently controllable sliding spool–type valves. These combinations are typically called **stack valves** (see Figure 6-26). Stack valves are commonly employed in portable earthmoving equipment.

With the use of electrical circuitry to activate hydraulic circuitry the potential for timers, lights, noises, heat, and other starting mechanisms

FIGURE 6-25

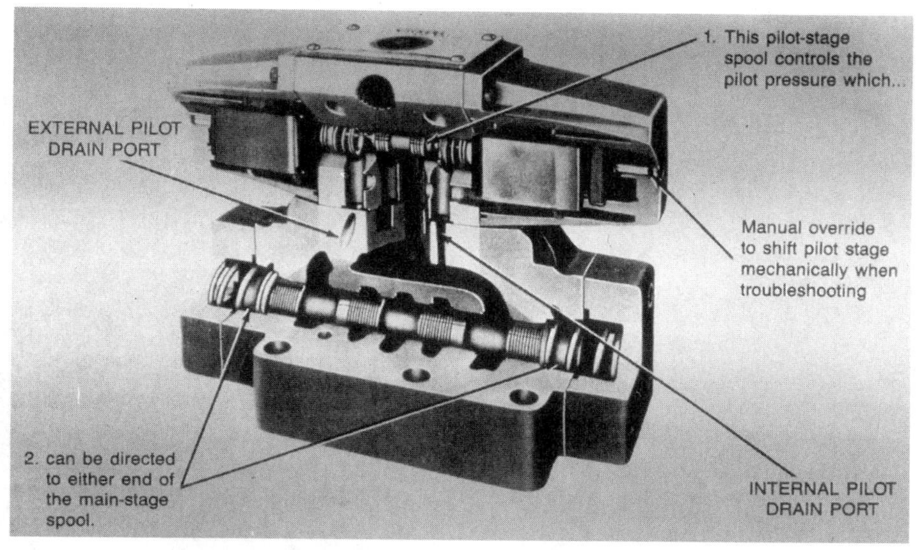

Piggy-back directional control package. (*Courtesy Vickers, Inc.*)

for machinery is achieved. The application of electronic sensors to machinery in combination with computer processors has produced a new generation of control in hydraulic circuitry.

6.6 SIZING DIRECTIONAL CONTROLS

Traditionally, directional controls have been sized by describing the nominal diameter of the fluid ports leading to and from the valve. This description demonstrates a fairly accurate method of sizing directional controls. However, conductors are normally sized based on pump delivery and ideal flow conditions within them.

By design, the directional control diverts the path of fluid flow, causing a resistance to that flow and pressure drop through the valve. Therefore, *directional controls are also sized by a rating that describes the maximum gpm {lpm} delivery that will produce no more than a 5% pressure drop between any two ports in the valve.* For example, a properly sized directional control that is operating in a circuit requiring 500 psi {*3,447.4 kPa*} and 3 gpm {*11.4 lpm*} on the outlet side of the valve may have up to 525 psi {*3,619.7 kPa*} evident at the inlet port location. If the circuit is not operating properly, the size (gpm) {*lpm*} rating should be increased to lessen the pressure drop through the valve. Above the proper rating, the relationship between pressure drop and flow rate is parabolic. Thus, if the flow rate in the valve described previously was increased to 9 gpm {*34.2 lpm*}, the pressure drop through the valve would be doubled.

FIGURE 6-26

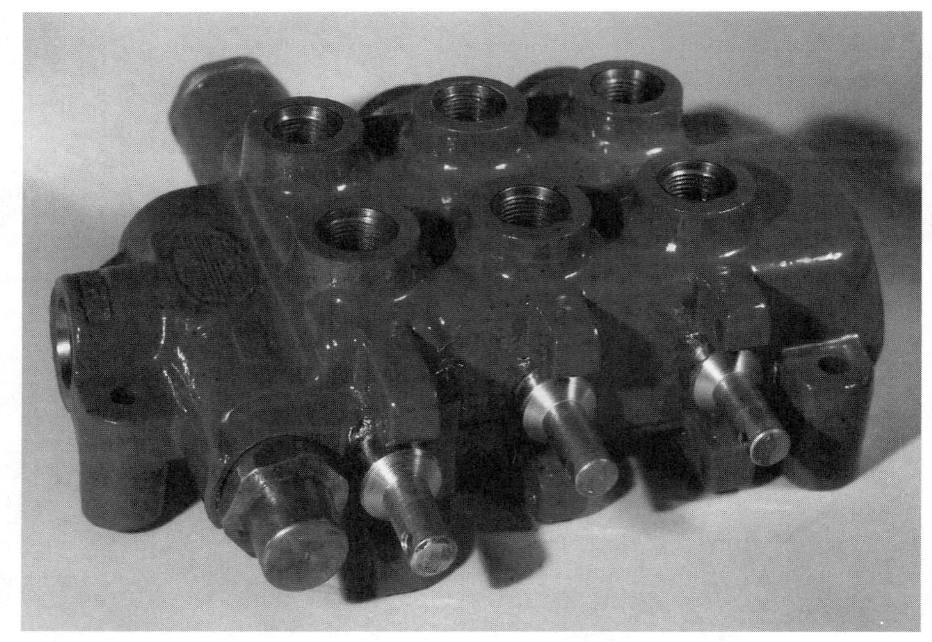

Stack valve package typically used on mobile excavation equipment. (*Photograph by the author.*)

Problem 6-10

A directional control rated at 5 gpm {*18.9 lpm*} is operating in a linear circuit using a 2-in. {*5.1-cm*} diameter cylinder having a 0.5-in. {*1.3-cm*} diameter rod and is being resisted by a 700-lb. {*3.1-kN*} load in both extension and retraction. Describe the maximum pressure evident at the directional control inlet.

P (psi) {*kPa*} = ________

[] [] [] 6.7 PIECE-RATE PARAMETER—POSITION PARAMETER DESIGN PROBLEM

In the shipping department for a metal-casting operation, the loading dock is located 3 ft. {*0.9 m*} above the plant floor and 4 ft. {*1.2 m*} away from the door through which trucks are loaded. Five castings are loaded onto a pallet and delivered to the shipping department. The total weight for the castings and pallet is 3,000 lbs. {*13.3 kN*}, and the sliding resistance for the castings and pallet has been calculated to be only 1,500 lbs. {*6.65 kN*}. Your assignment is to design a position parameter sequence circuit that will raise the castings and pallet and push them to the loading door at a rate of 2 cycles/min. Complete your assignment by specifying all components for the circuit in the following format and drawing an ANSI schematic of the circuit.

A Single End Rod, Double-Acting Cylinder—Raising Cylinder

PISTON		*ROD*			
Diameter	Area	Diameter	Area	Annulus Area	Stroke
______ in. {*cm*}	______ in.^{2} {*cm^2*}	______ in. {*cm*}	______ in.^{2} {*cm^2*}	______ in.^{2} {*cm^2*}	36in. {*91.4 cm*}

FORCE		*PRESSURE*		*SUPPLY*	
Extension	Retraction	Extension	Retraction	Extension	Retraction
3,000 lbs. {*13.3 kN*}	0 lbs. {*0 kN*}	_______ psi {*kPa*}	_______ psi {*kPa*}	________ in.^{3} {*cm^3*}	________ in.^{3} {*cm^3*}

			VELOCITY	
Total Supply	Rate	Supply Rate	Extension	Retraction
_______ in.^{3} {*cm^3*}	2 cyc/min.	_______ in.^{3}/min. {*cm^3/min.*}	_______ in./min.{*cm/min.*}	_______ in./min {*cm/min.*}

Port Size	Body Style	Mounting
_______ in.{*mm*}	__________	__________

B Single End Rod, Double-Acting Cylinder—Sliding Cylinder

PISTON		*ROD*			
Diameter	Area	Diameter	Area	Annulus Area	Stroke
______ in. {*cm*}	______ in.^{2} {*cm^2*}	______ in. {*cm*}	______ in.^{2} {*cm^2*}	______ in.^{2} {*cm^2*}	48in. {*120 cm*}

FORCE		*PRESSURE*		*SUPPLY*	
Extension	Retraction	Extension	Retraction	Extension	Retraction
1,500 lbs. {*6.6 kN*}	0 lbs. {*0 kN*}	_______ psi {*kPa*}	_______ psi {*kPa*}	________ in.^{3} {*cm^3*}	________ in.^{3} {*cm^3*}

			VELOCITY	
Total Supply	Rate	Supply Rate	Extension	Retraction
_______ in.^{3} {*cm^3*}	2 cyc/min.	_______ in.^{3}/min. {*cm^3/min.*}	_______ in./min{*cm/min*}	_______ in./min {*cm/min*}

B Single End Rod, Double-Acting Cylinder—Sliding Cylinder (*continued*)

Port Size	Body Style	Mounting
_______ in.{*mm*}	__________	__________

C Pump

Type	Design	Delivery	Drive Speed	Volumetric Efficiency
__________	__________	________ gpm{*lpm*}	__________ rpm	_______%

	PORT SIZE	
Displacement at V.E.	Inlet	Outlet
______ in.^{3}/rev. {cm^3/rev.}	________ in. {*mm*}	________ in. {*mm*}

D Master System Directional Control

Positions	Ways	Element	Activation	Port Size
________	_________	Sliding Spool	__________	_______ in. {*mm*}

E Position Parameter Sequencing Directional Control

Positions	Ways	Element	Activation	Port Size
________	________	Sliding Spool	__________	_______ in. {*mm*}

F Pressure Relief Valve

Type	Element	Setting	Mounting	Port Size
Simple	Poppet	_______ psi {kPa}	__________	_______ in. {*mm*}

G Filtration

		RATING		
Type	Element	Element	Flow	Port Size
Strainer	Wire	________ mesh	________ gpm {*lpm*}	______ in. {*mm*}

Description	Mounting
Full Flow	________

H Reservoir

CAPACITY	
Minimum	Maximum
______ gal. {*l*}	______ gal. {*l*}

I Electric Motor

Speed	Voltage	Phase	Power at V.E.
________ rpm	__________	__________	_________hp {*W*}

J Conductors

Inlet Type	Inlet Description	INLETS (I.D. DECIMAL) Minimum	Maximum	Working Type	Working Description
_______	_________	______ in. {*cm*}	______ in. {*cm*}	__________	_________

WORKINGS (I.D. DECIMAL)		LINES—NOMINAL	
Minimum	Maximum	Inlets	Workings
______ in. {*cm*}	______ in. {*cm*}	______ in. {*mm*}	______ in. {*mm*}

K Draw an accurate schematic diagram of the circuit described in the previous specification.

L The following are variations for the Piece-Rate Parameter—Position Parameter Design Problem (6.7):

Raising Cylinder: Double-Acting, Single End Rod Cylinder

	A	B	C	D	E
Stroke	40 in.	48 in.	24 in.	75 *cm*	1.5 *m*
Extension Force	3,500 lbs.	2,500 lbs.	4,300 lbs.	12 *kN*	18 *kN*
Retraction Force	0 lbs.	0 lbs.	0 lbs.	0 *kN*	0 *kN*
Cycle Rate	3 cyc/min.	4 cyc/min.	2 cyc/min.	2 cyc/min.	4 cyc/min.

Sliding Cylinder: Double-Acting, Single End Rod Cylinder

	A	B	C	D	E
Stroke	55 in.	60 in.	48 in.	90 *cm*	2.0 *cm*
Extension Force	1,700 lbs.	1,200 lbs.	2,500 lbs.	7 *kN*	10 *kN*
Retraction Force	0 lbs.	0 lbs.	0 lbs.	0 *kN*	0 *kN*
Cycle Rate	3 cyc/min.	4 cyc/min.	2 cyc/min.	2 cyc/min.	4 cyc/min.

6.8 SUGGESTED ACTIVITIES

1. Make arrangements to visit a construction site in your community during the excavation phase. Examine the hydraulic systems on equipment such as backhoes, front-end loaders, track-hoes, bulldozers, and the like. Locate the obvious directional controls and determine the number of ports and potential flow logic (ways, positions, etc.) in each valve, what type of elements they have, how the valves are activated, and how they are mounted.
2. Visit a local library and locate the section having books on automotive repair. Find the section on braking systems. After review of that section, determine where directional controls are located in the system. Describe the directional controls in terms of element types, flow logic, activation, and mounting.

3. Make arrangements to visit a local airport and interview aircraft maintenance personnel. Discuss hydraulic systems employed in the aircraft, paying special attention to directional controls. Write a brief report on your findings.
4. Disassemble a three-position, four-way, directional-control valve in your laboratory. Describe (either verbally or with sketches) the arrangements between external ports and the valving element that allow the described flow logic.
5. Fabricate a mock-up of the position parameter sequence circuit described in this chapter on a hydraulic trainer. Observe its operation and determine if it is operating as predicted by design parameters, such as pump delivery and cylinder sizes, when compared to actual times for extension and retraction strokes.

6.9 REVIEW QUESTIONS

1. What are the functions of a directional control?
2. Describe the operation of a check valve in terms of element, ways, positions, and activation.
3. Explain the inclusion of a check valve in the design for a cylinder fluid cushioning device.
4. List five methods of activating (shifting) directional controls.
5. Describe the design and operation of cartridge valves when used as directional controls in terms of element, spring, control port, signal port, and pilot port.
6. Describe the use of two-position, one-way directional controls in position parameter sequence circuitry.
7. If a two-cylinder position parameter sequence circuit is operating under a feed-rate parameter for delivery requirement and the cylinders are not the same size, why are the velocities on various strokes not identical?
8. Describe the operation of a two-position, four-way directional control in terms of "and/or" logic.
9. Draw a schematic of a three-position, four-way, closed-centered, normally-centered, solenoid-activated directional control.
10. Differentiate between the design and operation of sliding spool and rotary spool directional controls.
11. Describe advantages and disadvantages of closed, open, and tandem centers in three-position directional controls.
12. Discuss combination activation methods employed in special directional controls such as piggyback and stack valves.
13. Describe methods used to size directional controls.

[ ] [ ] [ ] [ ] 7 [ ]

Pressure Control

Imagine a house wired without fuses or circuit breakers and the potential resulting damage to the circuit and probably the entire building housing that circuit. Since pressure will develop to the point of overcoming the least resistant component, such safety devices are also required in hydraulic circuitry.

7.1 HYDROSTATIC OPERATION

In a closed hydraulic circuit, pressure results from a resistance to movement. Of course, it is a function of a load or force and the area acted upon by that force. However, pressure must develop in the circuit; that is, the necessary pressure to overcome a given force is not immediately available when the circuit is energized. When energized, pressure will be zero and will build, over time, until it overcomes the least resistant force–area combination. If this force–area combination represents the ultimate resistance in the circuit, then maximum circuit pressure will be defined. However, if other force–area combinations are evident in the circuit, pressure will continue to build in the circuit until all subsequent forces are overcome in order of resistance (see Figure 7-1).

This schematic represents a linear circuit using three ram-type cylinders having 2-, 3-, and 4-in. {*5.1-, 7.6-, 10.2-cm*} pistons. Each ram's extension is being resisted by a 1,000-lb. {*4,453.3-N*} load.

Problem 7-1

Describe the order of extension of the three cylinders. Which cylinder will extend:

FIGURE 7-1

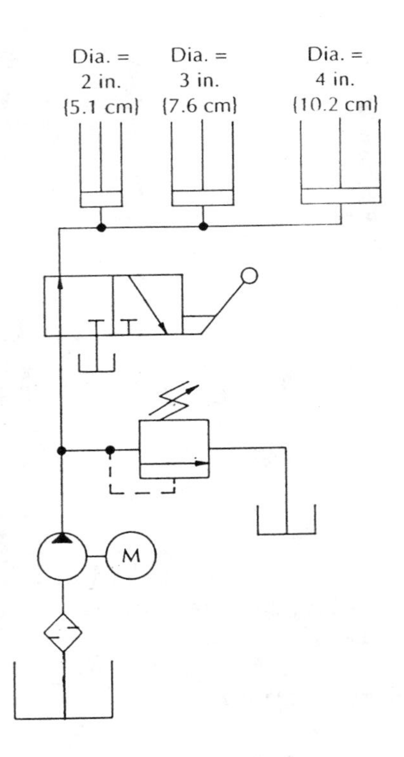

First? __________

Second? __________

Third? __________

Problem 7-2 What should be the setting on the pressure-relief valve to assure extension of all three cylinders?

Pressure (psi) {*kPa*} = __________

The circuit illustrated in Figure 7-1 results in a sequence of the three cylinders' extensions. The sequence circuit described in Chapter 6 used a directional control to cause sequence order. Regardless of the loads imposed on the cylinders in that sequence circuit, the order of sequence would be the same. The extensions of the cylinders in Figure 7-1 are not positively sequenced. If the loads on those cylinders varied, then the sequence order may be altered. However, other pressure-control devices may be incorporated to positively sequence the circuit in any required order or even to reduce pressure in the circuit. *The functions of pressure controls, therefore, are to cause actions to occur in the circuit brought about by changes in operating pressure, and to reduce pressure in a portion of a circuit as well as to limit maximum system pressure.* As such, **pressure-control valves** are typically named according to their usage, even though their physical constructions may be identical while performing different functions.

7.2 PRESSURE-RELIEF VALVES

Those pressure controls that limit maximum pressure in the circuit are called **relief valves**. Some relief valves may have a fixed setting determined by the force imposed by a biased spring. Others may achieve variable settings through the use of a screw-type device that causes different initiation forces to be encountered through compression of the spring.

7.2.1 Simple-Type Relief Valves

The **poppet relief valve**, described in circuitry previously presented in this book, is the most commonly used device to limit maximum pressure in a hydraulic circuit. In the context of this book, the poppet relief valve is referred to as the simple type of relief valve. However, its simplistic design may result in inaccurate control of circuit pressure. In operation, when system pressure becomes great enough to exert sufficient force over the valve's poppet to overcome the force exerted by the spring, the poppet will unseat and allow some flow to be directed to the tank. By design, this poppet movement will compress the spring, resulting in increasingly greater biased force as the poppet moves. The pressure necessary to cause primary relief to occur is known as **initiation** or **cracking pressure**. Of course, pressure will continue to build in the circuit until the poppet becomes fully unseated, allowing maximum flow through the valve to the tank. The pressure necessary to cause this secondary relief action is **operation** or **full-flow pressure** (see Figure 7-2).

The difference between operation and initiation pressures is called **pressure override**. Pressure override may be as great as 20% of system pressure in some poppet valve designs. This means that a circuit designed to require 1,000 psi {*6,895 kPa*} may develop up to 1,200 psi {*8,274 kPa*} when using a poppet relief valve. In some cases, pressure override may be detrimental to circuit operation. For example, a linear circuit having a poppet relief valve might be used as a clamp that must impose a specific force to hold a workpiece; but excessive force might cause damage to the workpiece. Still, poppet relief valves are superior safety valves because when these valves jam, they tend to stick in the open rather than the closed position. This assures that potentially damaging pressures will not occur in a well-designed circuit, even when the relief valve malfunctions.

FIGURE 7-2

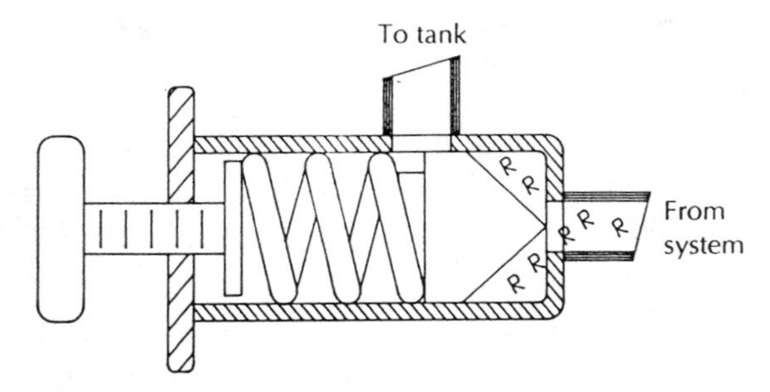

Below setting pressure

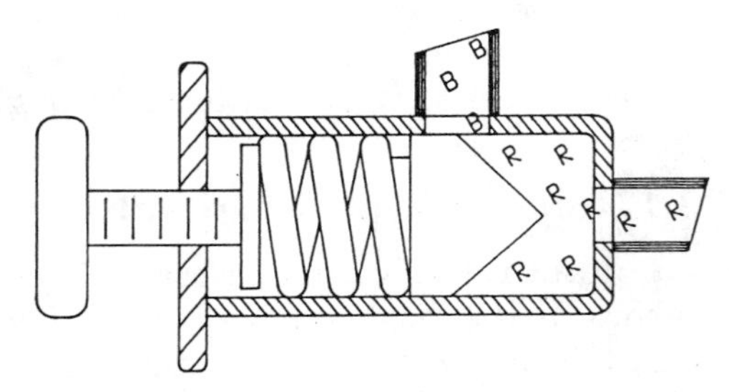

At initiation pressure

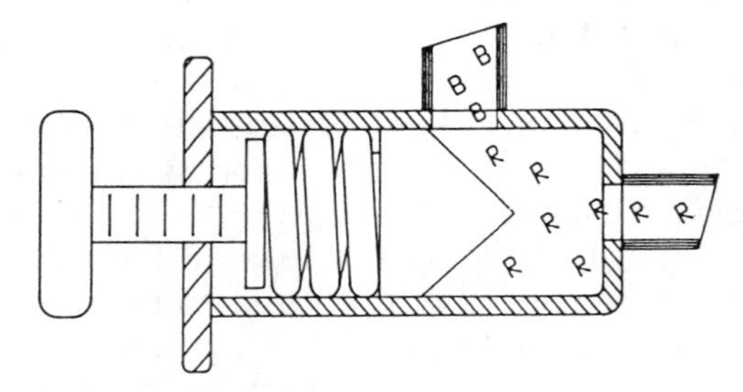

At operating pressure

7.2.2 Compound-Type Relief Valves

7.2.2.1 Piston element In circuitry that needs constantly and accurately controlled maximum system pressure, a relief valve that isolates and individually controls initiation and operation pressures is required. These characteristics are available in a **two-stage piston relief valve**. Pressure supplied to the piston relief valve operates on both stages simultaneously. The **primary stage** consists of a poppet-type relief section that senses system pressure through a small opening in the valve casing. Initiation pressure will cause the poppet to unseat and divert flow through a low-volume passageway in the center of the piston to the tank port. The **secondary stage** of the piston relief valve consists of a piston that also physically separates the inlet section from the poppet valve chamber. The small passageway that supplies fluid pressure to the poppet section also supplies that pressure to one side of the piston, while system pressure operates on the other side. Thus, the piston is designed to be in hydraulic balance. However, a light-duty spring forces the piston in a direction that normally closes a high-volume return from the valve inlet to the tank port (see Figures 7-3 and 7-4).

Pressure override in the piston relief valve can only develop to the equivalent of the force imposed by the light-duty spring that acts on the piston. This force is normally the equivalent of 20 to 30 psi {*138 to 207 kPa*} regardless of system pressure. However, the piston relief valve is not without fault. For example, if the small return passageway through the center of the valve to the tank would become clogged during initiation relief, system pressure would continue to rise. Neither stage would operate to relieve system pressure because pressure on both sides of the piston would be the same and the high-volume passageway would also remain closed. Two methods are typically used to overcome this fault of the piston relief valve. Each requires connection of supplementary valving through a vent port located in the primary stage of the piston relief valve (see Figure 7-5). The **automatic venting mechanism** using the poppet relief valve takes advantage of the nature of that valve to jam open and thus relieve poten-

FIGURE 7-3

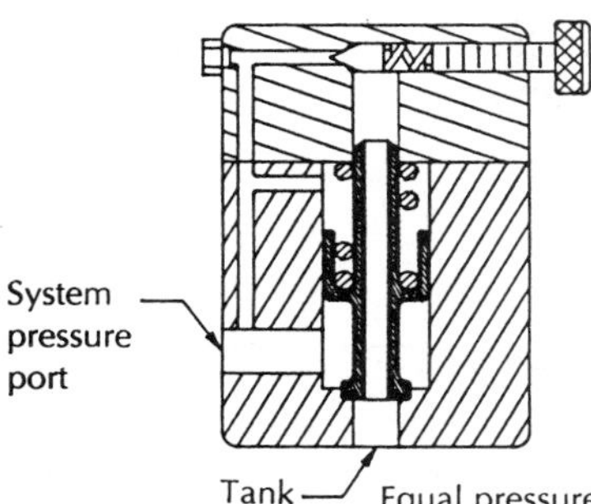

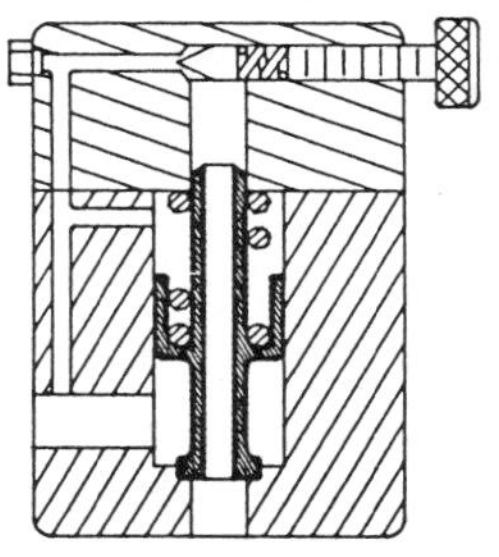

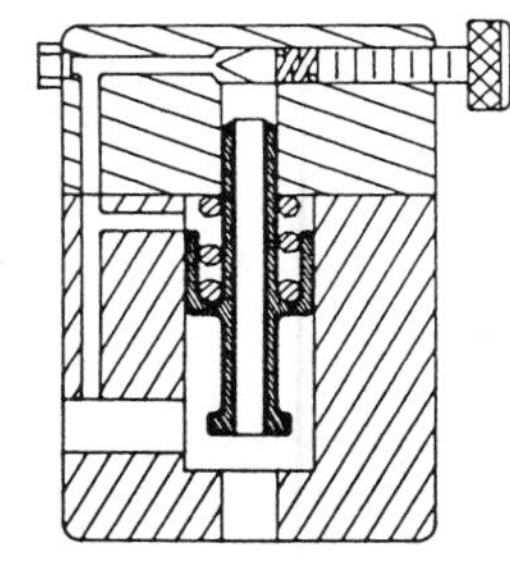

FIGURE 7-4

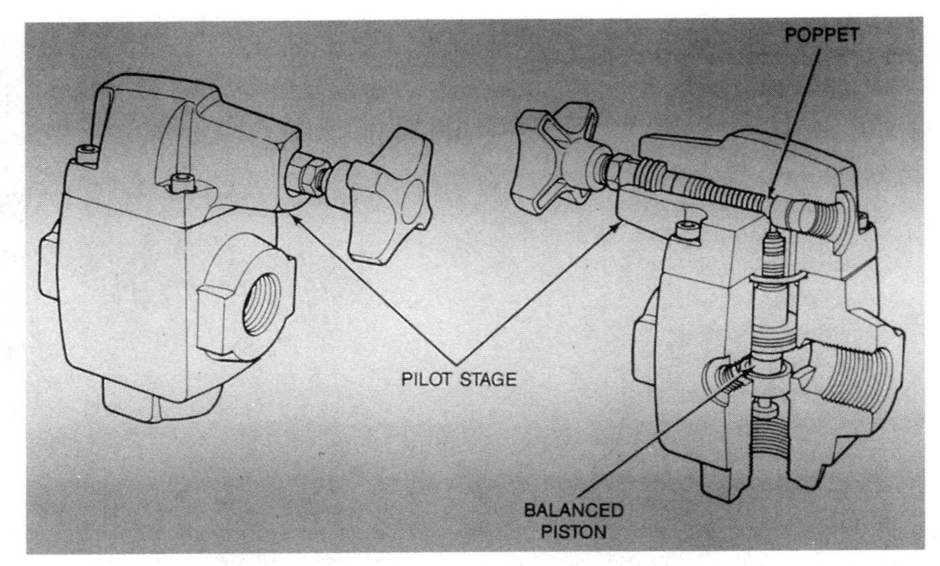

Two-stage, piston-element, pressure-relief valve. (*Courtesy Vickers, Inc.*)

tially damaging pressure, even if it fails to operate correctly. In this case, the poppet valve should be set to a minimum of 10% above operating pressure. *The venting mechanism using the two-position, one-way directional control is not automatic.* Rather than system control, which may also conceivably fail, this arrangement requires monitoring of a pressure gauge by an operator who would activate the directional control when excessive system pressure develops. In this sense, the venting mechanism should be considered the "kill" switch for the circuit. The operator could thereby shut down the system at any pressure if circuit damage were observed.

FIGURE 7-5

7.2.2.2 Spool element Another valve design that may be used to limit maximum system pressure is the **spool type**. *The construction of this valve uses three sections: the piloting section, the primary section, and the secondary section.* The **primary section** contains the valve inlet and the sensing mechanism for system pressure. The **secondary section** consists of the biased spring that determines valve shifting pressure and the porting for tank return. Separating these two sections is a sliding spool that normally closes the secondary return port. The **piloting section** allows system pressure to be exerted on the spool and shifts the spool when system pressure is great enough to overcome the biased spring (see Figures 7-6 and 7-7).

The design illustrated in Figure 7-6 is similar to a two-position, one-way, spring-détented, pilot-actuated, sliding spool directional control. However, *in this case, its function is to control pressure. Thus, it is named for that function, as are all pressure controls.*

Two features of this valve bear special notice. The passageway through the center of the spool allows any fluid that accumulates in the cavity

FIGURE 7-6

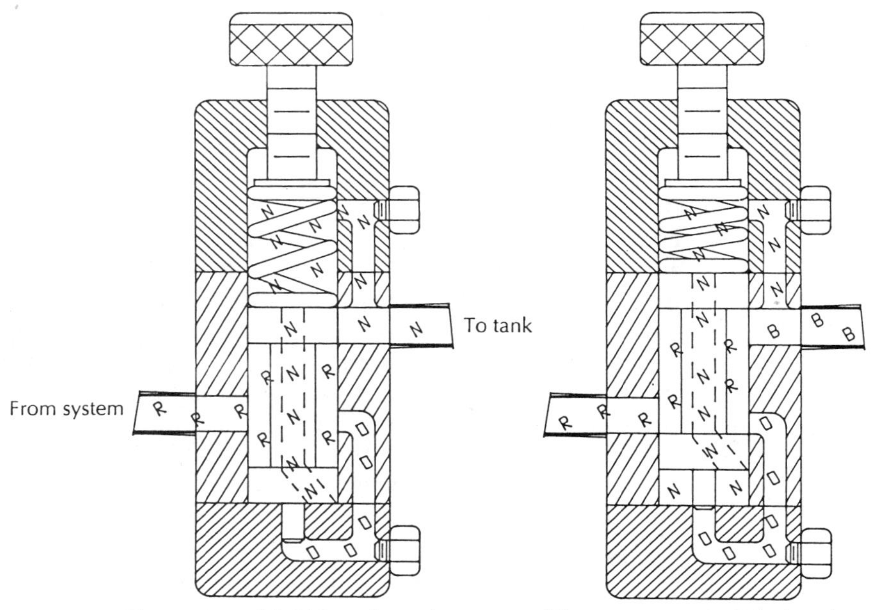

The passage of fluid from the system to the tank is normally excluded by the spool being forced downward by the biased spring.

When system pressure in the piloting section exerts sufficient force on the small piston extension to the spool to overcome the force on the biased spring, the spool will shift upward, diverting flow to the tank.

FIGURE 7-7

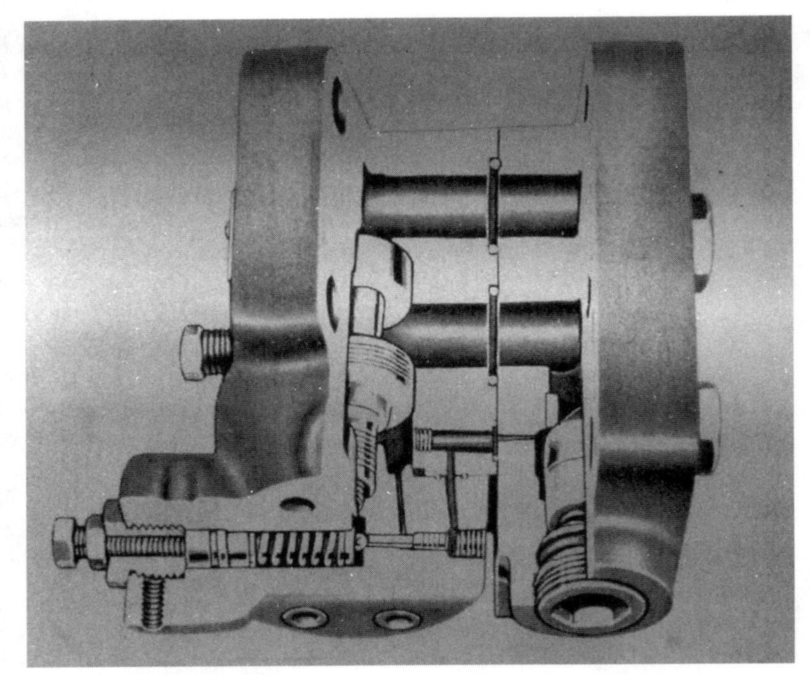

Multistage spool-element, presure-relief valve. (*Courtesy Vickers, Inc.*)

beneath the lower spool landing to drain back to the tank. Some seepage into this cavity is normal, since a small amount of fluid is allowed to lubricate the periphery of the landing. Without the **internal drain**, the spool might not shift back to the normally closed position, since the noncompressible fluid would stop spool movement. Some seepage is also allowed to lubricate the small **piston** attached to the spool in the piloting section. The spool-type pressure-relief valve would operate without this small piston with pressure acting over the entire surface of the lower spool landing. However, with this small piston attached to the large spool, shifting occurs more rapidly, since less fluid needs to be displaced to cause a large degree of linear spool movement. *The major advantage of the spool pressure control over other pressure-control designs is this rapid reaction.*

7.2.3 Cartridge-Type Pressure-Relief Valves

The cartridge valve previously described in Chapter 6 as a directional control may also be used as a pressure-relief valve. In this arrangement, the **cover cap** for the slide-in cartridge houses the primary-stage simple relief valve that controls initiation pressure. Most often, the cartridge-type relief valve relies on a compounded design. *It incorporates the cartridge element as the secondary stage used to control operating pressure and reduces pressure override to the pressure equivalent of the lightly biased spring incorporated into the cartridge package* (see Figure 7-8).

In operation, system pressure acts on both the signal port of the cartridge element and the poppet in the primary-stage cover. An interconnecting passageway in this cover also directs operating pressure to the control port and spring chamber of the cartridge valve. Because of the larger area in the control chamber and the cartridge spring exerting greater forces than the smaller area on the signal port, the cartridge remains closed and passage to the pilot port is excluded. When system pressure in the primary cover stage rises to the equivalent force exerted by the primary-stage spring (variably controllable), this poppet will open and allow flow to return to the tank through its spring chamber. At the same time, pressure in the interconnected control port area of the cartridge element rapidly drops, as fluid in this chamber is also exhausting to the tank through the primary stage. With pressure temporarily lower in the control port area than in the signal port area of the cartridge valve, its poppet will unseat and expose the pilot port area which is also open to tank return.

System pressure is adjusting throughout the valve at this time. When it reaches a level at which the pressure within the primary cover stage is "nearly equal" to the pressure in the cartridge valve's control port area, the cartridge will close once more. In this case the "nearly equal" is an

FIGURE 7-8

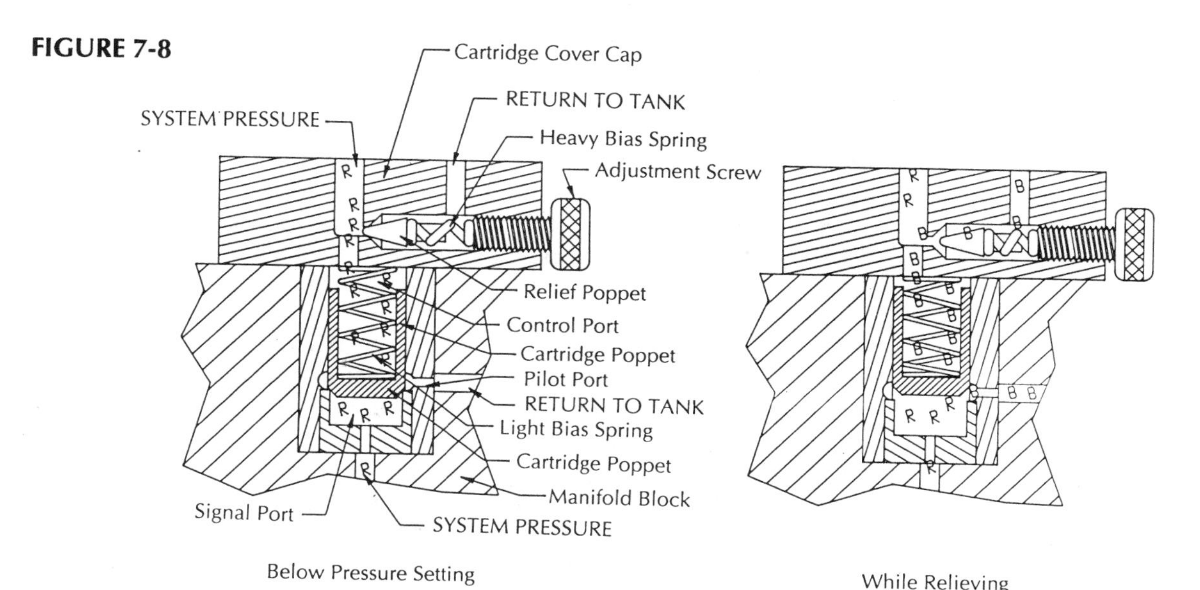

important concept because the lightly biased spring in the cartridge's control port area also holds it closed. So, any time that pressures at these two locations are within equivalent ranges in consideration of the light force imposed by the spring, the cartridge valve will close. Similar to the balanced-piston compound relief valve, this lightly biased spring in the control port area of the cartridge package limits the amount of pressure override to its (the spring's) equivalent pressure (normally a negligible value).

7.3 PRESSURE FUNCTIONAL VALVES

As previously described, pressure evident in the hydraulic circuit develops as a resistance to flow. The pressure-relief valve limited system pressure by directing flow to a nonresistant portion of the circuit (the tank) when a predetermined pressure was evidenced. As a pressure switching device, the normally closed pressure control may also be used to direct flow to another portion of the circuit that is resistant to fluid flow or to control flow outside the power supply to relieve pressure in a portion of the circuit. In linear circuitry, three such applications are common: **unloading valves, sequencing valves**, and **counterbalance valves**.

7.3.1 Unloading

In some linear circuits, pressure is evident due to operational design. A good example of this exists in the regenerative circuit. Due to design, a load is imposed during extension of the cylinder because system pressure is being imposed on the annulus area. Even though the rapid movement of the cylinder may be desirable, the resulting excessive pressure may be detrimental, especially if the major load is only evident at the end of the cylinder's extension stroke. For example, the operation of a log splitter would benefit from the impact of rapid movement until the log strikes the wedge. Thereafter, rapid movement is inconsequential, while back pressure reduces the force available to shear the wood. However, a circuit may be designed that allows regenerative extension until major loading is evident and afterward, affording simple linear extension using a pressure unloading valve (see Figure 7-9).

Notice that the **unloading valve** schematically represented in Figure 7-9 has **remote pilot** activation. Remote pilot activation in most spool-type pressure-control valves is achieved by simply rotating the pilot-section end cap 180 degrees (see Figure 7-10). During initial extension when the three-position, four-way directional control is shifted into the left position, flow is directed to the base end of the cylinder and exhaust from the head end of the cylinder is directed through check valve #1 and joins the pump supply in regeneration. When the extension encounters the

FIGURE 7-9

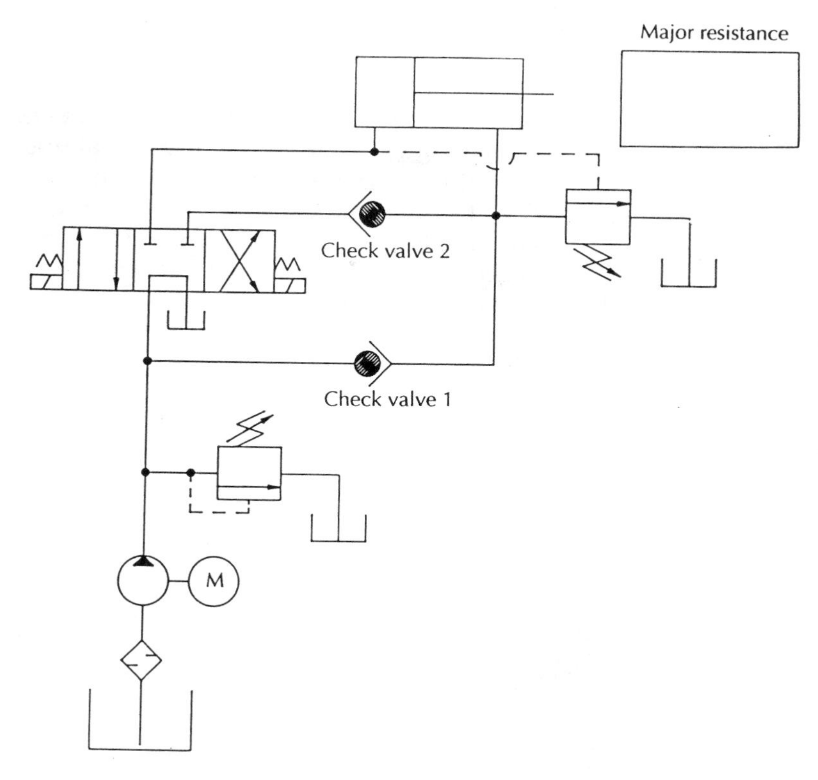

FIGURE 7-10

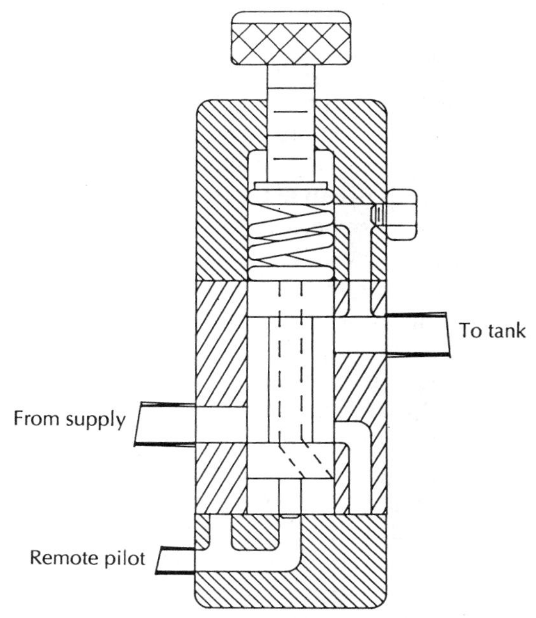

Assignment: Color-code the drawing.

major load, circuit pressure will rise. This increased pressure will be sensed at the unloading valve through the external pilot, opening the valve and directing cylinder exhaust to the tank without resistance. During retraction, when the three-position, four-way directional control is shifted to the right position, fluid flow is directed to the head end of the cylinder through check valve #2, while exhaust from the base end returns to tank through the directional control.

Problem 7-3

The circuit described in Figure 7-9 is being used to split logs with a maximum weight resistance of 500 lbs. {*2.2 kN*}. Shear resistance during splitting has been calculated to be 5,000 lbs. {*22.2 kN*}. The log splitter is equipped with a 5-in. {*12.7-cm*} diameter differential cylinder.

a. What pressure will develop during initial movement of the log?

P (psi) {*kPa*} = ________

b. What should be the setting on the unloading valve?

P (psi) {*kPa*} = ________

c. What should be the setting on the pressure-relief valve?

P (psi) {*kPa*} = ________

Another application of the unloading valve is illustrated by the **high–low circuitry** typically used in large hydraulic presses. Hydraulic press operation requires clearance between upper and lower platens for tooling setup and part removal. This clearance reduces productivity during operation because the platens must travel through a great deal of unnecessary space before performing the high-pressure pressing cycle. To increase the productivity of the press, high-displacement pumps are needed to cause rapid movement of the movable platen. However, high-displacement pumps normally cannot withstand high pressures without damage or excessive leakage and slippage. Ideally, a mechanism that incorporates a high-displacement pump during clearance movement and a low-displacement, high-pressure pump during pressing is needed in the high–low circuitry.

A **two-stage pumping mechanism** is used in the high–low circuit. During initial movement, both stages supply fluid to cause rapid cylinder movement. When high resistance is evident, during pressing, the pressure developed throughout the circuit affords a pilot pressure that unloads the high-displacement, low-pressure pump. The low-displacement, high-pressure pump is isolated from this unloading effect by a check valve. This stage continues delivering a supply to the cylinder while resisting the high pressure within the circuit (see Figure 7-11).

FIGURE 7-11

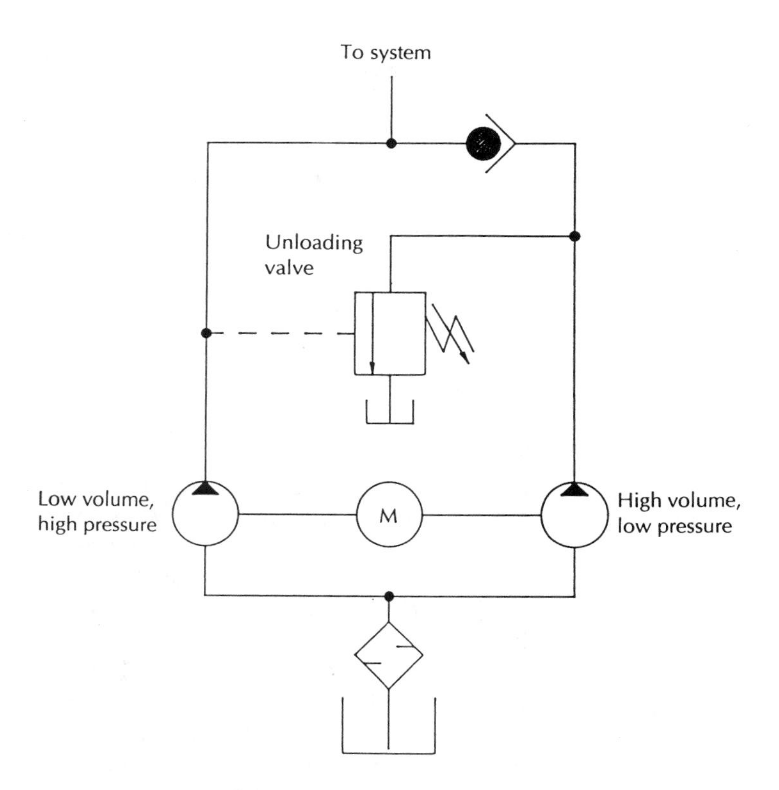

7.3.2 Counterbalancing

Up to this point, pressure controls have been used in combination with other hydraulic components to enhance circuit operation through allowance for increased speed. However, the operation of a hydraulic circuit may be enhanced through controlled decreased speed. While the unloading valve application in the regenerative circuit reacted to allow impact, *pressure control can also be used to reduce or eliminate impact.*

When cylinders are placed into a position that allows gravity to accelerate a heavy load, potentially damaging impact may occur at the end of that stroke. The degree of impact that may occur is a function of the size of the load and the distance traveled by the load. Any attempt to control a falling load through reducing flow or imposing a load on the inlet side of the cylinder will not reduce this effect. In this case, the cylinder will act as a pump displacing the exhausting fluid at a rate determined by acceleration due to gravity minus the resistance imposed by turbulence in that exhausting fluid. Pressure controls, called **counterbalance valves**, may be used to impose a load on the exhaust from a free-falling cylinder (see Figure 7-12).

FIGURE 7-12

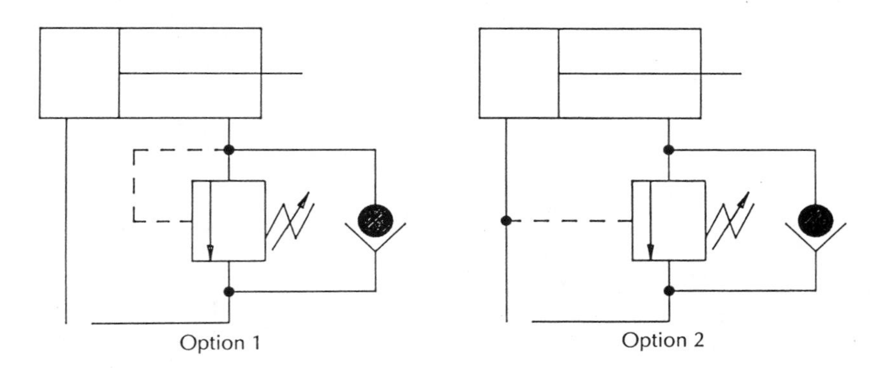

Counterbalancing

Notice that the circuit illustrated in Figure 7-12 may use counterbalance valves with either internal or external piloting. *Option 1 describes the design typically used for counterbalance circuits having relatively low loads*. In this option, the exhaust flow during extension is blocked until pressure on the annulus area of the cylinder becomes greater than the setting on the biased spring of the normally closed counterbalance valve. At this point, the valve opens and flow is returned through the directional control to the tank. When this occurs, pressure on the annulus area drops and the counterbalance valve closes. Of course, pressure will then rise again and reopen the counterbalance valve. It will open and close rapidly and slowly exhaust fluid in the head end of the cylinder until complete decompression of the cylinder is accomplished. *The setting on the counterbalance valve in Option 1 is determined by the annulus area of the cylinder and the load assisting movement*. Thus, a cylinder having a 3-sq. in. {*19.4-cm²*} annulus area and a 3,000-lb. {*13.3-kN*} load would develop 1,000 psi {*6,895 kPa*} pressure when the counterbalance valve is closed. The counterbalance valve, in this case, should be set to 10% above this value, or 1,100 psi {*7,585 kPa*}. When the directional control is shifted to the center position, the cylinder will be held stationary because the load itself will not exert enough force to develop sufficient pressure to open the counterbalance valve. Otherwise, the resistance to pump supply into the base end of the cylinder will result in additional pressure to operate the counterbalance valve.

In Option 2, the exhaust-blocking operation of the counterbalance valve is the same. However, with remote piloting from the supply side of the circuit, the counterbalance valve can be adjusted to virtually any setting regardless of the load. In this option, any pressure evident on the inlet side of the cylinder will overcome the spring setting on the counterbalance valve and will begin the oscillation of opening and closing that allows decompression. *This design is typically used on circuits having heavy loads*. Little waste of energy or horsepower is evident because low pressure develops, even with resistance to exhaust fluid.

7.3.3 Sequencing

Sequencing, as explained in Chapter 6, is simply the ordering of actuator actions. In that chapter, sequence circuits were designed from a position parameter, using a directional control that opened an alternate fluid path after an initial cylinder reached a specified position in its stroke. However, some circuit designs would benefit from sequencing based on a **pressure parameter** rather than a position parameter.

7.3.3.1 Sequence circuit descriptions If a two-cylinder sequence circuit were used in a clamping and shearing operation, a pressure parameter would be appropriate. The primary or clamping cylinder would need to exert a specific force on the piece part before the secondary shearing operation occurred. If the clamping and shearing operations were performed with cylinder extensions, the retraction of the cylinders would not necessarily need to be sequenced; and either the shear or the clamp cylinder may retract first. *A clamp and shear sequence circuit is illustrated in Figure 7-13 using a spool-type pressure-control preset to a value that is 10% higher than pressure needed to exert sufficient force with the clamping mechanism* (see also Figure 7-14).

FIGURE 7-13

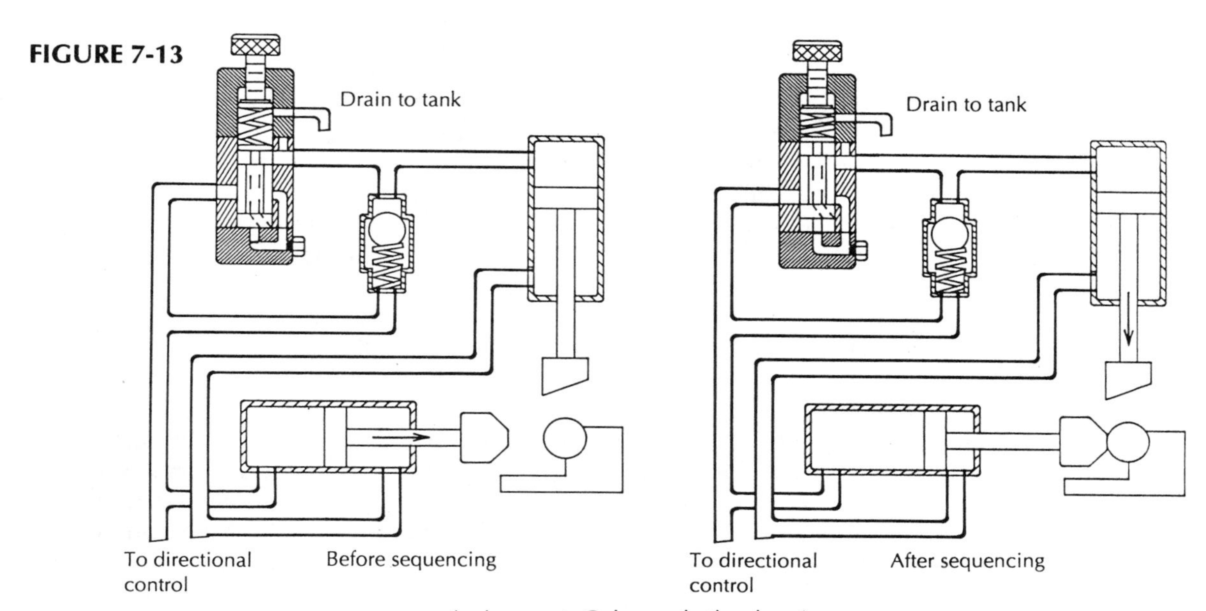

Assignment: Color-code the drawing.

FIGURE 7-14

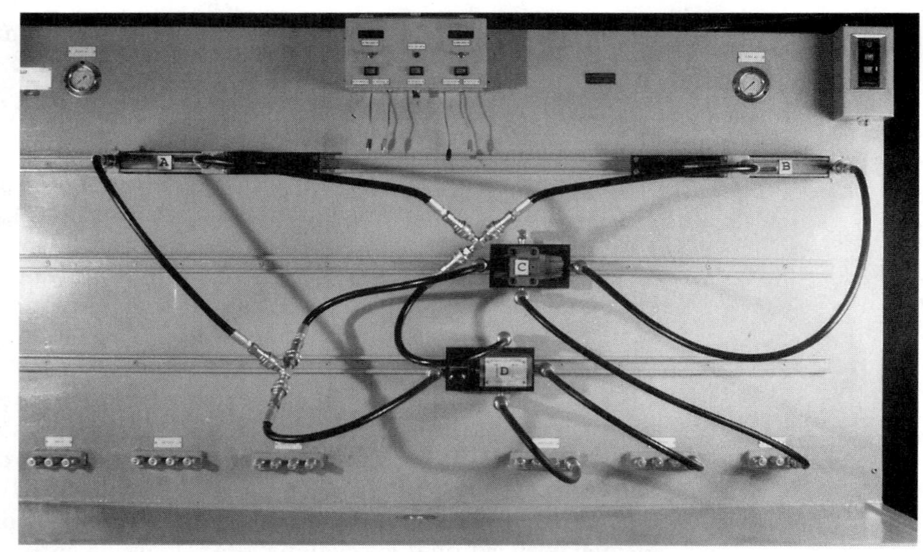

Mock-up of a pressure parameter sequence circuit during extension only of two cylinders, including **(A)** single end rod, double-acting cylinder (primary extension); **(B)** single end rod, double-acting cylinder (secondary extension); **(C)** sequence valve; and **(D)** three-position, four-way, directional-control valve. (*Photograph by the author.*)

Notice that the sequence valve illustrated in Figure 7-13 has the upper cap rotated 180 degrees from that position previously described in circuitry using the spool-type pressure control. Since the secondary port becomes a pressurized port after sequencing, drainage from the pocket beneath the spool cannot be directed to this port. Rotation of the upper cap accommodates **external drainage**, and a separate drain line must be incorporated to return drainage fluid to the tank. Also, this circuit includes a check valve that allows free flow from the secondary port to the primary section during shearing cylinder retraction. Without this check, the shearing cylinder would not retract because the pilot that causes the spool to shift is located in the primary section of the valve. During retraction, no pressure exists in the primary section; therefore, the spool would remain in its normally closed position. The check is often integrated into the design of the sequence valve.

7.3.3.2 Designing pressure parameter sequence circuits The methods used for designing pressure parameter sequence circuits are extensions of the techniques used to design position parameter sequence circuits. Pump delivery and the subsequent sizing of other components are still functions of either the number of cycles of the circuit in a given period of time or the necessary speed of operation of an actuator. The other major determinant of circuit design, system pressure, must accommodate descriptions of the sequence valve and another cylinder that will probably increase maximum operating pressure within the circuit.

For example, a pressure parameter sequence circuit might include two cylinders, each being resisted by a 1,000-lb. {*4.45-kN*} load during extension only. If the primary cylinder has a piston area of 2 sq. in. {*12.9* cm^2}, a pressure of 500 psi {*3,448 kPa*} will develop during extension of that cylinder. Standard design practices would call for the sequence valve, which separates the two cylinders, to be set to 10% above this pressure, or 550 psi {*3,792 kPa*}. If the secondary cylinder has a piston area of 1 sq. in. {*6.5* cm^2}, pressure will rise in the circuit to 1,000 psi {6,895 *kPa*} to overcome this force. Since no load is imposed on either cylinder during retraction, negligible pressure will develop. The pressure-relief valve for this circuit should be set to 10% above pressure to extend the secondary cylinder, or, in this case, to 1,100 psi {*7,585 kPa*}. In this circuit, ultimate system pressure is realized and controlled by the relief valve. Its setting will be a factor in specifications for other components that are described using pressure components such as the horsepower rating {*wattage*} for the drive component or motor. For this reason, *special attention should be paid to the setting on the pressure-relief valve to assure that adequate pressure will develop in the circuit to complete all functional operations.*

Figure 7-15 describes a pressure parameter sequence circuit having three cylinders positively sequence on both extension and retraction. Note that this operation requires four sequence valves, two each in extension and retraction mode. *A general rule of thumb states that the number of sequence valves required in a linear circuit is equal to the number of strokes to be positively sequenced minus one per stroke direction.* Thus, a circuit used to positively sequence four cylinders during extension only would require four minus one, or three, sequence valves. If that circuit also required positive sequencing of the retraction strokes, four minus one, or three, more sequence valves would be required—a total of six sequence valves in that circuit.

In the circuit shown in Figure 7-15, a load of 2,000 lbs. {*8,906.7 N*} is imposed on extension and retraction of all cylinders.

Problem 7-4

Describe the pressure setting for each sequence valve.

SV1. Pressure (psi) {*kPa*} = __________

SV2. Pressure (psi) {*kPa*} = __________

SV3. Pressure (psi) {*kPa*} = __________

SV4. Pressure (psi) {*kPa*} = __________

FIGURE 7-15

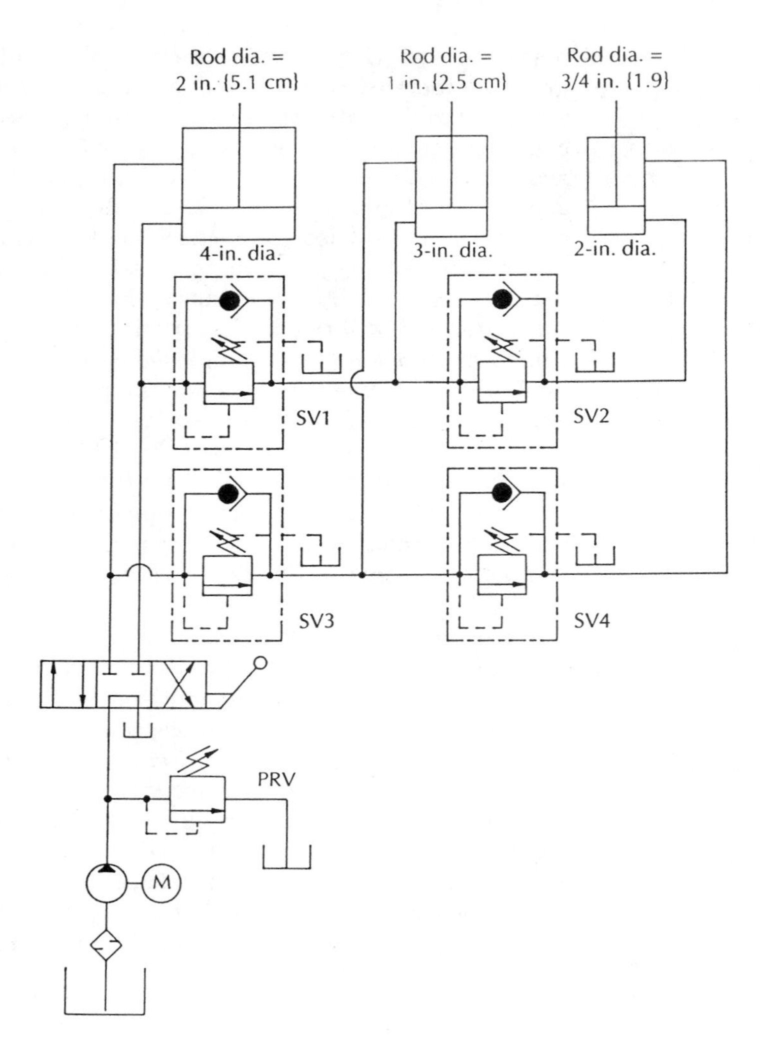

Problem 7-5

Describe the pressure setting for the relief valve.

PRV. Pressure (psi) {*kPa*} = __________

Most pressure-control valves do not operate below 250 psi {*1,724 kPa*}, and many industrial circuits operate below a maximum pressure of 1,200 psi {*8,274 kPa*}. Cylinders are generally sized to cause pressure in the circuit to remain within this range in consideration of loads imposed upon them. Also, cylinders may commonly be sized so that increasing pressure is required for sequential cylinder strokes. This would appear to negate the need for sequence valves. However, any variation in resistance might cause malfunction of the circuit. Therefore, sequence valves are included in such designs to assure a positive sequence of each cylinder.

7.4 PRESSURE-REDUCING VALVES

A final function of pressure controls is to limit pressure in a portion of a circuit. *Unlike other pressure controls, the pressure-reducing valve is a normally open hydrostat.* Also, rather than operating from an inlet-side pilot, *the pressure reducing valve is shifted using a pilot pressure taken from the outlet side of the valve* (see Figures 7-16 and 7-17). The pressure-reducing valve illustrated in Figure 7-16 is commonly referred to as a direct acting–type pressure-reducing valve. The terminology is probably due to the fact that its simplistic design, using only the spool element, typically requires no integral check valve to allow free flow from the outlet to the inlet. Other more complicated, more accurate, and more expensive pilot-operated types of pressure-reducing valves are available. Typically, the pilot-operated pressure-reducing valve must have an integral check for reverse flow from the outlet to the inlet and, thus, is considered a compound valve.

FIGURE 7-16

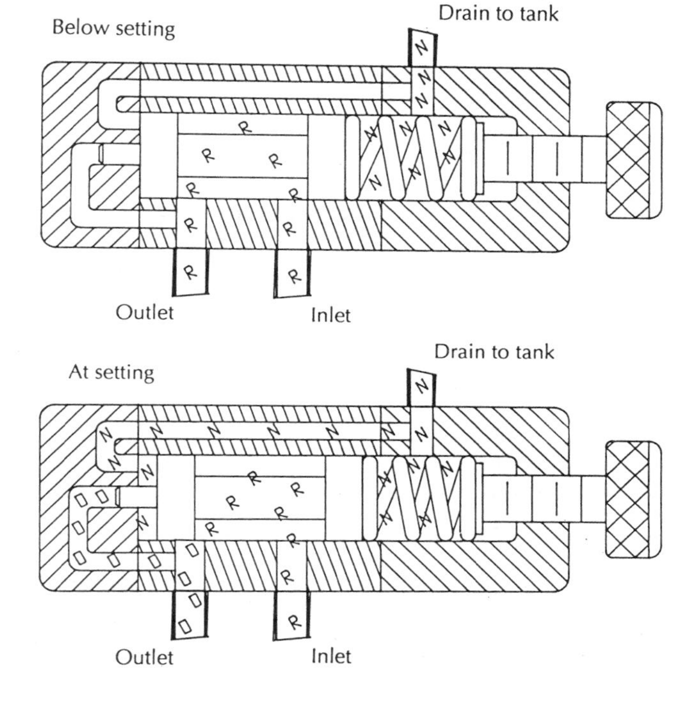

FIGURE 7-17

Pressure-reducing valve porting descriptions: including **(I)** inlet (high pressure), **(O)** outlet (reduced pressure), and **(D)** drain. (*Photograph by the author.*)

Pressure-reducing valves are included in hydraulic circuits to limit excessive pressure that might cause damage to or malfunction of machinery. That is, high pressure may be necessary to perform some portion of the cycle, while its existence would be detrimental to another function being performed in a separate portion of the cycle.

A clamping circuit used on light metals, plastics, or woods may require pressure reduction. During extension of the heavy clamping frame, a certain pressure would develop as determined by the size of the cylinder and the load imposed by the frame. During retraction, the heavy clamping frame would also be a resistance. However, with the reduced annulus area, higher pressure would be required to move the same clamp. Subsequently, the pressure-relief valve would need to be set 10% above the pressure necessary for retraction. This pressure would also be evident at the end of the extension clamping stroke. This excessive pressure might cause damage to piece parts. The schematic representation in Figure 7-18 describes a clamping circuit that may be used to limit the maximum pressure during clamping to a value equal to that required to move the clamping frame and a slight additional force for holding (see also Figure 7-19).

FIGURE 7-18

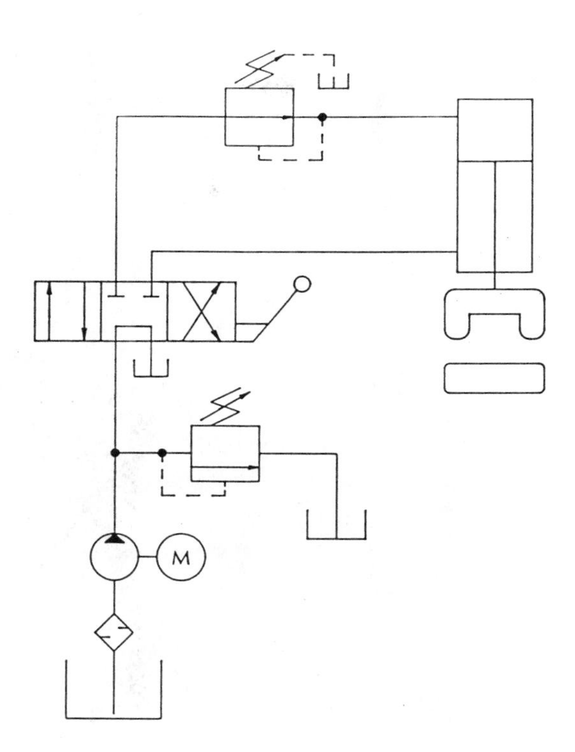

This clamping circuit is being used to move a 1,000-lb. {*4,453.3-N*} clamping frame and then impose an additional 200-lb. {*890.7-N*} force to hold an aluminum piece part. The cylinder in this circuit has a 2-in. {*5.1-cm*} diameter piston and a 1½-in. {*3.8-cm*} diameter rod.

Problem 7-6

What should be the setting on the pressure-reducing valve?

Pressure (psi) {*kPa*} = __________

Problem 7-7

What sould be the setting on the pressure-relief valve?

Pressure (psi) {*kPa*} = __________

Beyond the benefits derived from limiting clamping pressure in the circuit design shown in Figure 7-18, the pressure-reducing valve might play another important role. Figure 7-20 describes a sequence circuit using a pressure-reducing valve to limit pressure on the secondary extension (see also Figure 7-21).

FIGURE 7-19

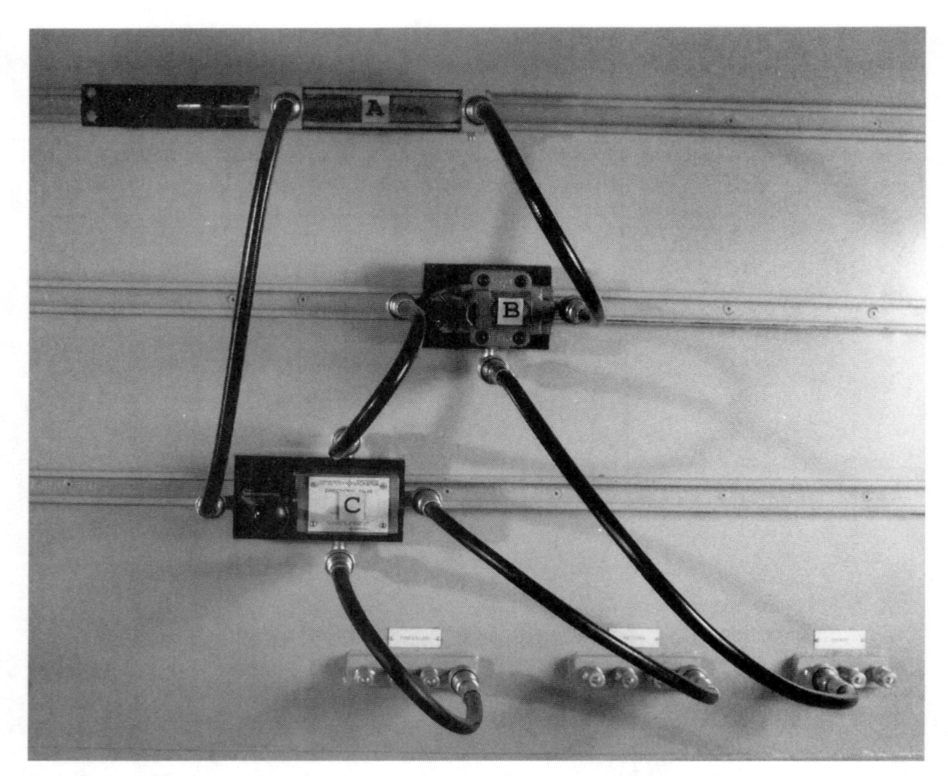

Mock-up of linear circuit having pressure-reducing capabilities during extension of the cylinder. Components include **(A)** single-end rod, double-acting cylinder; **(B)** pressure-reducing valve; and **(C)** three-position, four-way, directional-control valve. (*Photograph by the author.*)

The sequence circuit shown in Figure 7-20 is used to clamp a workpiece with 5,000 lbs. {*22.2 kN*} of force and then insert alignment pins into it, with the arbor cylinder also requiring 5,000 lbs. {*22.2 kN*} of force.

Problem 7-8 What should be the setting on the sequence valve?

Pressure (psi) {*kPa*} = __________

Problem 7-9 What should be the setting on the pressure-reducing valve?

Pressure (psi) {*kPa*} = __________

Problem 7-10 What should be the setting on the pressure-relief valve?

Pressure (psi) {*kPa*} = __________

FIGURE 7-20

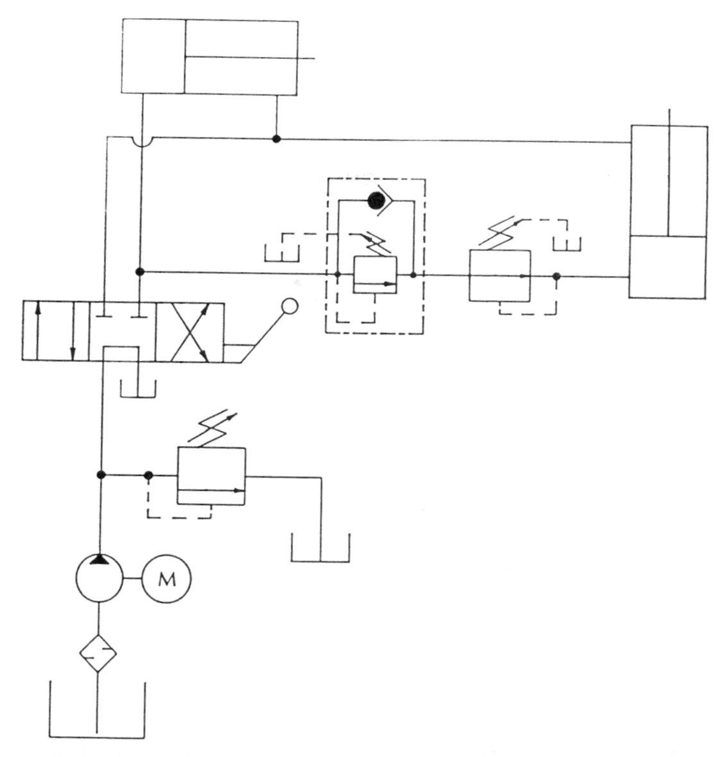

FIGURE 7-21

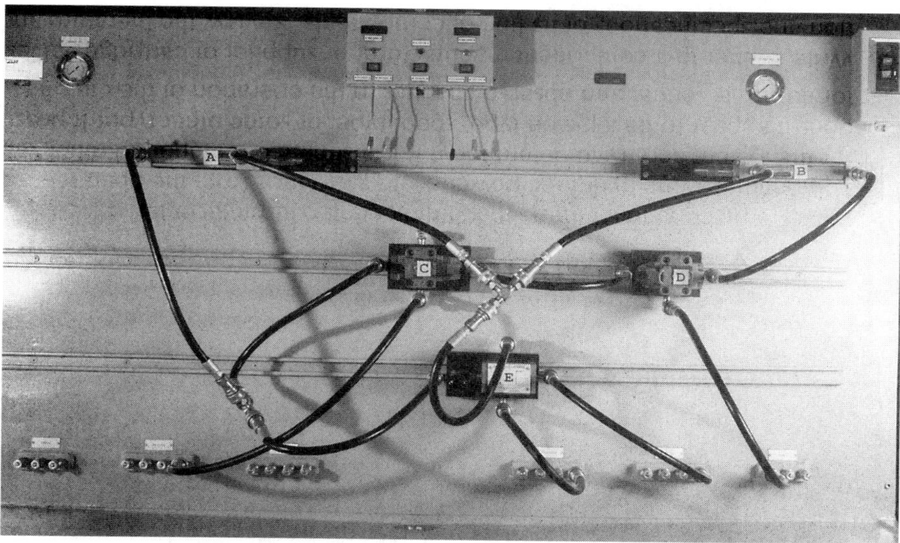

Mock-up of pressure parameter sequence circuit having pressure-reducing capabilities during extension of the secondary cylinder. Components include **(A)** single end rod, double-acting cylinder (primary extension); **(B)** single end rod, double-acting cylinder (secondary extension); **(C)** sequence valve; **(D)** pressure-reducing valve; and **(E)** three-position, four-way, directional-control valve. (*Photograph by the author.*)

Without the pressure-reducing valve, system pressure would drop throughout the circuit during secondary cylinder extension. This would cause the sequence valve to close and immediately open again when system pressure then rose to overcome its setting. The fluctuation of opening and closing the sequence valve would cause erratic movement of a secondary cylinder. Since the secondary cylinder in this case might be controlling a machining operation, damage to the workpiece and the machine tool is likely to occur. With the pressure-reducing valve included in the circuit design, smooth operation of the secondary cylinder's extension is assured. Only that pressure required to operate that cylinder is allowed beyond the pressure-reducing valve, while higher system pressure is maintained at all points before it. The sequence valve will, therefore, be held constantly open without fluctuations.

7.5 PRACTICAL DESIGN PROBLEM

A plastics manufacturing operation has requested that you supply them specifications for hydraulic components within the design of the circuit illustrated in Figure 7-22. This circuit is to be used in a resin-transfer system currently being considered for molding large body panels for diesel trucks. As you can see, the circuit is basically a **three-cylinder sequence circuit**.

The first cylinder (**loading cylinder**) is used to advance and retract the mold from the transfer press through a distance of 12 in. {*30.5 cm*}. The second cylinder (**clamp cylinder**) is used to hold the mold halves tightly together during resin transfer. This clamp cylinder has a maximum 6-in. {*15.25-cm*} stroke. The third cylinder (**thrust cylinder**) is used to transfer a charge of plastic resin through a distance of 20 in. {*50.8 cm*} from a storage and preparation tank to the mold.

Resistance to movement of the **loading cylinder** has been calculated to be 300 lbs. {*1.3 kN*} during both extension and retraction. The mold clamping frame has a sliding resistance of 1,000 lbs. {*4.45 kN*} during retraction; and the total clamp force, including frame weight, during extension needs to be 10,000 lbs. {*44.5 kN*}. The thrust cylinder will need to overcome a resistance of 5,000 lbs. {*22.2 kN*} to pump resin through the nozzle and sprue-gate assembly into the mold cavity. *During retraction, the resistance to thrust cylinder movement will be negligible.* Your specifications should be made to accommodate two full molding cycles per minute. *Feel free to include and specify any other components that you deem necessary or eliminate any unnecessary components. However, you should include with your report your reasons for these actions.*

FIGURE 7-22

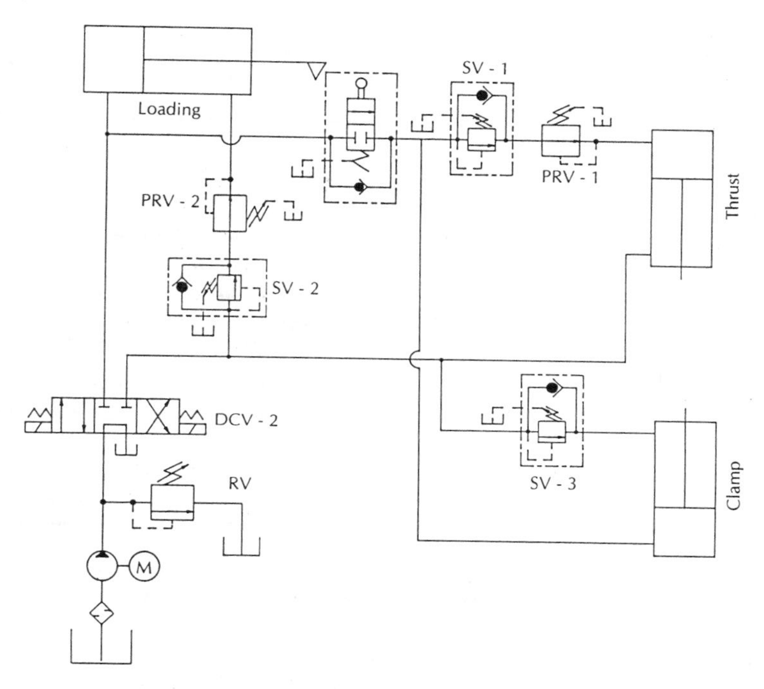

A Single End Rod, Double-Acting Cylinder—Loading Cylinder

PISTON		*ROD*			
Diameter	Area	Diameter	Area	Annulus Area	Stroke
______ in. {*cm*}	______ in.² {*cm²*}	______ in. {*cm*}	______ in.² {*cm²*}	______ in.² {*cm²*}	12 in. {*30.5 cm*}

FORCE		*PRESSURE*		*SUPPLY*	
Extension	Retraction	Extension	Retraction	Extension	Retraction
300 lbs. {*1.3 kN*}	300 lbs. {*1.3 kN*}	______ psi {*kPa*}	______ psi {*kPa*}	______ in.³ {*cm³*}	______ in.³ {*cm³*}

			VELOCITY	
Total Supply	Rate	Supply Rate	Extension	Retraction
______ in.³ {*cm³*}	2 cyc/min.	______ in.³/min. {*cm³/min.*}	______ in./min{*cm/min*}	______ in./min {*cm/min*}

Port Size	Body Style	Mounting
______ in.{*mm*}	______	______

B Single End Rod, Double-Acting Cylinder—Clamping Cylinder

PISTON		*ROD*			
Diameter	Area	Diameter	Area	Annulus Area	Stroke
______ in. {*cm*}	______ in.^{2} {*cm*^{2}}	______ in. {*cm*}	______ in.^{2} {*cm*^{2}}	______ in.^{2} {*cm*^{2}}	48in. {*120 cm*}

FORCE		*PRESSURE*		*SUPPLY*	
Extension	Retraction	Extension	Retraction	Extension	Retraction
10,000 lbs. {*44.5 kN*}	1,0000 lbs. {*4.45 kN*}	______ psi {*kPa*}	______ psi {*kPa*}	______ in.^{3} {*cm*^{3}}	______ in.^{3} {*cm*^{3}}

			VELOCITY	
Total Supply	Rate	Supply Rate	Extension	Retraction
______ in.^{3} {*cm*^{3}}	2 cyc/min.	______ in.^{3}/min. {*cm*^{3}/*min.*}	______ in./min{*cm/min*}	______ in./min {*cm/min*}

Port Size	Body Style	Mounting
______ in.{*mm*}	______	______

C Single End Rod, Double-Acting Cylinder—Thrust Cylinder

PISTON		*ROD*			
Diameter	Area	Diameter	Area	Annulus Area	Stroke
______ in. {*cm*}	______ in.^{2} {*cm*^{2}}	______ in. {*cm*}	______ in.^{2} {*cm*^{2}}	______ in.^{2} {*cm*^{2}}	20in. {*50.8 cm*}

FORCE		*PRESSURE*		*SUPPLY*	
Extension	Retraction	Extension	Retraction	Extension	Retraction
5,000 lbs. {*22.2 kN*}	0 lbs. {*0 kN*}	______ psi {*kPa*}	______ psi {*kPa*}	______ in.^{3} {*cm*^{3}}	______ in.^{3} {*cm*^{3}}

			VELOCITY	
Total Supply	Rate	Supply Rate	Extension	Retraction
______ in.^{3} {*cm*^{3}}	2 cyc/min.	______ in.^{3}/min. {*cm*^{3}/*min.*}	______ in./min{*cm/min*}	______ in./min {*cm/min*}

C Single End Rod, Double-Acting Cylinder—Thrust Cylinder (*continued*)

Port Size	Body Style	Mounting
_______ in.{*mm*}	__________	__________

D Pump

Type	Design	Delivery	Drive Speed	Volumetric Efficiency
________	________	________ gpm{*lpm*}	_______ rpm	_________ %

Displacement at V.E.	*PORT SIZE* Inlet	*PORT SIZE* Outlet
__________ in.³/rev{*cm³/rev*}	________ in.{*mm*}	________ in.{*mm*}

E Reservoir

CAPACITY Minimum	*CAPACITY* Maximum
______ gal. {*l*}	______ gal. {*l*}

F Filtration

Type	Element	*RATING* Element	*RATING* Flow	Port Size
Strainer	Wire	_________ mesh	________ gpm {*lpm*}	________ in. {*mm*}

Description	Mounting
Full Flow	__________

G Master System Directional Control

Positions	Ways	Element	Activation	Port Size
________	________	Sliding Spool	__________	________ in. {*mm*}

H Position Parameter Sequencing Directional Control

Positions	Ways	Element	Activation	Port Size
________	________	Sliding Spool	__________	________ in. {*mm*}

I Pressure Parameter Sequence Valve—1

Type	Element	Setting	Mounting	Port Size
__________	__________	_______ psi {kPa}	__________	_______ in. {*mm*}

J Pressure Parameter Sequence Valve—2

Type	Element	Setting	Mounting	Port Size
__________	__________	_______ psi {kPa}	__________	_______ in. {*mm*}

K Pressure Parameter Sequence Valve—3

Type	Element	Setting	Mounting	Port Size
__________	__________	_______ psi {kPa}	__________	_______ in. {*mm*}

L Pressure-Reducing Valve—1 (If Required)

Type	Element	Setting	Mounting	Port Size
__________	__________	_______ psi {kPa}	__________	_______ in. {*mm*}

M Pressure-Reducing Valve—2 (If Required)

Type	Element	Setting	Mounting	Port Size
__________	__________	_______ psi {kPa}	__________	_______ in. {*mm*}

N Pressure-Relief Valve

Type	Element	Setting	Mounting	Port Size
______	______	______ psi {kPa}	______	______ in. {*mm*}

O Electric Motor

Speed	Voltage	Phase	Power at V.E.
______ rpm	______	______	______hp {*W*}

P Conductors

Inlet Type	Inlet Description	*INLETS (I.D. DECIMAL)* Minimum	Maximum	Working Type	Working Description
______	______	______ in. {*cm*}	______ in. {*cm*}	______	______

WORKINGS (I.D. DECIMAL) Minimum	Maximum	*LINES—NOMINAL* Inlets	Workings
______ in. {*cm*}	______ in. {*cm*}	______ in. {*mm*}	______ in. {*mm*}

Q Report any changes made in the circuit or specifications (i.e., elimination of or relocation of pressure reducing valves) that would alter the schematic that was provided.

7.6 SUGGESTED ACTIVITIES

1. Locate a piece of machinery (machining center, turning center, etc.) or portable excavation equipment having hydraulic power. With assistance from the operator, locate the relief valve for the system and try to determine if it is a simple or compound relief valve. If compound, does it appear to be a piston type, spool type, or some other type? If possible, have the operator adjust its setting and observe changes in the operation of the circuitry.
2. Write a letter to a component manufacturer or supplier of cartridge valves asking for information and specifications of its line of components. Review the literature and write a brief report on this line of cartridge valves, paying special attention to design differences, operation, specifications, and sizing of their components.
3. Set up a linear circuit on a hydraulic trainer, operate the circuit, and then adjust the pressure-relief valve. Observe differences in the system's operation, paying special attention to pressure gauges. Alter the circuit with appropriate valving to achieve circuitry that would perform the following functions:
   a. pressure reduction during extension of the cylinder
   b. pressure reduction during retraction of the cylinder
   c. counterbalancing during extension of the cylinder
   d. counterbalancing during retraction of the cylinder
4. Draw a schematic of a pressure parameter circuit that would clamp a workpiece during the extension of a single end rod, double-acting cylinder; then shear the workpiece with the extension of another single end rod, double-acting cylinder. *Note*: Retractions need not be sequenced in this circuit because it does not matter if the workpiece is unclamped or if the shear repositions first.
5. Fabricate a mock-up of the circuit described in Activity 4 on a laboratory hydraulic trainer. Run the system and observe its operation, taking pressure readings (with gauges) at appropriate locations.

7.7 REVIEW QUESTIONS

1. What are the three functions of pressure-control valves?
2. Discuss pressure override in the simple relief valve and how it is compensated for in the designs of compound spool-type, piston-type, and cartridge-type valves.
3. Name three applications of a normally closed pressure-control valve.
4. Describe the application of the spool-type pressure control when used in unloading and high–low pumping operations.
5. Differentiate between "near" and "remote" piloting in counterbalance circuitry.
6. Why must the pressure parameter sequence valve be externally drained rather than internally drained?
7. How is a pressure-reducing valve different from a pressure-relief valve?

8

Flow Control

Another major control factor (other than direction and pressure) available within hydraulic circuitry is flow control. *Flow-control valves are used to reduce the volume of fluid to or from a portion of the operating circuitry during a period of time.* Thus, they modulate effective delivery. This results in a reduction in operating speed of actuators such as cylinders.

8.1 FIXED-ORIFICE FLOW CONTROLS

In Chapter 5, a linear circuit designed to operate a single end rod, double-acting cylinder at a specific rate during extension was shown to have a resulting retraction velocity greater than that of extension. The rate at which velocity was specified or calculated was shown to be a function of cross-sectional cylinder areas and pump delivery. The necessary delivery to produce a specific cylinder velocity can be calculated using the following equation:

$$\text{DEL (gpm) } \{lpm\} = \frac{\text{VEL (ft./sec.) } \{m/sec.\} * \text{A (in.}^2\text{) } \{cm^2\}}{0.3208\ \{0.1667\}}$$

A cylinder having a 2-in. {*5.08-cm*} diameter piston requiring an extension velocity of 0.5 ft./sec. {*0.15 m/sec.*} would require a pump delivery as calculated:

$$\text{DEL (gpm) } \{lpm\} = \frac{0.5 \text{ (ft./sec.) } \{0.15\ m/sec.\} * 3.14 \text{ (in.}^2\text{) } \{20.26\ cm^2\}}{0.3208\ \{0.1667\}}$$

$$\text{DEL (gpm) } \{lpm\} = \frac{1.57\ \{3.04\}}{0.3208\ \{0.1667\}}$$

$$\text{DEL (gpm) } \{lpm\} = 4.89\ \{18.23\}$$

If that cylinder has a 1-in. {*2.54-cm*} diameter rod, the annulus area operating during retraction would be 2.36 sq. in. {*15.20 cm²*}. The necessary delivery to produce a retraction velocity of 0.5 ft./sec. {*0.15 m/sec.*} would be calculated to be the following:

$$\text{DEL (gpm)}\ \{lpm\} = \frac{0.5\ (\text{ft./sec.})\ \{0.15\ m/sec.\} * 2.36\ (\text{in.}^2)\ \{15.20\ cm^2\}}{0.3208\ \{0.1667\}}$$

$$\text{DEL (gpm)}\ \{lpm\} = \frac{1.18\ \{2.28\}}{0.3208\ \{0.1667\}}$$

$$\text{DEL (gpm)}\ \{lpm\} = 3.68\ \{13.68\}$$

To achieve this effective delivery to the cylinder during retraction, the pump delivery specified for extension must be reduced. One method of achieving this is to place a restriction in the supply line between the directional control and the head end of the cylinder.

Figure 8-1 illustrates a circuit using a check valve that allows unobstructed supply to the head end of the cylinder during retraction. When exhaust is directed from the head end of the cylinder, during extension, flow is allowed to pass through a small orifice (hole) in the check valve's

FIGURE 8-1

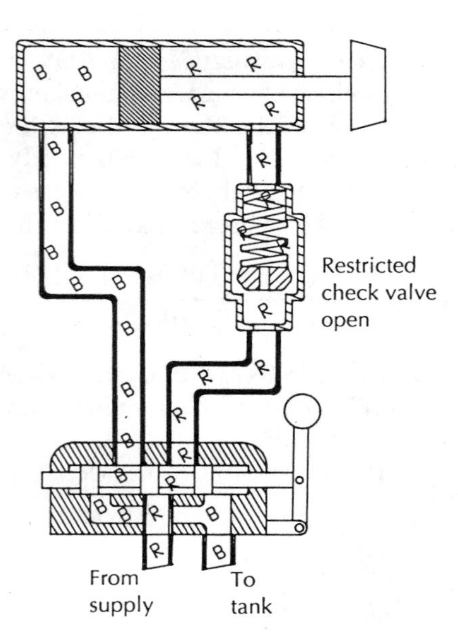

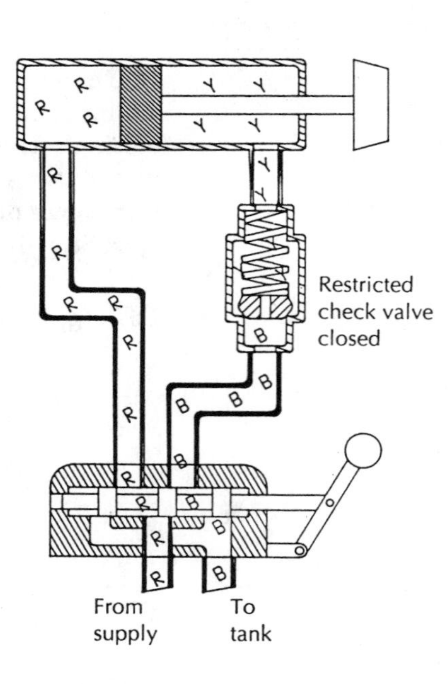

poppet. The hole may be sized to allow passage of only a specific limited amount of fluid within a period of time.

The **fixed-orifice check valve** used in the circuitry shown in Figure 8-1 may also be used to reduce the retraction speed of a linear circuit by restricting the exhaust flow during retraction. In this application, the circuit will be able to allow a vertically loaded ram-type cylinder to slowly decompress during retraction. The rate at which decompression will occur is a function of the orifice size in the check valve (see Figure 8-2).

FIGURE 8-2

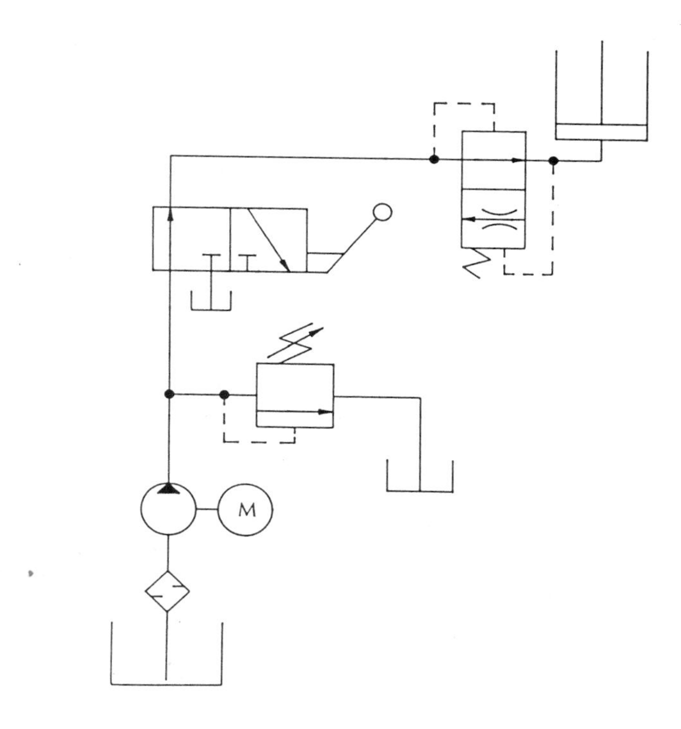

The ram in this circuit has a piston diameter of 4 in. {*10.2 cm*}, and it is desired that the ram decompress at a constant rate of 2 in. {*5.1 cm*} per second.

Problem 8-1

What rate of fluid flow should be allowed to pass through the restricted check valve during retraction?

DEL (gpm) {*lpm*} = __________

8.2 VARIABLE-FLOW CONTROLS

The fixed-orifice flow control, although simplistic in design, is often inaccurate when a specific actuator speed is desired. If pump supply varies, no adjustment can be made to accommodate this fluctuation. Portable hydraulic devices that operate under internal combustion engine drives typically have a wide variation in pump supply because of the difficulty of accurately controlling engine speed. Therefore, circuitry such as this typically includes **variable-flow controls** to adjust actuator speed.

8.2.1 Noncompensated Flow Controls

The most common designs of variable-flow controls include **gate, globe, and needle valves** (see Figures 8-3 and 8-4). Even simple variable-flow controls such as those in Figure 8-3 typically include a reverse-free flow check valve when used in hydraulic circuits. The needle valve commonly is located in a fixed-orifice **venturi** (restriction) so that even when the needle is fully retracted, some reduction in flow is evident. *Gate and globe valves are infinitely variable* and may be set to allow any supply between full pump delivery and total exclusion of pump delivery.

If the flow control is fully open when used in a linear circuit, negligible pressure differential will occur from one side to the other; and the pressure will be determined by the load on the cylinder and the appropriate

FIGURE 8-3

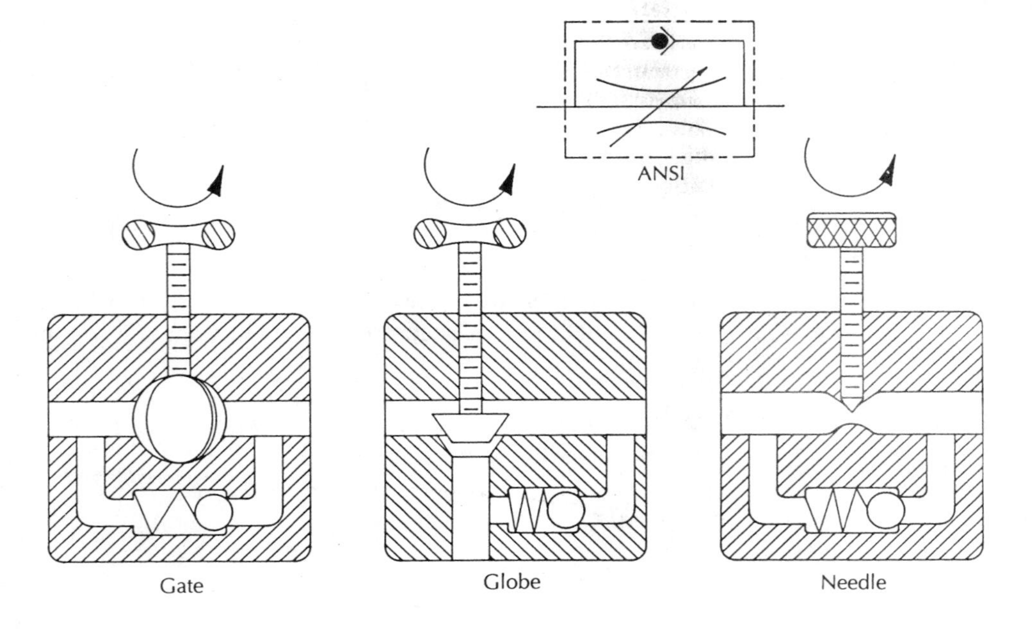

FIGURE 8-4

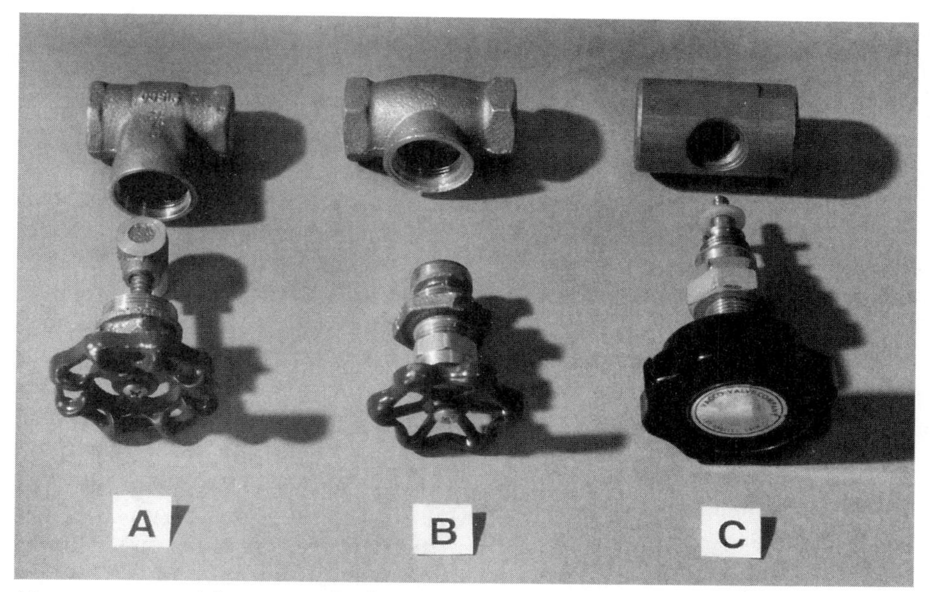

Noncompensated flow-control valves: **(A)** gate valve, **(B)** globe valve, and **(C)** needle valve. (*Photograph by the author.*)

cylinder area. When fully closed, pressure will rise to the relief valve setting on one side of the flow control; and no pressure will exist on the other side. However, at any degree of restriction between these extremes, different pressures will be evident on each side of the flow control. In other words, *a pressure drop is evident between the inlet and outlet of the flow-control valve during cylinder movement.*

When a high degree of restriction is encountered, flow from the pump will be greatly reduced; and a large degree of pressure drop is evident. Inversely, low restriction yields little reduction in supply flow and a small degree of pressure drop. After the cylinder caps out, pressure will equalize on both sides of the flow control and attain the value established by the circuit's pressure-relief valve.

8.2.2 Pressure-Compensated Flow Controls

To compensate for pressure differential between the inlet and outlet sides of the flow control, a compound valve is required. By combining the **throttle** (flow-restricting device) with a pressure control that is pilot activated from the valve's outlet side, the pressure differential can be adjusted to a negligible value.

The throttle control, normally adjusted by a rotary knob, may be a simple notch in the periphery of a spool that gradually increases and decreases in cross-sectional area to allow limited variable restriction. These are typically called single-range flow controls. Other throttle designs use a notched-end spool that exposes and closes many small passageways between the valve's inlet and outlet through multiple rotations of the spool. These valves are known as multirange flow controls (see Figures 8-5 and 8-6).

One design of pressure-compensated flow control incorporates a normally closed pressure compensator. The compensator shunts between the flow control's inlet and outlet, bridging the valve's throttle. The compensator normally closes a passageway between the inlet and a tank exhaust port using a light-duty biased spring on the outlet side of the valve. Pilot pressure from the outlet side assists the spring in holding this passageway closed. Whenever pressure on the inlet side of the valve is greater than the combined outlet side pilot pressure and spring pressure, the compensator spool will shift, opening the path to tank, and relieve inlet pressure to a value nearly equal to that on the outlet side. Regardless of the restriction of the throttle, pressure at the inlet will equalize that of the outlet side within the pressure value of the light-duty biased spring (normally 10 to 20 psi {*69 to 138 kPa*}). An illustration of the **pressure-relief compensated flow control** appears in Figure 8-7.

FIGURE 8-5

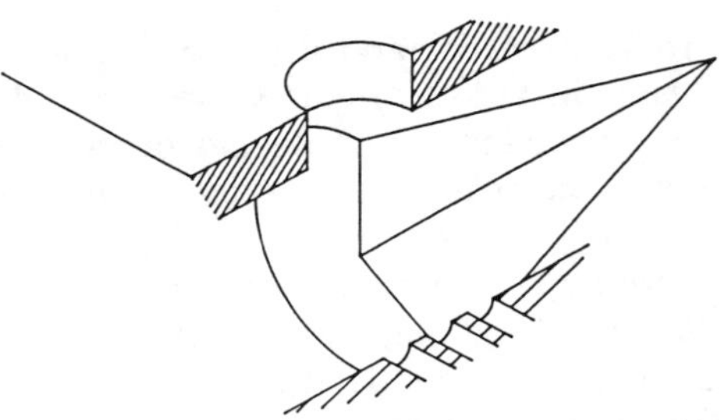

Rotation of the throttle spool opens or closes more banks of passageways.

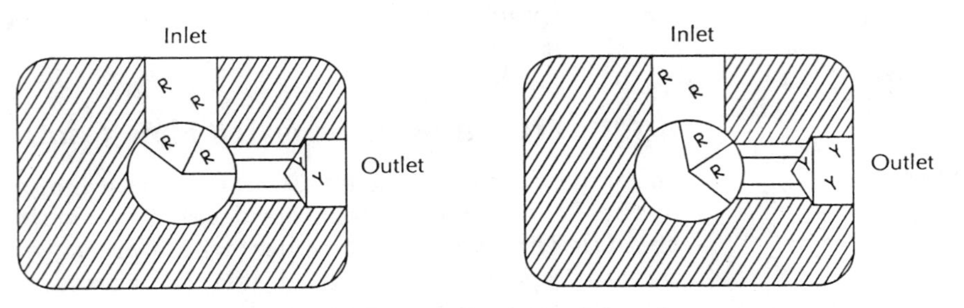

Rotation of the throttle spool opens or closes more layers of passageways.

FIGURE 8-6

Multirange, pressure-compensated, flow-control valve. Porting designations include **(I)** inlet and **(O)** outlet. This valve has an integral check that allows reverse-free flow from the "O" port to the "I" port. Variably adjustable flow control occurs when flow is from the "I" port to the "O" port. (*Photograph by the author.*)

FIGURE 8-7

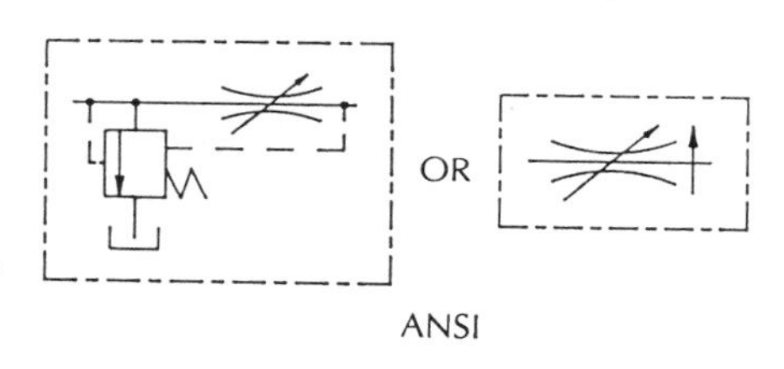

In this pressure-compensated flow control, outlet-side pilot pressure in combination with a light-duty spring holds the spool so that fluid passage to the tank port (T) is normally blocked.

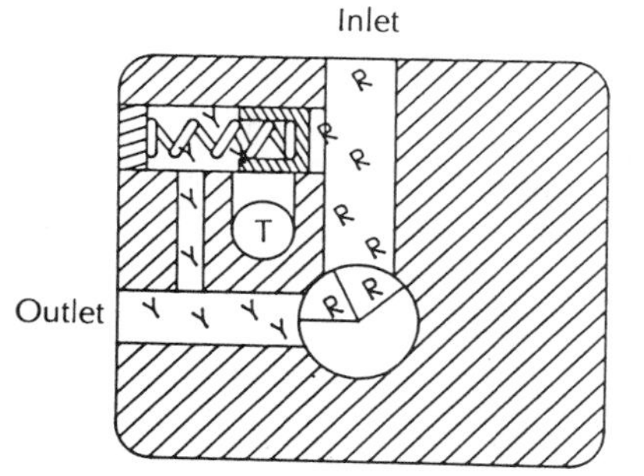

A pressure drop is evident through the throttle that makes pressure on the outlet side less than pressure at the inlet side.

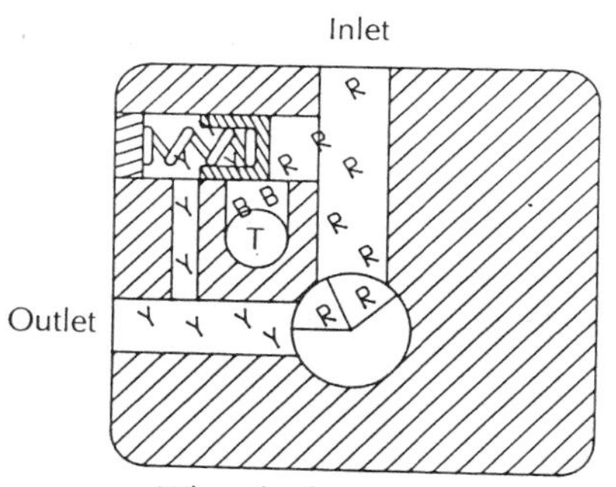

When the force components of pressure differential between inlet and outlet become great enough, the spool will shift. This limits inlet pressure to a value no greater than the combined pressure of the spring and outlet-side pressure.

Another design of pressure-compensated flow control incorporates a normally open pressure compensator. The compensator shunt in this design includes the inlet passage. The light-duty biased spring, in combination with outlet-side pilot pressure, holds the compensator spool in a position that normally allows unobstructed passage of fluid through the valve inlet. Inlet pilot pressure greater than those combined pressures on the outlet side of the compensator will shift the spool, closing the inlet passage. The pressure through the throttle will be reduced to a value equivalent to outlet pressure within the pressure rating of the light-duty biased spring. An illustration of the **pressure-reducing compensated flow control** appears in Figure 8-8.

FIGURE 8-8

In this pressure-compensated flow control, outlet-side pilot pressure in combination with a light-duty spring holds the spool so that fluid passage through the inlet is normally open. An inlet-side pilot pressure acts on the end of the spool opposite the biased spring.

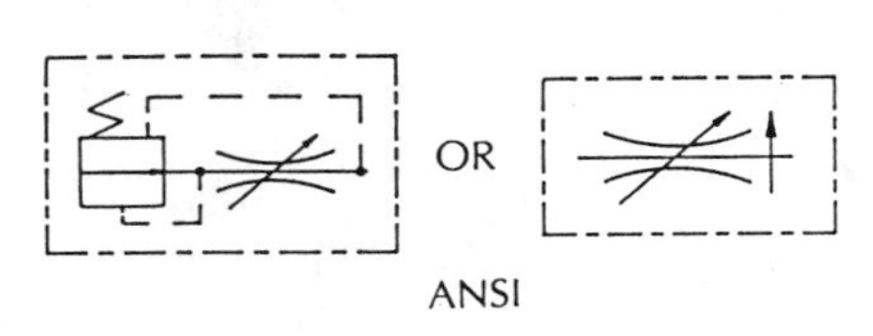

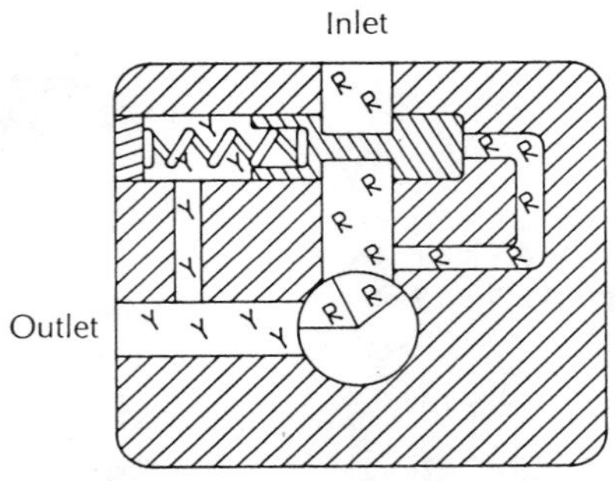

Pressure drop is evident through the throttle, making pressure on the outlet side less than pressure at the inlet side.

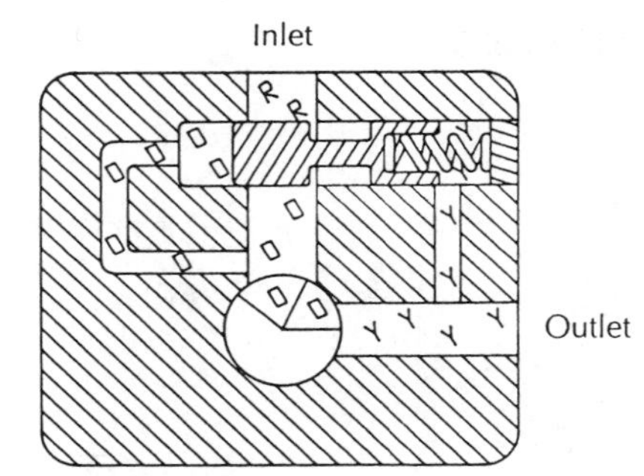

When the force components of pressure differential between inlet and outlet become great enough, the spool will shift. This closes down the inlet and reduces pressure through the throttle to a value no greater than the combined pressure of the spring and outlet-side pressure.

8.2.3 Characteristics of Pressure-Compensated Flow Controls

The pressure-relief compensated flow control will only operate as designed when a resistance resulting in pressure on the outlet side is evident. Otherwise, the pressure at the inlet will overcome the light-duty biased spring and divert all flow to the tank port. No fluid will be available to pass through the throttle. *The pressure-reducing compensated flow control will operate as designed, regardless of whether pressure exists at the valve outlet.* Yet, the pressure-relief type is generally considered to be the more accurate device.

Without adaptations, the throttling effect will occur regardless of the direction of flow through the valve. However, when reverse flow is directed through the valve, the pressure compensator will malfunction. Therefore, pressure-compensated flow controls are further compounded with an integral check valve to allow unrestricted reverse flow (see Figure 8-9).

FIGURE 8-9

Pressure-relieving compensation

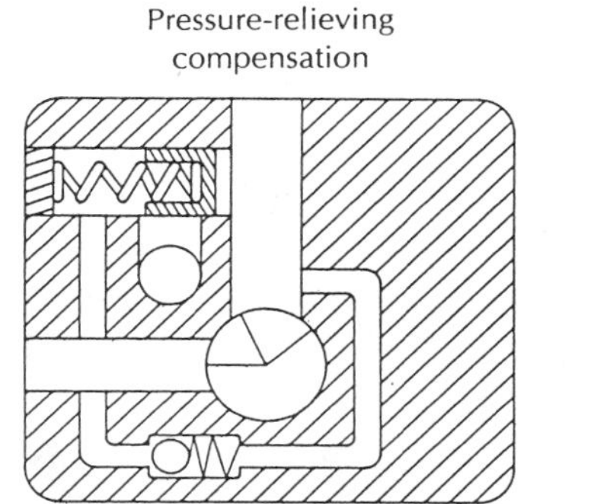

Pressure-reducing compensation

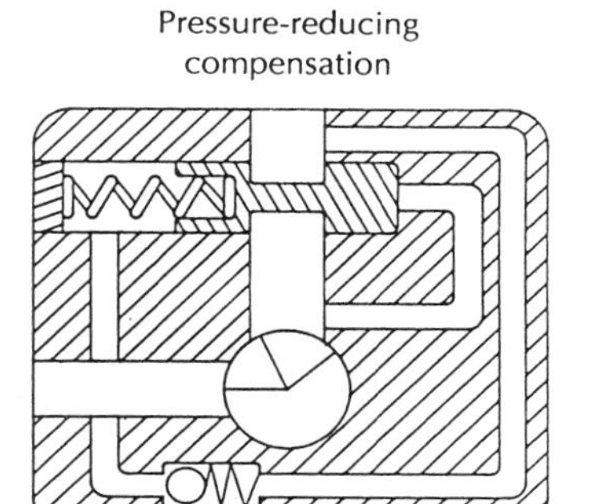

Assignment: Color-code the drawings for reverse-free flow.

8.3 FLOW-CONTROL CIRCUITRY

The method of flow control used in a linear hydraulic circuit is determined by its operation. Three options are common: **meter-in, meter-out, and bleed-off**.

8.3.1 Meter-In Circuitry

When the load on the cylinder constantly resists its movement, meter-in circuitry is appropriate. In meter-in circuitry, pressurized flow from the directional control is directed to the inlet port of the flow control. After passing through the flow control, flow is then directed to the appropriate port on the cylinder. If this flow is directed to the base end of the cylinder, the circuitry is described as **meter-in on extension**. Inversely, flow-controlled fluid directed to the head end of the cylinder results in **meter-in on retraction**. In both cases, either type of pressure compensation may be used; however, the more accurate pressure-relief type is typically selected (see Figures 8-10 and 8-11).

FIGURE 8-10

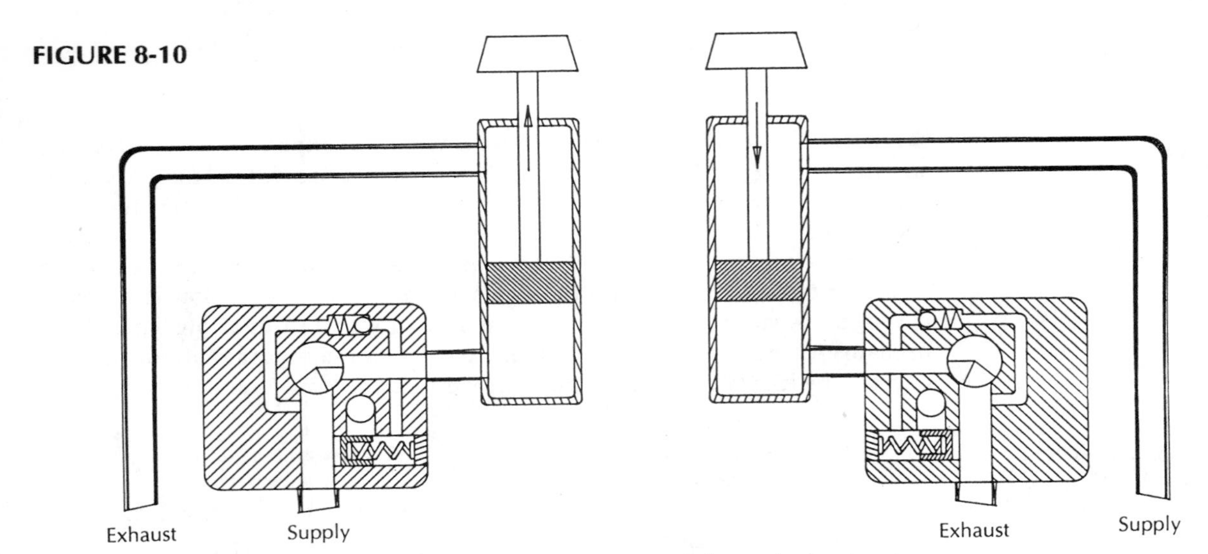

Assignment: Color-code the drawing.

FIGURE 8-11

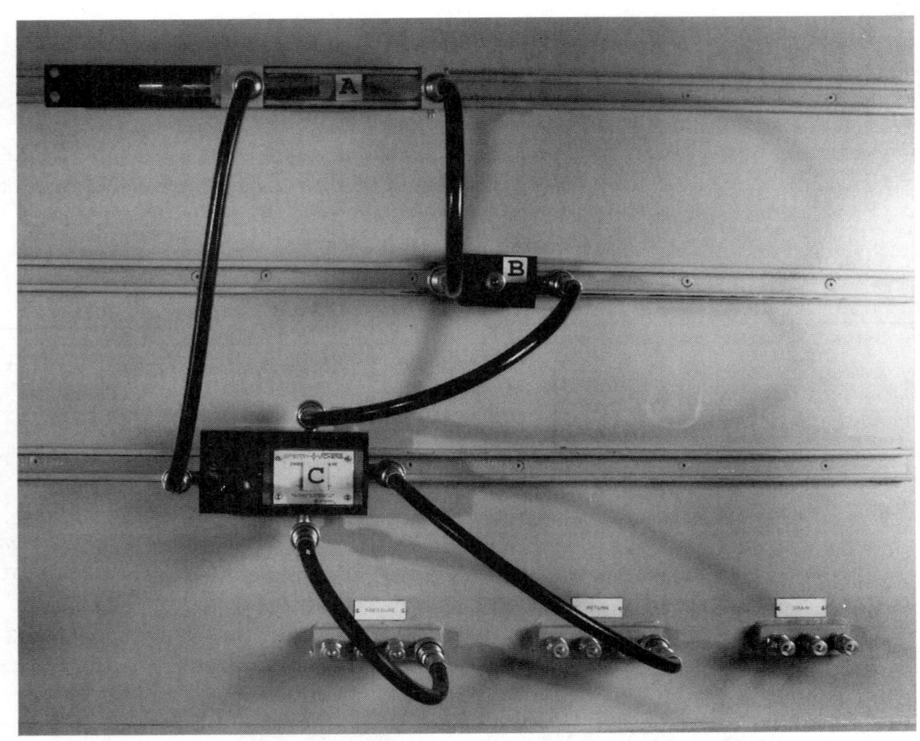

Mock-up of a linear circuit having meter-in flow-control capabilities during extension of the cylinder. Components include **(A)** single end rod, double-acting cylinder; **(B)** flow-control valve; and **(C)** three-position, four-way, lever-activated, directional-control valve. (*Photograph by the author.*)

8.3.2 Meter-Out Circuitry

When a load may tend to assist the movement of the cylinder, meter-out circuitry is appropriate. Such may be the case in a vertically mounted cylinder, under load during retraction, if gravity tends to assist that stroke. In meter-out circuitry, exhaust from the cylinder is directed to the inlet side of the flow control. After passing through the restriction of the throttle, flow is returned to the tank, as normal, through the directional control. *Meter-out circuitry may be applied on extension, retraction, or both.* Because no pressure is evident on the outlet side of the valve, a pressure-reducing compensated flow control is used in meter-out circuitry (see Figures 8-12 and 8-13).

FIGURE 8-12

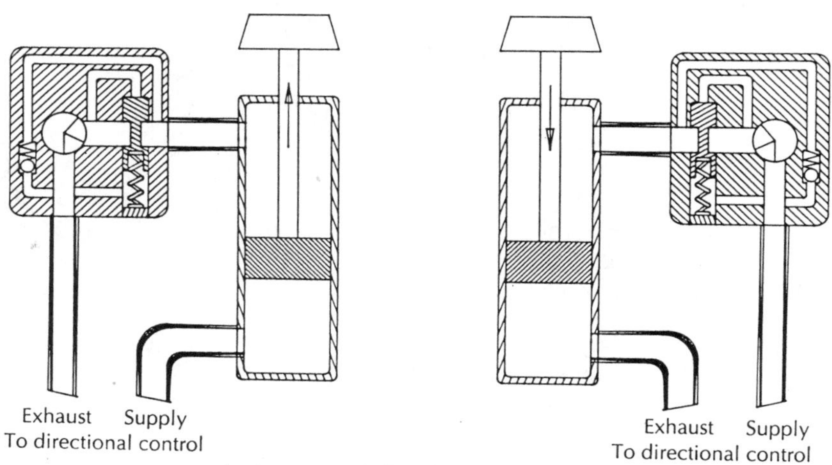

Assignment: Color-code the drawings.

The functional operation of meter-out flow-control circuitry is similar to counterbalancing, described in Chapter 7. However, the counterbalance circuit was a pressure parameter design; and as such, the valve setting at which decompression occurred varied as the load varied. When using the meter-out circuit, cylinder decompression will remain relatively constant, regardless of varying loads when using a pressure-compensated flow control. In general, counterbalancing is used when the major concern is impact at the end of the cylinder stroke and rate of cylinder movement is of minor consequence. *Meter-out circuitry is selected when a controlled rate of cylinder movement is required.*

FIGURE 8-13

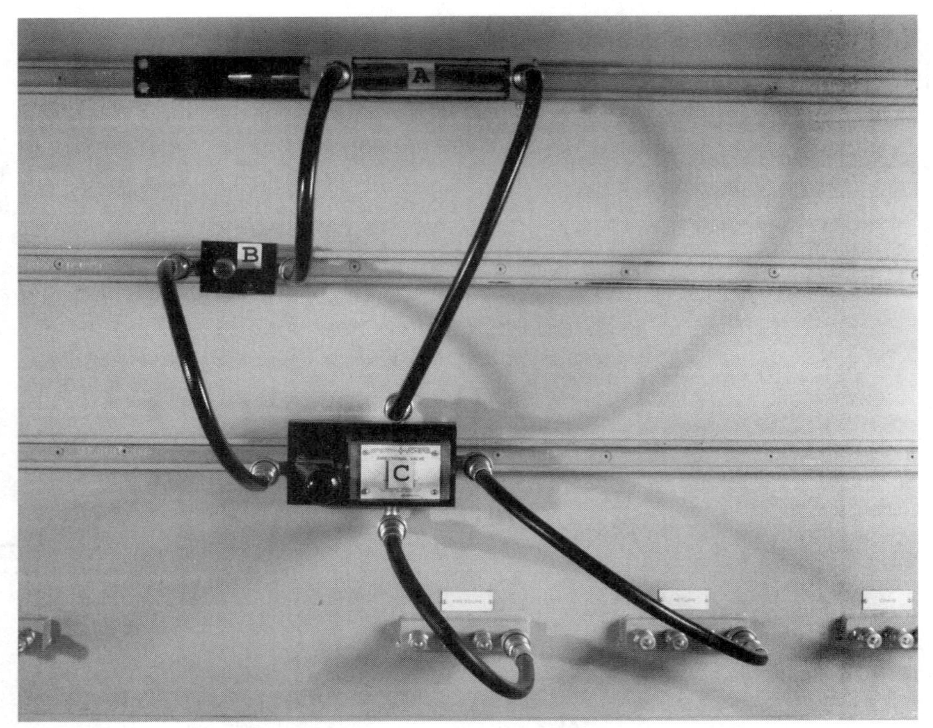

Mock-up of a linear circuit having meter-out flow-control capabilities during extension of the cylinder. Components include **(A)** single end rod, double-acting cylinder; **(B)** flow-control valve; and **(C)** three-position, four-way, lever-activated, directional-control valve. (*Photograph by the author.*)

8.3.3 Bleed-Off Circuitry

When simple speed control is required but the actual rate of cylinder movement is not critical, bleed-off circuitry is appropriate. In bleed-off circuitry, supply flow is "teed" off a pressure line and directed to the inlet side of the flow control. After passing through the valve, flow is directed unconditionally back to the tank rather than through the directional control. *The bleed-off design may be used on extension, retraction, or both.* Because no pressure is evident at the valve outlet, pressure-reducing compensated flow controls should be used. However, bleed-off circuitry typically uses nonpressure-compensated valves because critical control is not normally necessary. Flow will never be directed to the valve's outlet; thus, no concern for reverse-free flow need be made and integral checks are unnecessary (see Figures 8-14 and 8-15).

FIGURE 8-14

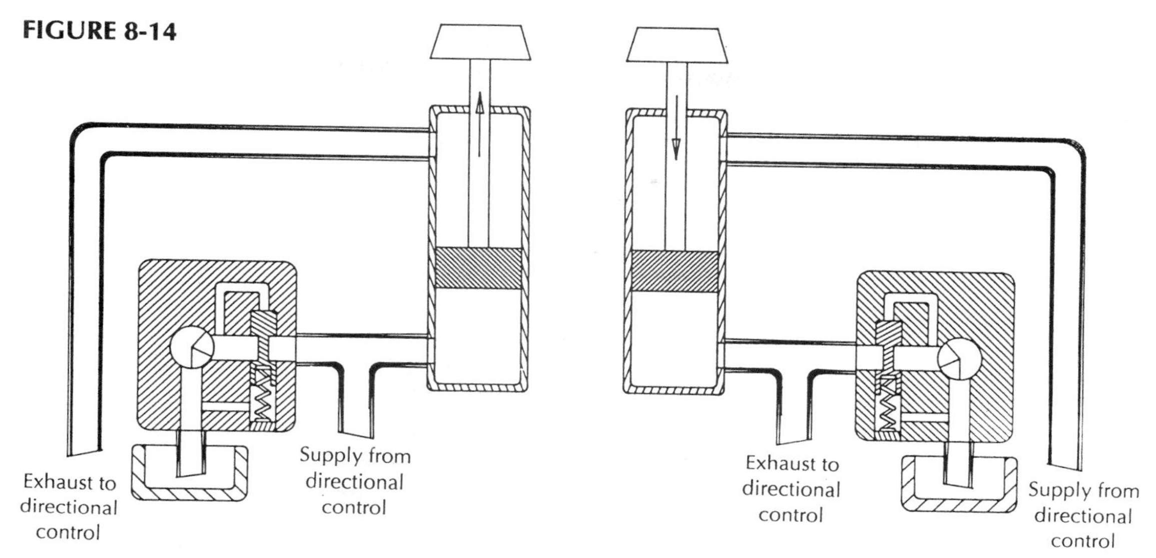

Assignment: Color-code the drawings.

FIGURE 8-15

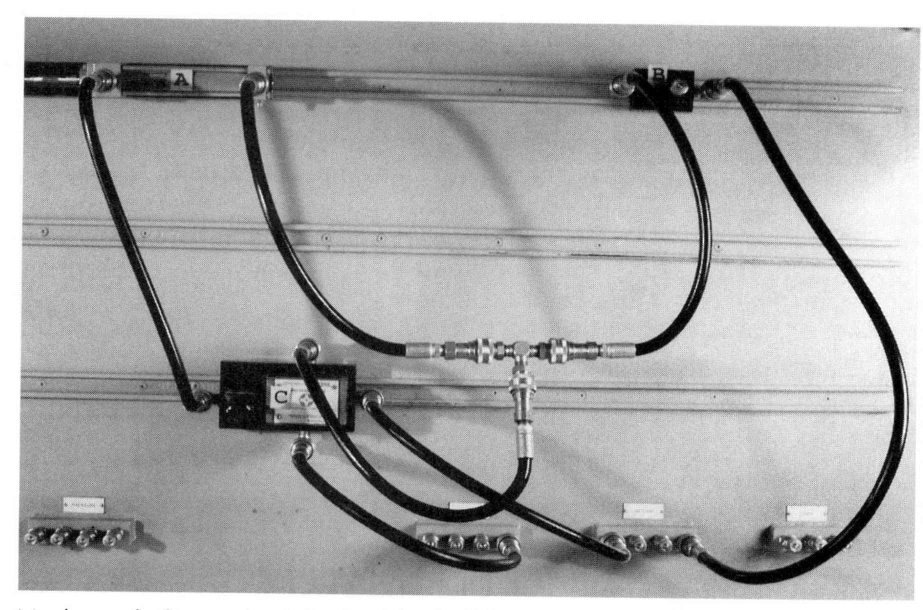

Mock-up of a linear circuit having bleed-off flow-control capabilities during extension of the cylinder. Components include **(A)** single end rod, double-acting cylinder; **(B)** flow-control valve; and **(C)** three-position, four-way, lever-activated, directional-control valve. (*Photograph by the author.*)

When using bleed-off flow control, circuit pressure will rise only to the level required to cause cylinder movement. Unlike other flow-control methods, excessive pressure is not necessary to cause operation; and minimal wasted energy is needed to bypass fluid over the relief valve. In operation, the flow control in bleed-off circuitry appears to operate in reverse. *To decrease cylinder speed, the flow-control restriction is decreased.* Increasing the restriction increases cylinder speed. This is opposite to adjustments for meter-in and meter-out flow control.

8.4 DESIGN PARAMETERS FOR FLOW-CONTROL CIRCUITS

Flow-control valves are sized according to port diameter and maximum flow rate (delivery) through the valve. Most flow controls may be varied from that maximum rating to a minimum that allows no flow through the valve. In general, a flow control is specified for use in a given circuit that will allow the required flow rate near mid-range. For example, in a situation that requires 3 gal./min. {*11.15 l/min.*} to pass through the flow control, it would be best to have a flow control rated from 0 to 6 gal./min. {*0 to 22.30 l/min.*}. This would allow a great degree of latitude when setting the valve to mid-range.

In circuit designs other than bleed-off, the pressure-relief valve setting, as compared to the pressure necessary to overcome the cylinder load, defines the pressure drop across the flow control. Therefore, the pressure-relief setting, in theory, could be made to any value greater than that required to overcome the load or loads and have minimal effect on circuit operation. This would be true if it were not for turbulent flow. There is a relationship between pressure drop across the flow-control throttle and the velocity of the fluid flowing through it. Within the operating pressure range of most industrial hydraulic circuits, *the square root of the pressure drop is equal to the velocity of fluid flow in ft./sec.*:

$$\sqrt{\Delta P \text{ (psi)}} = \text{VEL (ft./sec.)}$$

Recall that velocity through any pressurized fluid conductor should be maintained at a level between 7 and 20 ft./sec. {*2.1 and 6.1 m/sec.*} to eliminate turbulent flow. These factors, therefore, establish the typical pressure drop range allowed over a flow control.

Problem 8-2

What will be the pressure drop over a throttle that has fluid flowing through it at a velocity of 7 ft./sec. {*2.1 m/sec.*}?

ΔP (psi) {kPa} = __________

Problem 8-3

What will be the pressure drop over a throttle that has fluid flowing through it at a velocity of 20 ft./sec. {*6.1 m/sec.*}?

ΔP (psi) {kPa} = __________

As a result, the addition of any pressure between 49 and 400 psi {*338 and 2,758 kPa*} to the pressure needed to overcome cylinder force will be a permissible value for the pressure-relief valve setting. This variability may be useful when using flow controls in pressure parameter sequence circuits and other multiple-actuator circuits.

8.4.1 Meter-In Circuit Design

In the meter-in circuit, the pressure at the cylinder inlet will be determined by the load and the appropriate cross-sectional cylinder area. Pressure at the cylinder outlet will, of course, be 0. The pressure setting at the relief valve is determined by that necessary to do the cylinder operation and an allowance for pressure drop across the flow control. Delivery to the cylinder inlet from the flow control may be determined by the necessary cylinder velocity and appropriate area:

$$\text{DEL (gpm)}\ \{lpm\} = \frac{\text{VEL (ft./sec.)}\ \{m/sec.\} * \text{A (in.}^2\text{)}\ \{cm^2\}}{0.3208\ \{0.1667\}}$$

Any pump delivery may be specified, and a resultant return velocity for the cylinder may then be determined.

Fluid conductor size in the meter-in circuit is a function of standard fluid velocity in each portion where varied deliveries are encountered. In the single-cylinder meter-in circuit, three areas of varied delivery exist:

1. In the power supply (pump delivery)
2. Between the power supply and the cylinder, including cylinder exhaust (cylinder delivery)
3. Exhaust from the relief valve (pump delivery – cylinder delivery)

In each case, the range of inside diameters for fluid lines may be calculated by applying the extreme fluid velocity values (7 and 20 ft./sec.) {*2.1 and 6.1 m/sec.*} and the delivery value in that section to the following formula:

$$\text{I.D. (in.)}\ \{cm\} = \sqrt{\frac{\text{DEL (gpm)}\ \{lpm\} * 0.3208\ \{0.1667\}}{\text{VEL (ft./sec.)}\ \{m/sec.\} * 0.7854}}$$

Before specifying a nominal size for working lines in the meter-in flow-control circuit, it is important to consider that all these lines will at some time operate to pass full pump delivery. For example, the flow-controlled delivery line will operate as a cylinder exhaust line during the opposite cylinder stroke; and the tank exhaust line will need to pass the full pump delivery at the end of the cylinder strokes. Specification of the nominal working line size should, therefore, be made so that no turbulent flow exists in any situation. *One value that falls within all three permissible size ranges should be selected for all working lines, if possible.*

Problem 8-4

A meter-in flow-control linear circuit is operating with a pump delivery of 6 gal./min. {*22.7 l/min.*} and a cylinder delivery of 4 gal./min. {*15.1 l/min.*}. Describe the permissible range of working line sizes for each portion of the circuit.

Pump Delivery

a. Maximum Velocity: I.D. (in.) {*cm*} = __________

b. Minimum Velocity: I.D. (in.) {*cm*} = __________

Cylinder Delivery

a. Maximum Velocity: I.D. (in.) {*cm*} = __________

b. Minimum Velocity: I.D. (in.) {*cm*} = __________

Relief Exhaust Delivery

a. Maximum Velocity: I.D. (in.) {*cm*} = __________

b. Minimum Velocity: I.D. (in.) {*cm*} = __________

Problem 8-5

Specify one permissible size for all working lines in this circuit (use Schedule 40 {*STD*} sizing). Express this size nominally (see Appendix A).

Nominal Size: I.D. (in.) {*mm*} = __________

The inlet line (from the reservoir to the pump) size range may be calculated using the extreme permissible velocity values (2 and 5 ft./sec.) {*0.6 and 1.5 m/sec.*} and the pump delivery adjusted for volumetric efficiency for the flow-control circuit. Other design specifications for meter-in circuits may be made using the procedures common to simple linear circuit design.

8.4.2 Meter-Out Circuit Design

Pressure that develops within the meter-out circuit is essentially a matter of balance. When the flow control restricts exhaust flow, a back pressure develops that resists cylinder movement along with the major load (force) on the cylinder. The pressure that develops from flow restriction is a matter of choice. Recall that the permissible pressure drop across a throttle to eliminate turbulent flow may range between 49 and 400 psi. Any value may be specified and the meter-out flow control circuit will operate. However, *for the sake of design, the maximum value of 400 psi {2,758 kPa} should be used for further pressure calculations.* Resulting system pressure may then be adjusted downward with other controls. The pressure-relief valve setting will then become the determining factor for system pressure.

The first consideration in determining system pressure is to calculate the total force to be overcome with the cylinder. Primarily, the circuit will develop an **internal force (F_1)** equal to the **pressure drop (P_1)** across the flow control (400 psi {*2,758 kPa*} maximum) times the appropriate **cylinder area (A_1)**:

$$F_1 \text{ (lbs.) } \{N\} = P_1 \text{ (psi) } \{Pa\} * A_1 \text{ (in.}^2\text{) } \{m^2\}$$

If the circuit is meter-out on extension, A_1 will be the annulus area; and in meter-out on retraction, A_1 will be the piston area. The **total force (F_t)** to be overcome by the cylinder is a summation of the internal force (F_1) and the **external load** applied to the cylinder **(F_2)**:

$$F_t \text{ (lbs.) } \{N\} = F_1 \text{ (lbs.) } \{N\} + F_2 \text{ (lbs.) } \{N\}$$

System pressure (P_t) required to overcome this total resistance is calculated using the **opposite** appropriate **cylinder area (A_2)**:

$$P_t \text{ (psi) } \{Pa\} = \frac{F_t \text{ (lbs.) } \{N\}}{A_2 \text{ (in.}^2\text{) } \{m^2\}}$$

In meter-out on extension circuitry, A_2 is the piston area, while A_2 will be the cylinder's annulus area in meter-out on retraction circuitry. The calculated system pressure is the maximum setting for the pressure-relief valve. Also of interest may be the minimum pressure-relief valve setting. This value may be achieved by substituting the minimum permissible pressure drop (P_1) rate of 49 psi {*338 kPa*} into the previous formulation where appropriate (see Figure 8-16).

FIGURE 8-16

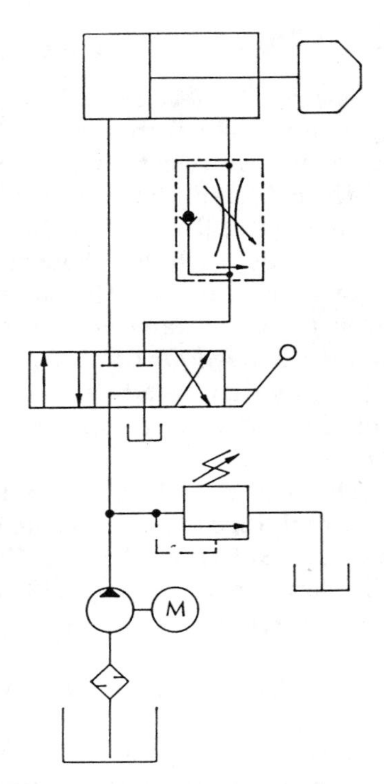

This ANSI schematic represents a meter-out on extension flow-control circuit using a cylinder having a 3-in. {*7.6-cm*} diameter piston and a 1-in. {*2.5-cm*} diameter rod. The major external load resisting extension is 5,000 lbs. {*22.3 kN*}.

Problem 8-6

Describe the setting for the pressure-relief valve that will allow this circuit to operate correctly.

Maximum $\Delta P - P_t$ (psi) {*kPa*} = __________

Minimum $\Delta P - P_t$ (psi) {*kPa*} = __________

The delivery passed through the flow control to produce a specified cylinder speed is a function of the opposite-area component of the cylinder. **In meter-out on extension flow-control circuitry, the following formula may be used, incorporating the cylinder's annulus area:**

$$\text{DEL (gpm)}\ \{lpm\} = \frac{\text{VEL (in./min.)}\ \{cm/min.\} * \text{A (in.}^2\text{)}\ \{cm^2\}}{231\ \text{(in.}^3\text{/gal.)}\ \{1{,}000\ cm^3/l\}}$$

In the meter-out on retraction circuit, the delivery through the flow control may be calculated using the same formula. *In this case, substitute the cylinder's piston area for the previously used annulus area.*

Pump delivery in the meter-out circuit may be any value greater than that required through the flow control. Delivery passed through the pressure-relief valve may be calculated by subtracting the delivery through the flow control from the pump delivery.

Specifications for conductor sizes and other components follow along the lines established for meter-in circuitry.

8.4.3 Bleed-Off Circuit Design

The bleed-off flow control serves as a diversion device. In essence, *circuit pressure during bleed-off will rise to the level necessary to displace the load in consideration of the appropriate cylinder area.* Any pressure in excess of this value will be negated by fluid return to the tank through the flow control. Therefore, no pressure drop is evident through the relief valve during bleed-off. The relief valve should be set to a value of 10% above system pressure during the bleed-off stroke. However, if bleed-off occurs during cylinder extension and the load is also imposed on retraction, the pressure-relief valve setting should be made in consideration of pressure that will develop during retraction.

The bleed-off flow control also diverts flow. *Delivery to the cylinder is determined by the necessary velocity for the cylinder stroke and the appropriate cylinder area (piston or annulus).* Pump delivery may be set to any value above that necessary for cylinder velocity. The flow control is "tuned" to pass any excess delivery to the tank.

Other components in the flow-control circuit are specified as in linear circuitry and other flow-control circuits.

8.5 BIDIRECTIONAL LINEAR FLOW CONTROL

Flow control for a single end rod, double-acting cylinder is somewhat complicated. Further complications are evident if accurately variable volume control is required in both stroke directions. The integral check valve could be eliminated from the valve, resulting in meter-in flow control in one stroke direction and meter-out in the other. However, cylinder speed would then only be variably specific in one direction. Cylinder speed in the other stroke direction would be a resultant of that setting. Therefore, single end rod, double-acting cylinder circuits typically use two flow-control valves if accurately controllable, variable cylinder speed is needed. With this arrangement, the method of control (meter-in, meter-out, bleed-off) and the necessary settings can be made independently (see Figure 8-17).

FIGURE 8-17

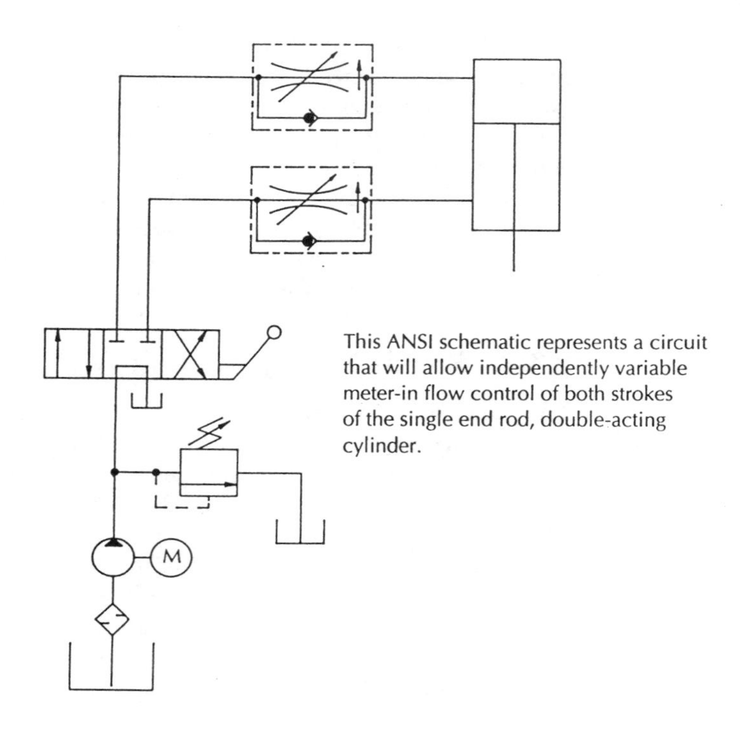

This ANSI schematic represents a circuit that will allow independently variable meter-in flow control of both strokes of the single end rod, double-acting cylinder.

When double end rod, double-acting cylinders are used, the design for flow control is greatly simplified. Since these cylinders have identical effective areas on either side of the piston, necessary supply to achieve a specific cylinder speed will be identical in both stroke directions. Hence, the exhaust from the opposite end of the cylinder will also be identical. In these circuits, only one flow-control valve is necessary to achieve a matched variable cylinder speed in both stroke directions. The location of the flow-control valve within the circuit will determine the type of flow control achieved. Figure 8-18 illustrates three circuit designs that allow matched-type flow control in both stroke directions of a double end rod, double-acting cylinder.

Notice that in each circuit design illustrated in Figure 8-18, no integral checks are located in the flow controls. During normal operation, no flow will ever be directed to the outlet sides of the flow controls. Therefore, checks are unnecessary. Mixed-type control may also be achieved by placing the unchecked flow-control valve in the appropriate location between the directional control and the cylinder. This is only appropriate when the load resists cylinder movement in one stroke direction while assisting it in the other.

Design and specifications for components in the double end rod, double-acting flow-control circuit will follow similar procedures as appropriate for single end rod, double-acting flow-control circuits. The one exception is that annulus areas only are used in formulation.

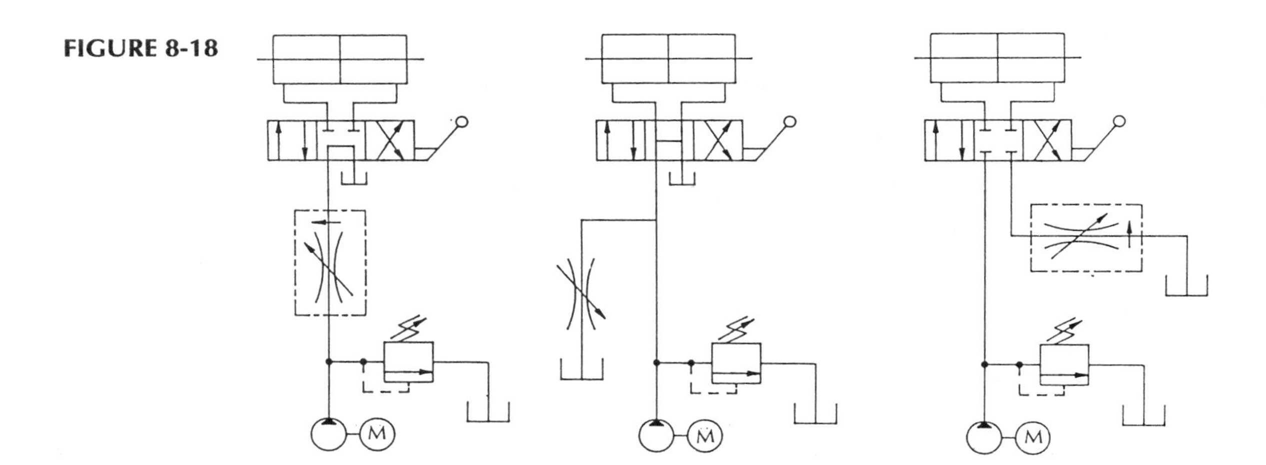

FIGURE 8-18

8.6 TEMPERATURE COMPENSATION

Constant or repeated flow of fluid through a restriction results in friction and shear of the fluid. Even if laminar flow is maintained, some increase in fluid temperature will result. This will decrease the viscosity of the fluid by thinning it. The thinned fluid will flow through the throttle with less resistance, resulting in increased cylinder velocity. In rapid-cycling circuitry, some compensation for this is required. The most common device employed for temperature compensation incorporates a spring made from a metal with a high coefficient of expansion. When heated, this spring expands and forces the throttle to close more of the passageways (see Figure 8-19). Of course, the further restriction of the flow in the **temperature compensator** will aggravate the temperature problem. Therefore, temperature-compensated flow controls should only be used when extremely precise and consistent flow control is required. Even then, other arrangements to cool the fluid in the reservoir are necessary. The temperature compensator will then act to accommodate minor fluctuations in system temperature.

8.7 INTERMITTENT OR PARTIAL-STROKE FLOW CONTROL

Many hydraulically controlled mechanisms require a reduced velocity only during a portion of the cylinder stroke. The rapid traverse on the table feed of a milling machine is an example of such a mechanism. By shunting around the flow control with a directional control, flow can be diverted to exclude or include the flow-control valve in the operating phase of the circuitry. Since hydraulic fluid will flow in the direction of least resistance, a directional control having an open position will allow flow through the shunt rather than the resistant flow control. When shifted, the direc-

FIGURE 8-19

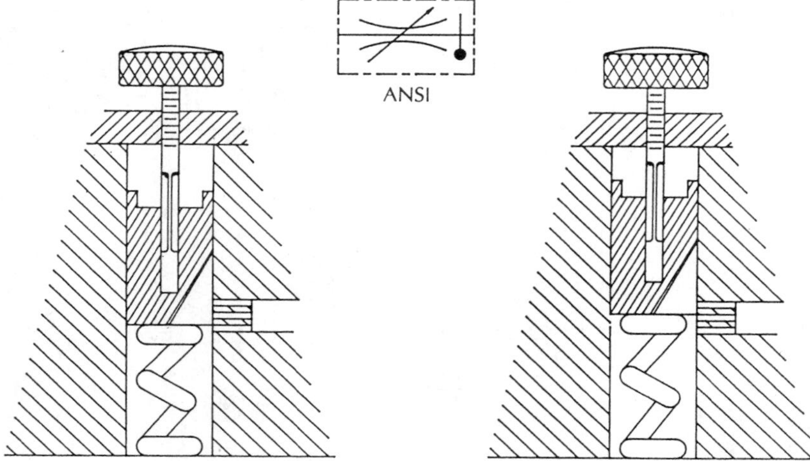

The throttle is splined into a rotating shaft that sets the major flow adjustment and compresses the bimetallic spring.

When temperature increases, the bimetallic spring expands and slides the throttle back on its splined shaft, closing off outlet passageways and reducing flow through the valve.

tional control will close the shunt path and flow must be directed through the flow control. Figure 8-20 describes a traverse circuit that allows rapid cylinder movement upon demand through manual activation of a two-position, one-way directional control.

FIGURE 8-20

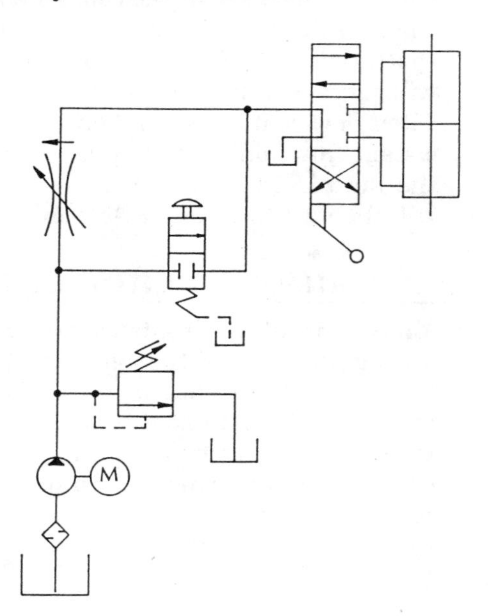

Problem 8-7

In Figure 8-20, verbally describe the method of flow control used in this traverse circuit.

____

____

Problem 8-8

Verbally describe when the flow-controlled (reduced-speed) cylinder movement will occur.

____

____

Problem 8-9

Verbally describe when the traverse (rapid-speed) cylinder movement will occur.

____

____

Machinery typically requires clearance of the tooling from the workpiece for setup and safety purposes. Movement of the tool from a station point to the touch point (tool–workpiece interface) is not resisted. When operating, this movement represents wasted time in the production cycle. Therefore, this distance should be traversed as rapidly as possible. Once the tool reaches a specified position, a controlled feed rate should be attained to move the tool through the workpiece. In other cases, controlled feed rate is required at the beginning of a cylinder stroke, followed by rapid movement for clearance purposes. A position parameter **feed and traverse circuit** that affords variability in both activation position and feed rate is illustrated in Figure 8-21.

Some variability may be achieved by changing the relationship between the cylinder and the directional control. The stroke of the cylinder should be specified to describe the linear distance of traverse or feed and overall stroke.

Other similar circuit designs use a fixed orifice in the normally closed position of the directional control and eliminate the flow control. These **decelerating valves**, as they are called, do not allow variability in cylinder speed.

With the design illustrated in Figure 8-21, variability in cylinder speed (meter-out) may be achieved through throttle adjustment on the flow-control valve. The integral check included in the flow control allows unrestricted flow to the cylinder during retraction. Of course, *meter-in traverse and feed circuitry is possible by arranging the shunt and flow control into the line supplying the base end of the cylinder* (see Figure 8-22).

FIGURE 8-21

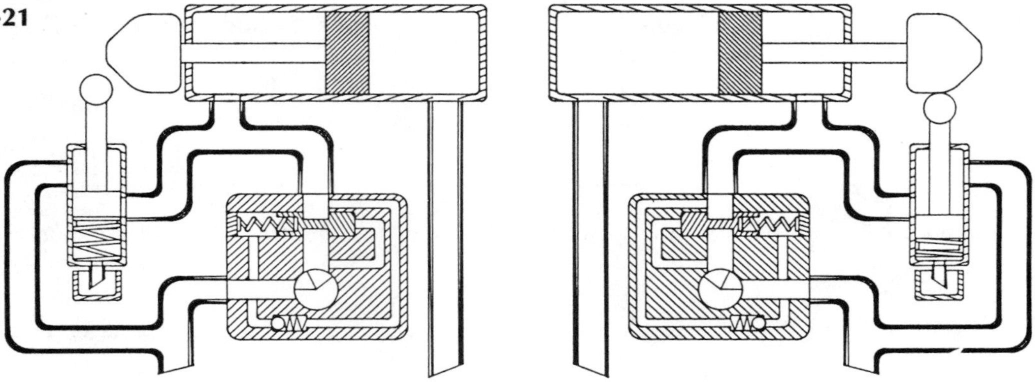

In the feed and traverse circuit, the two-position, one-way directional control is mechanically shifted. The cam attached to the cylinder rod forces the valve's plunger and shifts its spool from its normally closed position. This opens the shunt path and diverts flow to allow full flow and restore full delivery velocity. By changing the directional control to a normally open valve, traverse and feed circuitry can be achieved.

FIGURE 8-22

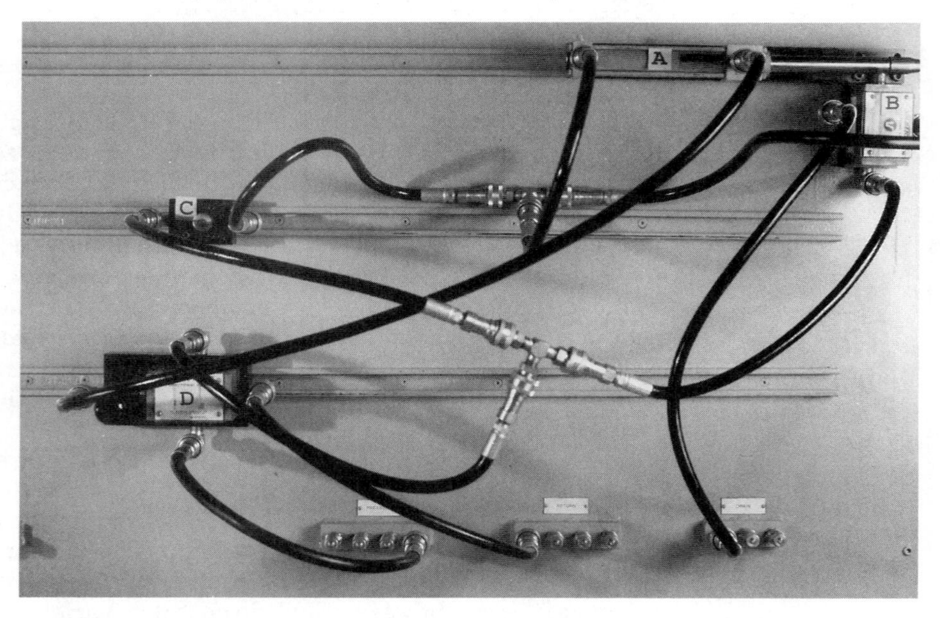

Mock-up of a linear circuit having traverse and feed meter-in flow-control capabilities during extension of the cylinder. Components include **(A)** single end rod, double-acting cylinder; **(B)** two-position, one-way, normally open, plunger-activated, directional-control valve; **(C)** flow-control valve; and **(D)** three-position four-way lever-activated, direction-control valve. (*Photograph by the author.*)

8.8 OTHER FLOW-CONTROL DEVICES AND SYSTEMS

Counterbalance linear circuitry and braking rotary circuitry have the net effect of reducing speed, but both are primarily concerned with preventing damage caused by excessive pressure or uncontrollable loads. It is possible, but not often feasible, to modulate speed of an actuator by varying the displacement of the system's pump or reducing the drive speed of the pumping package. Because the pump package in the power supply is generally remote to other controls in an operating circuit, it may not be convenient to vary actuator speed by these methods.

8.8.1 Proportional Flow Controls

Operational convenience is becoming increasingly important in modern hydraulic circuitry. This is true even in flow-control devices. Many of the innovations in these devices are marriages of hydraulic and electrical devices in a field known as electrohydraulics. The most important electrohydraulic flow-control device is the **proportional valve**. In two of its more common forms, the proportional valve is no more than a pressure-compensated flow control having sophisticated electrical or electronic devices that perform remote throttle adjustment.

The **proportional solenoid** type relies on a controllable advance and retract solenoid coupled to the throttle. Linear movement of the throttle then variably opens and closes the passageways to control flow. Because of design characteristics, these valves typically operate with end-notched throttle spools.

Peripherally notched throttle spools rely on a stepper-type motor to provide incremental adjustment through rotating the throttle. The **torque-motor proportional flow control** operates through electronic linkages between the throttle and motor and adjusts with variations in input current.

Programmable control is available in either design of proportional flow control through linkages to relay or digital logic controllers.

8.8.2 Flow Dividers

One of the more challenging problems in hydraulic control is to cause a large flow stream to segment into many small ones. Simply placing a "T" or a "Y" in a conductor will not assure that flow will evenly divide at that location. Because hydraulic fluid flowing in a system takes the path of least resistance, the flow will typically follow only one of the optional paths. This, of course, is dependent on which path requires the least resistance to overcome loads encountered. **Flow dividers** are designed to split one input flow into two or more output flows in an even or proportional ratio.

In its simplest form, a flow divider may be no more than two adjustable throttles in a branching network. Fine adjustment of each throttle dependent on loading conditions downstream from the valve or valves will cause the flow stream to divide equally when the paths are identically resistant. More sophisticated versions of the **adjustable-throttle flow divider** employ pressure compensators and proportional solenoids on each throttle in an interacting feedback loop that constantly adjust throttle openings to balance the two branches. This works well when only two paths are needed; however, adjustment becomes cumbersome when more alternatives are required.

The rotary-motor flow divider is capable of dividing input flow into an infinite number (theoretically) of output flows. This design typically employs a series of gear-type hydraulic motors mounted in tandem on a single power-takeoff shaft. Since all operate in conjunction, the constant rotating speed of one will cause intake of equivalent volumes of hydraulic fluid in all motors. During operation, each motor is simultaneously discharging an identical amount of fluid from its exhaust side. This exhaust may then be collected and used in multiple branch paths. (*For further discussion of design and operation of gear motors, see Chapter 9.*)

As complex as flow dividers appear, it is still important to closely monitor output requirements from the system. Great differences in loads in alternate paths can cause even the most sophisticated flow divider to malfunction.

[] [] [] 8.9 PRACTICAL DESIGN PROBLEM

MOFOCO produces rough and partially finished castings to supply engine blocks for Rumbling Diesel. In one finishing operation, a gang-drill head is used to produce twenty-four blind holes. Simultaneously, the gang-drill head rough reams the four cylinder walls. Thus, some operations of the gang drill produce "through" configurations, while other are "bottomed out." Currently, a jig borer is being used in this operation. However, MOFOCO has been having problems with chip removal and location indexing for the parts. For these reasons, MOFOCO has decided to produce its own machine to perform this operation. The concept that the production engineers have envisioned is to use a horizontal boring machine that will load, index, and clamp the castings, then sequentially feed the gang-drill head to perform the hole-producing operations. This will facilitate chip removal and allow the installation of a laser sensor to locate and index each casting.

The hydraulic circuit called for is a two-cylinder sequence circuit. It will load and index each 2,600-lb. {*11.6-kN*} casting through a 22-in. {*55.9-cm*} stroke into a fixture and then apply an additional 1,400 lbs. {*6.2 kN*} of force to clamp the casting. After the clamping and positioning force is realized, a pressure parameter sequence will occur that will allow the 2,200-lb. {*9.8-kN*} gang-drill head to advance rapidly through 11 in. {*27.9 cm*} of traverse stroke at a rate of 5 in./sec. {*12.7 cm/sec.*} to a "touch point." At this point, the drilling and reaming operation will occur through an additional 9 in. {*22.9 cm*} of feed stroke at a rate of 180 in./ min. {*457.2 cm/min.*}. An additional external force of 500 lbs. {*2.2 kN*} is anticipated during the machining phase of this **traverse and feed** cycle. After the machining operation is completed, the operator will reverse the cycle and retract the gang-drill head. *Anytime after 9 in. {22.9 cm} of retraction stroke of the drill head,* a position parameter sequence should occur to allow the clamp cylinder to retract and unload the casting. The schematic that appears in Figure 8-23 is provided as a reference for you to use in specifying all components to the enclosed, formatted guide. *Feel free to make changes in the basic design, if necessary, but thoroughly specify all additional components (or explain deletions in writing) and alter the schematic accordingly.*

FIGURE 8-23

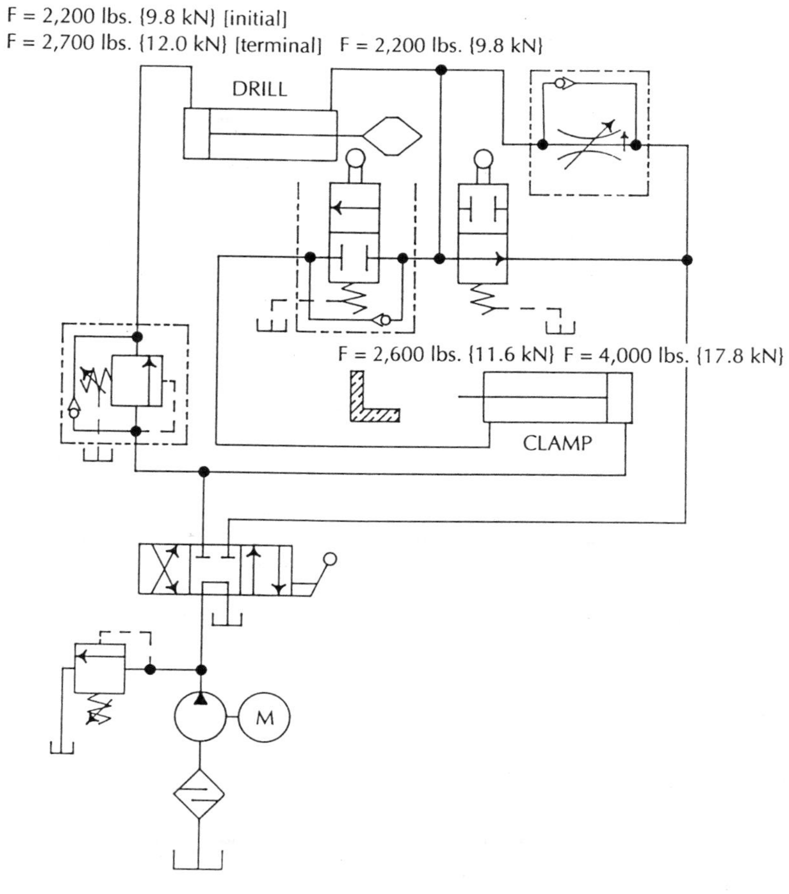

A Loading, Indexing, and Clamping Cylinder

PISTON		*ROD*			
Diameter	Area	Diameter	Area	Annulus Area	Stroke
______ in. {*cm*}	______ in.^{2} {*cm*^{2}}	______ in. {*cm*}	______ in.^{2} {*cm*^{2}}	______ in.^{2} {*cm*^{2}}	22in. {*55.9 cm*}

FORCE		*PRESSURE*	
Extension	Retraction	Extension	Retraction
4,000 lbs. {*17.8 kN*}	2,600 lbs. {*11.6 kN*}	_______ psi {*kPa*}	_______ psi {*kPa*}

VELOCITY		*TIME*	
Extension	Retraction	Extension	Retraction
_______ in./sec. {*cm/sec.*}	_______ in./sec. {*cm/sec.*}	______ sec.	______ sec.

Port Size	Body Style	Mounting
_______ in.{*mm*}	__________	__________

B Drill-Head Cylinder

PISTON		*ROD*		
Diameter	Area	Diameter	Area	Annulus Area
______ in. {*cm*}	______ in.^{2} {*cm*^{2}}	______ in. {*cm*}	______ in.^{2} {*cm*^{2}}	______ in.^{2} {*cm*^{2}}

FORCE		*PRESSURE*		
Extension	Retraction	Extension	Retraction	Stroke
2,200 lbs. {*9.8 kN*}(init.)	2,200 lbs. {*9.8 kN*}	_______ psi {*kPa*}	_______ psi {*kPa*}	11 in. {*27.9 cm*} traverse
2,700 lbs. {*12 kN*}(term.)		_______ psi {*kPa*}		9 in. {*22.9 cm*} feed 20 in. {*51.8 cm*} total

B Drill-Head Cylinder *(continued)*

FEED RATE		
Extension	Retraction	Supply Rate
5 in./sec. {*12.7 cm/sec.*} (traverse) 180 in./min. {*457.2 cm/min.*} (feed)	_______ in./sec. {*cm/sec.*}	_______ in.^{3}/min. {*cm^3/min.*}

TIME				
Extension	Retraction	Port Size	Body Style	Mounting
_______ sec.	_______ sec.	______ in.{*mm*}	__________	__________

C Pump

Type	Design	Delivery	Drive Speed	Volumetric Efficiency
________	________	________ gpm{*lpm*}	_______ rpm	________ %

	PORT SIZE	
Displacement at V.E.	Inlet	Outlet
__________ in. ^{3}/rev{*cm^3/rev*}	________ in.{*mm*}	________ in.{*mm*}

D Flow-Control Valve

Method	Type	Control Setting	Delivery through Valve at Flow-Control Setting
__________	___________	180 in./min. {*457.2 cm/min.*}	__________ gal./min. {*l/min.*}

Allowed Pressure Drop	Pressure Compensation Design	Delivery Range	Port Size
__________ psi {*kPa*}	__________	0 to ______ gpm {*lpm*}	______ in. {*mm*}

E Reservoir

CAPACITY	
Minimum	Maximum
______ gal. {*l*}	______ gal. {*l*}

F Filtration

		RATING		
Type	Element	Element	Flow	Port Size
Strainer	Wire	________ mesh	________ gpm {*lpm*}	________ in. {*mm*}

Description	Mounting
Full Flow	________

G Master System Directional Control

Positions	Ways	Element	Activation	Port Size
________	________	________	________	________ in. {*mm*}

H Position Parameter Sequencing Directional Control

Positions	Ways	Element	Activation	Port Size
________	________	________	________	________ in. {*mm*}

I Feed Shunt Directional Control

Positions	Ways	Element	Activation	Port Size
________	________	________	________	________ in. {*mm*}

J Pressure Parameter Sequence Valve

Type	Element	Setting	Mounting	Port Size
________	________	________ psi {kPa}	________	________ in. {*mm*}

K Pressure Relief Valve

Type	Element	Setting	Mounting	Port Size
________	________	________ psi {kPa}	________	________ in. {*mm*}

L Electric Motor

Speed	Voltage	Phase	Power at V.E.
________ rpm	________	________	________hp {*W*}

M Conductors

Inlet Type	Inlet Description	Working Type	Working Description	*INLETS (I.D. DECIMAL)*	
				Minimum	Maximum
________	________	________ in. {*cm*}	________ in. {*cm*}	________	________

	WORKINGS (I.D. DECIMAL)		*LINES—NOMINAL*	
	Minimum	Maximum	Inlets	Workings
Pump	______ in. {*cm*}	______ in. {*cm*}	______ in. {*mm*}	______ in. {*mm*}
Cylinder	______ in. {*cm*}	______ in. {*cm*}		
Relief	______ in. {*cm*}	______ in. {*cm*}		

N Report any changes made in the circuit or specifications that would alter the schematic that was provided.

O The following are variations for the Practical Design Problem (8.9):

Component: Indexing Cylinder—Double-Acting, Single-End Rod Cylinder

	A	B	C	D
Stroke	20 in.	18 in.	60 *cm*	45 *cm*
Extension Force—Total	3,800 lbs.	4,500 lbs.	16 *kN*	18 *kN*
Retraction Force	2,300 lbs.	2,800 lbs.	11.5 *kN*	12 *kN*

Component: Drill Cylinder—Double-Acting Single End Rod Cylinder

	A	B	C	D
Stroke—Traverse	12 in.	14 in.	38 *cm*	35 *cm*
Stroke—Feed	8 in.	10 in.	25 *cm*	21 *cm*
Extension Force—Initial	2,400 lbs.	2,600 lbs.	12 *kN*	15 *kN*
Extension Force—Terminal	2,800 lbs.	3,000 lbs.	14.5 *kN*	17.3
Retraction Force	2,400 lbs.	2,600 lbs.	12 *kN*	15 *kN*
Traverse Rate	4 in./sec.	3 in./sec.	15 *cm/sec.*	13 *cm/sec.*
Feed Rate	120 in./min.	75 in./min.	430 *cm/min.*	400 *cm/min.*

8.10 SUGGESTED ACTIVITIES

1. Using a garden hose, a watch, and a 5-gal. bucket, determine the delivery in gallons per minute {liters per minute} of your water supply when the faucet is wide open.
2. Examine the faucet that you used in Activity #1 and determine the number of turns of the faucet handle that it takes to go from closed to wide open. Open the faucet halfway and repeat your determination of the delivery for your water supply.
3. Make arrangements to visit a local commercial machine shop or school laboratory that has a hydraulically powered machining center. Have the operator or instructor show you the adjustments for feed and speed control on the machine. Have the operator run the machine and make adjustments to speed or feed. If that feed or speed is calibrated, record its value at a certain point. Find out what the cylinder size is for that portion of the machining center and determine the delivery though the flow control at various settings.
4. Visit a local plumbing supply store and have the clerk show you gate, globe, ball, and needle valves used in household plumbing systems. If possible, request the costs for the various valves and determine why they vary in cost.
5. Disassemble a pressure-compensated flow-control valve in your laboratory. Describe either verbally or with sketches the type of throttle employed and the type of pressure compensator used in your valve.
6. Fabricate a mock-up of various flow-control circuits such as meter-in, meter-out, bleed-off, traverse and feed, etc. Observe the operation of the various circuits and adjust their speeds. If you know the delivery of the pump and the size of cylinder you are using, determine (by timing) the delivery through the flow control at various settings on these circuits. Check these calculations with actual readings from a flow gauge.
7. Visit a local or school library. Look up books and reference materials listings for Osborne Reynolds, Daniel and Johann Bernoulli, and Evangelist Toricelli. Write a report on their contributions to fluid power and especially the flow conditions of fluids.

8.11 REVIEW QUESTIONS

1. What are the typical results of a reduction of effective delivery to a portion of a hydraulic circuit?
2. For what purpose are fixed-orifice check valves normally used?
3. Discriminate among gate, globe, and needle valve throttles used in noncompensated flow controls.
4. Justify the use of pressure compensation in flow controls used in hydraulic circuitry.
5. Describe the pressure-relief compensated flow control, its design, operation, characteristics, and uses.

6. Describe the pressure-reducing compensated flow control, its design, operation, characteristics, and uses.
7. Discriminate among meter-in, meter-out, and bleed-off methods of flow control.
8. What is the generally accepted range of pressure drop allowed across a flow control?
9. Explain why the pressure-relief valve operating in a meter-in or meter-out circuit actually determines the amount of pressure drop that will occur across the flow control.
10. If a meter-in flow-control method is employed in a linear hydraulic circuit, and the flow control is plumbed into the line between the pump and the directional control, what effect may this have on cylinder operation in both stroke directions? Why?
11. Why is temperature compensation used in a flow control, and how is it effected in the design of a flow-control valve?
12. Draw an accurate schematic of a traverse and feed flow-control linear circuit having meter-out flow control on the termination of the extension stroke only.
13. Differentiate between proportional solenoid flow controls and torque-motor proportional flow controls in design and operation.
14. Describe the adjustable-throttle flow divider.
15. Describe the rotary-motor flow divider.

9

Rotary Circuitry

Not all hydraulic circuitry serves the purpose of producing linear motion. A great deal of hydraulic circuitry serves the purpose of controlling and supplying fluid to produce rotary motion. The concepts that govern linear motion—including displacement, direction, and pressure control—also affect the operation of rotary circuits. Nearly all components included in the simple linear circuit may also be found in the rotary circuit. The major exception is the inclusion of a rotary actuator, called a **motor**, in place of the linear actuator or cylinder.

9.1 WORK IN THE ROTARY ACTUATOR

As in linear circuitry, work is, by definition, force times distance. However, *in a rotary actuator, distance is circular displacement and the cycle is a rotation*. The force is applied tangentially to this circular path; and this, in combination with the distance to the center of rotation, develops **torque**. *Torque may be calculated by multiplying the force applied times its directed distance from the center of the circular rotor* (see Figure 9-1).

Torque will exist in the operating hydraulic motor regardless of whether rotation is evident. The net effect of increased torque may be exemplified by considering the operation of removing an overly tightened nut from a bolt. The force applied by pulling on the end of the wrench handle may not be sufficient to dislodge the nut. By sliding a pipe over the handle, the distance from the center of the nut to the point at which the load is applied is increased and the nut may be easily dislodged.

FIGURE 9-1

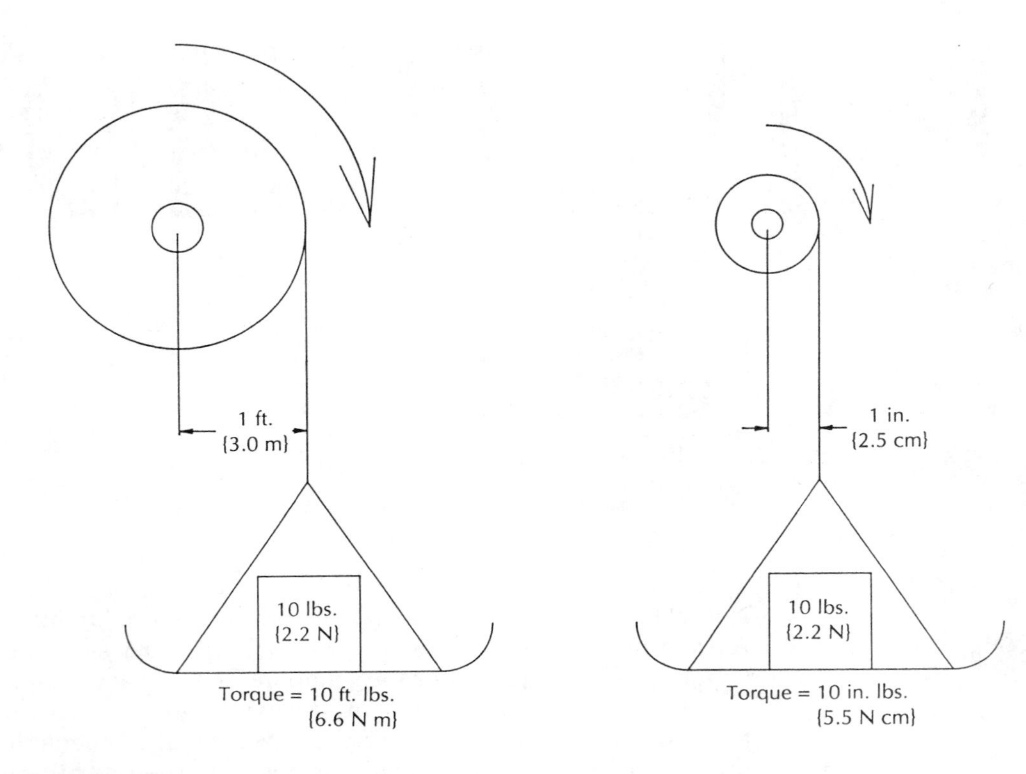

9.2 TORQUE DEVELOPMENT IN MOTORS

Resistance to rotation in the motor results in pressure developing in the rotary circuit. The pressure will operate on the internal construction of the motor to produce a rotary force. When that force becomes greater than the external load imposed on the motor, rotation will begin. Since the internal construction of the motor does not change, the control variable to initiate rotation is pressure. If allowed to rise, system pressure will build until sufficient torque (turning force) is available to produce rotation. The theoretical pressure required is a function of the volume of fluid inside the motor acting on its construction. It may be calculated using the following formula:

$$\text{P (psi) } \{Pa\} = \frac{\text{Torque (in. lbs.) } \{N\ m\} * 2\pi}{\text{DIS (in.}^3\text{/rev.) } \{m^3/rev.\}}$$

All motors require an initial pressure higher than that calculated to overcome inertia. In some cases, a 50% override may be necessary. However, *in this presentation, override will be standardized to 10% for design purposes*. Therefore, any calculated value should be multiplied by 1.1 to determine the pressure necessary to start the motor.

Problem 9-1

A circuit requires 500 in. lbs. {*56.5 Nm*} of torque to operate a hydraulic motor having a displacement of 10 in.^{3}/rev. {*0.00016 m^3/rev.*}. What pressure should be designed for to start the motor?

P (psi) {*kPa*} = __________

Another, and often more appropriate, description of a motor is its load capacity. Since the construction of any given motor is standard (if the motor is not variable displacement), its displacement will be constant. The capacity of that motor to produce torque will increase in direct proportion to an increase in operating pressure. Typically, motors will be rated for torque capacity per 100 psi {*1,000 kPa*}. The **torque rating (T.R.)** for a motor may be calculated by dividing its theoretical torque by its operating pressure and multiplying this answer by 100 {*1,000*}.

Problem 9-2

What will be the torque rating for the motor described in Problem 9-1?

T.R. (in. lbs./100 psi) {*N m/1,000 kPa*} = ____________

The torque rating is beneficial when trying to rapidly calculate or approximate operating pressure needed by a motor to overcome varying loads. This may be accomplished by dividing the necessary torque by the torque rating and, in turn, multiplying that answer by 100 {*1,000*}. Thus, a motor having a torque rating of 300 in. lbs./100 psi, would require approximately 500 psi to overcome a 1,500-in. lb. torque load. A motor having a torque rating of 300 *Nm*/1,000 *kPa* would require approximately 5,000 *kPa* to overcome a 1,500-*Nm* torque load.

9.3 SPEED DEVELOPMENT IN MOTORS

The operating speed of a motor is rated in revolutions per minute, similar to the cycles per minute of a cylinder. Also similar to the cylinder, *the number of revolutions per minute of the motor is a function of the delivery to the motor and its volumetric capacity or displacement*. Typically, a motor's displacement will be known; and a necessary operating speed will be specified. Therefore, the pump delivery to produce that speed will need to be described. Hydraulic motors are similar—and, in some cases, identical—in design and operation to hydraulic pumps. Slippage and leakage in

the motor will necessitate an adjustment, based on the motor's volumetric efficiency, in calculation for the pump delivery:

$$\text{DEL (gpm) \{lpm\}} = \frac{\text{DIS (in.}^3\text{/rev.) } \{cm^3/rev.\} * \text{SPeed (rev./min.)}}{[231 \text{ (in.}^3\text{/gal.)] [1,000 } \{cm^3/l\}] * \text{V.E.(\%)}}$$

Problem 9-3

A hydraulic motor has a displacement of 1.5 in.^{3}/rev. {*24.58 cm^3/rev.*}. What delivery will be necessary to operate this motor at 1,000 rev./min., assuming 80% volumetric efficiency?

DEL (gpm) {*lpm*} = _________

9.4 POWER IN THE ROTARY CIRCUIT

If the delivery to a motor and the pressure under which it is operating are known, horsepower may be calculated by using the same formula that was used to calculate horsepower in the linear circuit:

$$\text{hp} = \frac{\text{P (psi) } \{Pa\} * \text{DEL (gpm) } \{lpm\} * 0.000583 \ \{0.02235\}}{\text{V.E.(\%)}}$$

[Note: For the sake of clarity, discussion of metric equivalents of horsepower have been omitted from the following section since this rating is not common in the SI. However, calculation of wattage requirements for hydraulic circuitry may be made by multiplying answers derived from the previous formulation by 746 (there are 746 watts in each horsepower).]

However, a more simplified method of calculating horsepower requirement exists for rotary circuits. This method uses the design parameters generally used to describe motor operation:

$$\text{hp} = \frac{\text{T (in. lbs.) } * \text{SP (rev./min.) } * 0.0000158}{\text{V.E.(\%)}}$$

Problem 9-4

An 80% volumetrically efficient motor running at 1,000 rev./min. is operating under a 2,400-in. lb. torque load. What horsepower is needed to be supplied to the circuit to operate this motor?

hp = _________

9.5 TYPES OF MOTORS

Hydraulic motor designs are similar to hydraulic pumps. Therefore, three classifications are common:

1. Gear types
2. Piston types
3. Vane types

In the first two cases, designs may be identical to the pump designs. However, *in vane motors, special design features must be incorporated.* Therefore, the following presentation will give major emphasis to those special design features of vane motors. Only the basic operating principles of piston and gear motors will be discussed.

9.5.1 Gear-Type Motors

As would be expected, gear motors are either **internal** or **external.** *They typically have the lowest volumetric efficiency of hydraulic motors, ranging from 70% to 80%.* **External-gear motors** may be either a **gear-on-gear** or a **lobe** design. The pressured fluid acting on the exposed face of the gear chamber at the inlet creates a rotary force that is transmitted to the power takeoff creating torque (see Figure 9-2).

FIGURE 9-2

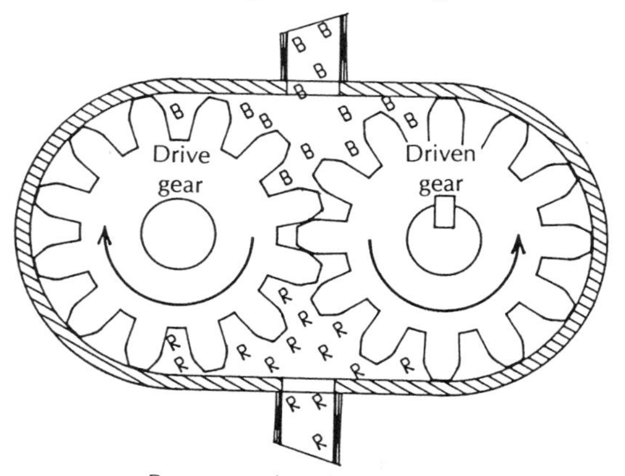

On closer inspection, it is realized that the pressure acting on the gear face of two gear chambers combines to create torque. The involute curves of the gear face produce an area upon which the pressure operates to create a force. The distance from the pressure point (the point at which the two gears mesh) to the center of the shaft, in combination with the force, produces torque. The actual torque produced is nearly twice this value, since an identical situation exists on the other gear.

The direction of rotation in the gear motor may be changed by simply reversing pressurized flow and exhaust flow at the ports. Of course, this may easily be accomplished by shifting a directional control.

Internal-gear motors will be either of a **gear-in-gear** or **gerotor** design. In the **gerotor motor**, torque develops as pressure acts on the gear face of the one gear chamber at maximum exposure. This occurs at the rotation interface between the pressure and exhaust sides of the motor and causes rotation (see Figure 9-3).

FIGURE 9-3

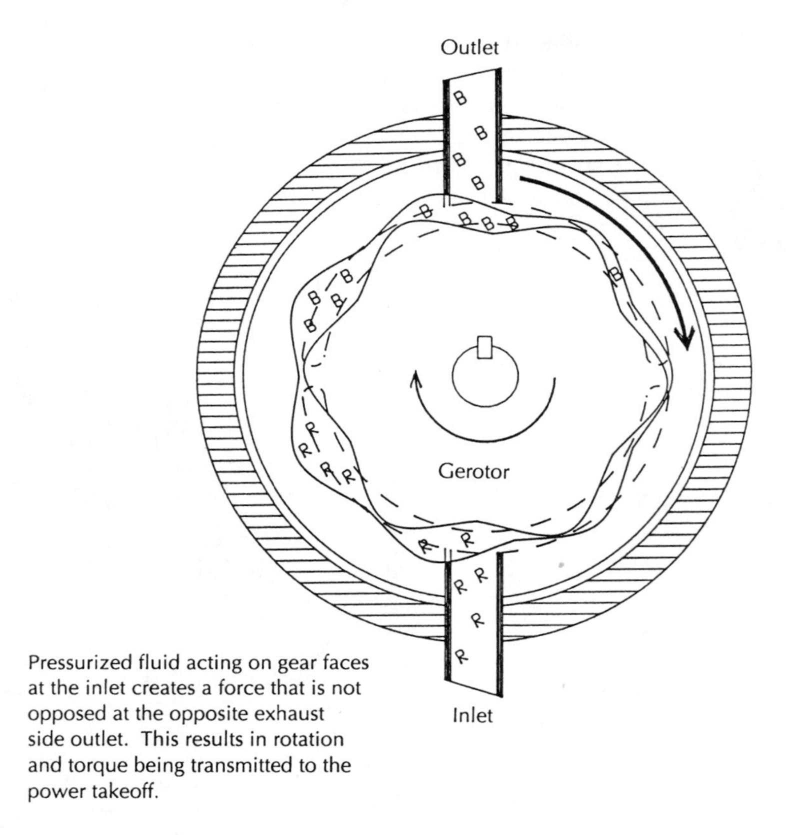

Pressurized fluid acting on gear faces at the inlet creates a force that is not opposed at the opposite exhaust side outlet. This results in rotation and torque being transmitted to the power takeoff.

9.5.2 Piston-Type Motors

Piston motors, like piston pumps, will be of either the radial or axial design. In each case, a pressure exerted on the face of the piston will create a force that is transmitted through a distance to the power takeoff creating torque. *Piston motors have the potential to be the most volumetrically efficient types, with ratings ranging from 85% to 95%.*

The **axial-piston motor** typically has one less than half of its odd number of pistons creating force during rotation. This combined force is transmitted to the power takeoff through a distance equal to one-half the piston circle diameter creating torque (see Figure 9-4).

FIGURE 9-4

Pressurized fluid acting on the pistons forces them downward, sliding the piston shoe down the incline of the swash plate. This turns the cylinder block and drives the power takeoff.

As pistons travel up the incline of the swash plate, fluid in these cylinders is exhausted.

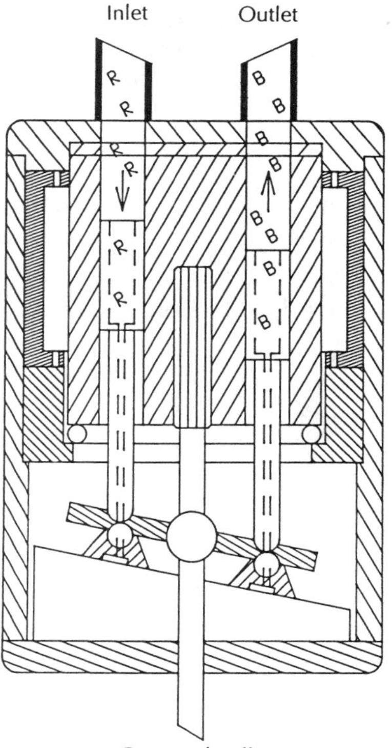

Operating speed in the axial-piston motor is a function of the total volume potential in all cylinders and the delivery to the motor. Axial-piston motors are capable of operating at higher speeds and pressures than other hydraulic motors. Therefore, *they will typically be found in high-efficiency, fast-operating, high-pressure circuitry.*

9.5.3 Vane-Type Motors

Unlike piston and gear types, *vane motors differ in design from vane pumps.* In each case, chambers are created to cause pumping action or torque development. In vane pumps, centrifugal force created by an external drive cause the vanes to extend from their slots and seal against the pump case, creating chambers. However, in motor applications, no external drive exists to cause pre-extension of the vanes before pressure is evident in the motor chambers. Therefore, design alternations are required to cause a basic vane pump to operate as a vane motor. In essence, *all vane motors will operate as pumps; but all vane pumps will not operate as vane motors.*

9.5.3.1 Mechanical vane extension One method of vane pre-extension in a motor is through the use of springs that mechanically force the vanes from their slots. Leaf-type springs are housed in pockets machined into the face of the rotor. This leaf spring is then pinned on a center pivot that applies a force to the rear tips of two vanes (see Figure 9-5).

9.5.3.2 Hydraulic vane extension The concept of using hydraulic pressure to cause vane extension was presented in a previous discussion of vane pumps (see Chapter 4). In that pump, passageways were machined into the pump housing from the outlet side to direct pressurized

FIGURE 9-5

fluid through the rotor to chambers located behind each vane. Two problems exist when applying this design feature in the vane motor.

Vane motors, like most other hydraulic motors, are designed to operate in both directions. Therefore, the pressure locations will change according to rotation direction. Simply tapping both locations to achieve a pilot pressure that consolidates at the rotor does not solve the problem. This consolidation would create a straight shunt between the pressure and exhaust sides of the motor. An exclusionary device that allows pressure to be directed to the rotor while shutting off the exhaust side is required. The **shuttle valve** performs this function, regardless of rotation direction in the motor (see Figure 9-6).

FIGURE 9-6

Pressure from the inlet side is directed to the shuttle valve, regardless of rotation direction. The shuttle valve directs this flow to the pintle through passageways and forces the vanes to extend. The shuttle valve also blocks this flow from the exhaust side of the motor. When shifted, the shuttle valve reverses its position and performs the same function in the opposite rotation direction.

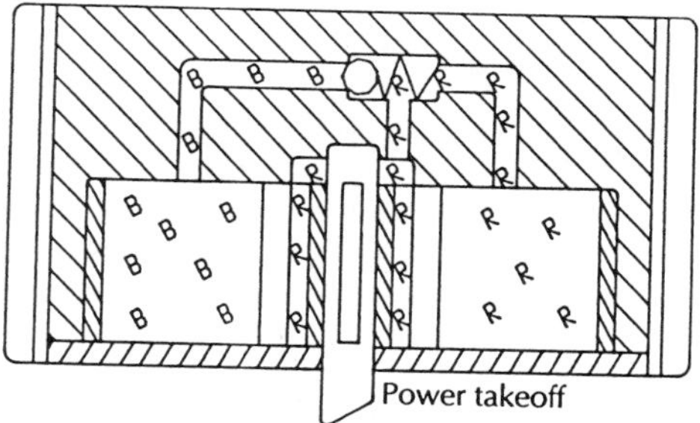

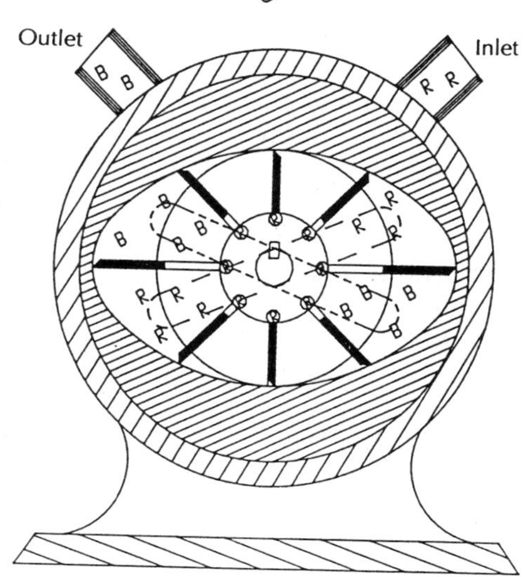

One other design feature may be incorporated to assure vane pre-extension. When pressurized fluid is directed to the motor, the same value of pressure will be provided in the pump chambers, as is evident at the rotor supplying pressure behind the vanes. Since the effective areas represented on both ends of the vane are identical, the forces that develop in front of and behind the vanes are equal. In this balanced condition, the vanes may not pre-extend. Either greater pressure or greater area is required behind the vanes to assure pre-extension. If the shuttle-valve arrangement is used, increased cross-sectional area on the backside of the vanes is necessary (see Figure 9-7).

FIGURE 9-7

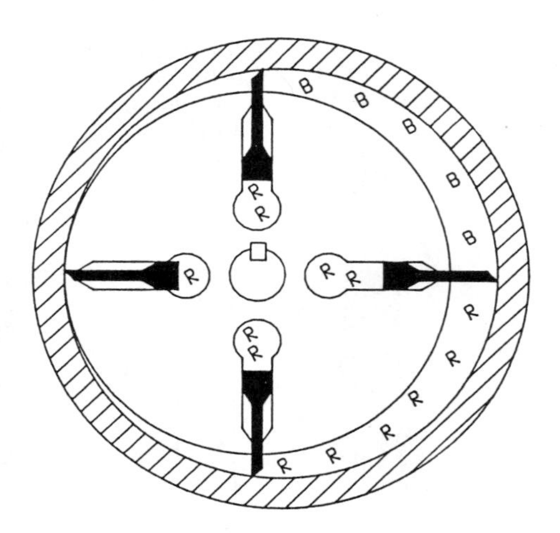

The shuttle-valve mechanism may be eliminated by directing an intensified pressure to the rotor. A check valve located between the pump and the directional control creates a slight pressure drop throughout the rotary circuit outside the power supply. A slightly higher pilot pressure, taken from the power supply, pressurizes the rotor and overcomes the lower pressure evident in the pump chamber to pre-extend the vanes (see Figure 9-8).

9.5.3.3 Torque development in the vane motor In the **unbalanced vane motor**, when one vane is at maximum extension, it will create a chamber that has system pressure on one side of the vane. The opposite side of that vane will be connected to the exhaust port, resulting in no pressure on that side. The cross-sectional area of that face of the vane exposed to system pressure creates a force that is concentrated upon the center of that vane. In combination with the distance to the center of the power takeoff, this force creates torque in the vane motor (see Figure 9-9).

FIGURE 9-8

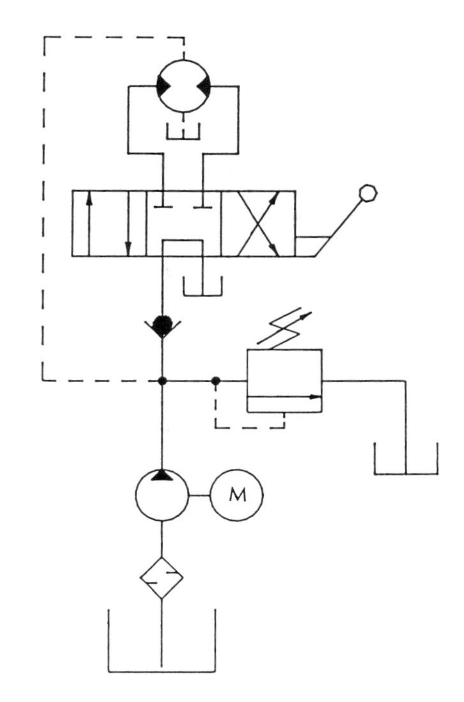

FIGURE 9-9

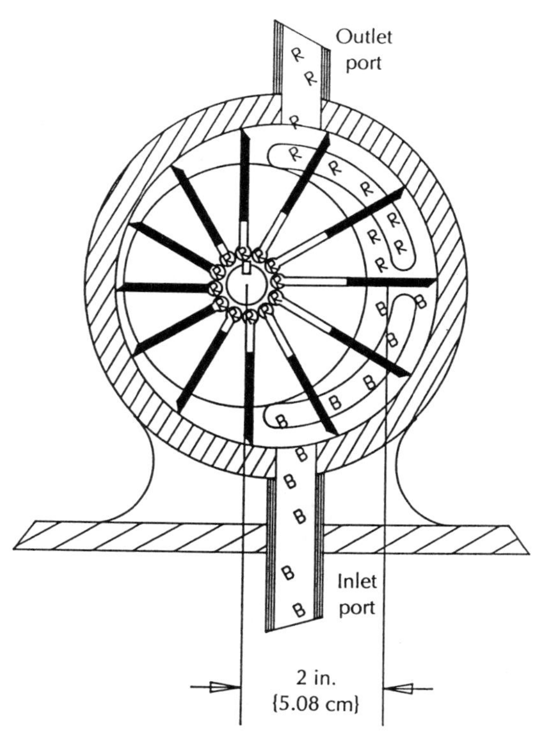

In Figure 9-9, a system pressure of 500 psi {*3,447.4 kPa*} is acting on this motor. The maximum area of vane exposure is 1.5 in.^{2} {*9.7 cm^2*}.

Problem 9-5

What turning force will develop in this motor?

T (in. lb.) {*N cm*} = __________

9.6 ROTARY CIRCUITS

The simple rotary circuit, like the simple linear circuit, is composed of a power supply, directional control, and an actuator (in this case a motor). Figure 9-10 describes probably the simplest bidirectional rotary circuit using a two-position, four-way directional control.

FIGURE 9-10

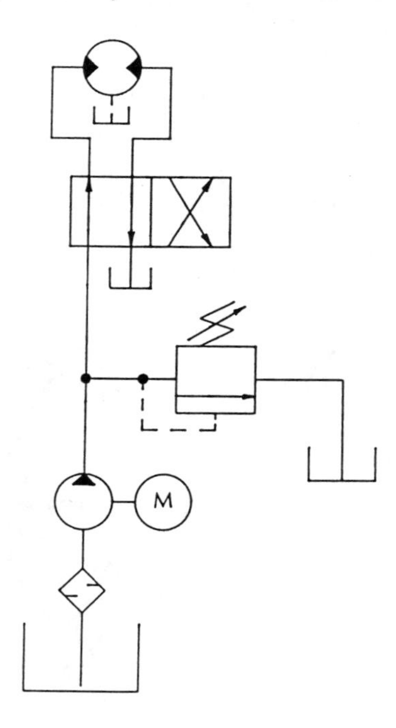

This ANSI schematic describes a simple rotary circuit requiring 750 in. lb. {*412.5 N cm*} of torque and operating at 500 rpm, using a motor having a 2 in.^{3}/rev. {*32.8 cm^3/rev.*} displacement.

Problem 9-6

What should be the setting on the pressure-relief valve?

P (psi) {*kPa*} = __________

Problem 9-7 What delivery is needed from the pump if the motor is rated at 80% volumetric efficiency?

DEL (gpm) {*lpm*} = __________

Although relatively simplistic, the circuit illustrated in Figure 9-10 describes one of the major advantages of rotary hydraulic circuits. Under low inertial loads, the direction of rotation can easily be reversed by simply shifting the directional control. Although there will be obvious losses of power comparing the electric motor drive to the potential power output of the hydraulic motor, this ease of reversibility sometimes more than compensates for these losses. Furthermore, the physical size of a hydraulic motor, rated at a certain horsepower level, compared to the same size electrical motor is much smaller. This allows for a great deal of design flexibility when locating and positioning the output rotary drive within a mechanism.

The simplest rotary circuit also has deficiencies. For example, no contingency is made to neutralize rotary motion. Rapid reversal of rotation direction is accommodated, but stopping may only be accomplished by stalling or stopping the electric motor rotation. Mechanical brakes may be used to halt hydraulic motor rotation. This would cause increased pressure and overcome the relief valve (an obvious advantage over electrical motors, since this would not cause damage to the hydraulic circuit). However, various reactions to inertial rotation may be accomplished by using a three-position, four-way directional control. Figure 9-11 illustrates three potential center logics used in rotary circuits.

FIGURE 9-11

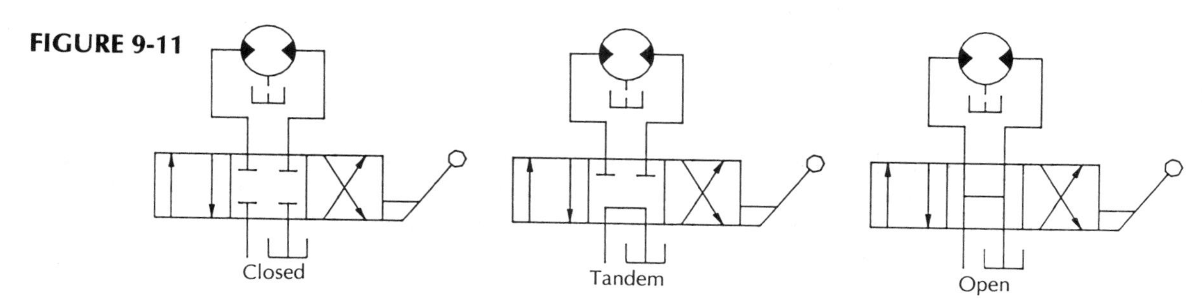

The rotary circuit using a closed-center directional control has the advantage of rapidly halting rotation in either direction when centered. However, when stalled for extended periods of time, a great deal of energy is wasted because pressure will develop in the power supply to the level of the relief valve setting. *The rotary circuit using a tandem-center position*

will overcome this deficiency. However, it may not be useful in more complicated, combination circuitry because all flow will be diverted to the tank in the center position, excluding it from any other branch. Either of these circuits may be damaged when shifted to the center position if high inertial loads are present.

Because the motor is designed like a pump, it will act like one when driven by inertia. This pseudo pumping action will create a vacuum condition at the inlet side of the motor and a pressure condition at the outlet. The vacuum will cause **cavitation** in the inlet lines. Of more potential damage is the pressure created in the outlet line. The motor branch is isolated from the power supply when the directional control is centered. Therefore, no pressure-relief capability is evident in the motor branch. Pressure will develop there until it overcomes some circuit component. Hopefully, this will cause the motor to stall. However, if inertia is great enough, pressure could develop to a high enough level to burst a line or connection or damage the motor.

The rotary circuit using an open-center directional control will not isolate the power supply from the motor branch. Therefore, pressure and vacuum development in inertial operation will not occur. Exhaust will be directed to the tank, and supply fluid will be available to charge the inlet side of the motor. For this reason, the open-center design will normally include a check valve in the tank line from the directional control to exclude reverse flow from the tank while the motor is coasting. Yet, the open center will not slow or stop motor rotation under inertial loading conditions. Other circuit design features are needed in the rotary circuit to accommodate deceleration and exclude the pseudo pumping nature of the coasting motor.

9.6.1 Rotary Braking Circuitry

The circuit designed to retard rotation of the coasting motor uses a normally closed pressure control **(brake valve)** and a closed- or tandem-centered directional control to absorb the energy created through inertia. A shunt between the outlet and inlet of the motor includes the brake valve. Piloted from the exhaust side of the motor, this brake valve will open at a preset pressure and return the exhaust flow to the inlet. This reduces pressure on the exhaust side, and the brake valve closes. The brake valve will fluctuate open and closed, absorbing the inertial energy while resupplying the inlet side to eliminate cavitation. In the bidirectional rotary circuit, two brake valves are used to allow independent control. *Each valve is typically set to a pressure of 10% above that needed to operate the circuit in the opposite rotation direction* (see Figure 9-12).

FIGURE 9-12

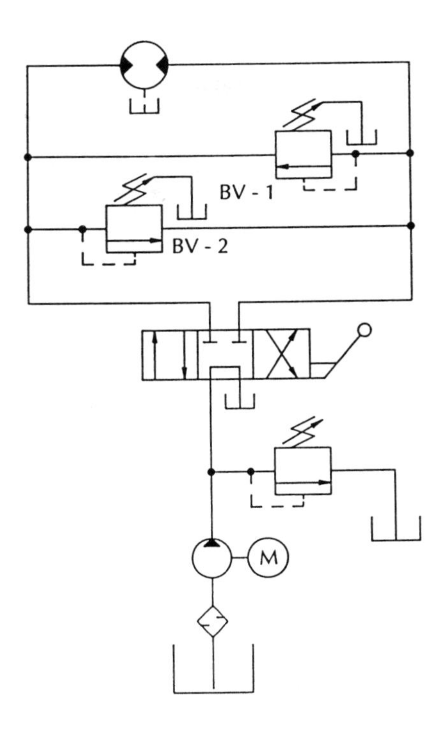

This bidirectional rotary circuit has a motor with a 1.5-in.^{3}/rev. {*24.6-cm^3/rev.*} displacement. When it is rotating clockwise, 150 in. lbs. {*82.5 N cm*} of torque will develop. When it is rotating counterclockwise, 500 in. lbs. {*275.0 N cm*} of torque will develop.

Problem 9-8

Describe the settings necessary on each brake valve.

BV-1. P (psi) {*kPa*} = __________

BV-2. P (psi) {*kPa*} = __________

9.6.2 Flow-Controlled Rotary Circuits

In rotary circuitry, flow control will use the meter-in, meter-out, or bleed-off method. Since the constructional design of a hydraulic motor is nearly identical regardless of rotation direction, *flow controls are normally plumbed into the power supply before the directional control.* Using a bidirectional motor, this affords equal rotation speed in both directions using only one flow control. This type of arrangement is referred to as dependent application, since the speed of one rotation direction is dependent upon the setting used for the other direction's rotation. If independently controllable speed is required, two flow controls may be plumbed into the circuit between the directional control and the motor. Figure 9-13 illustrates some variations of flow-controlled rotary circuits.

FIGURE 9-13

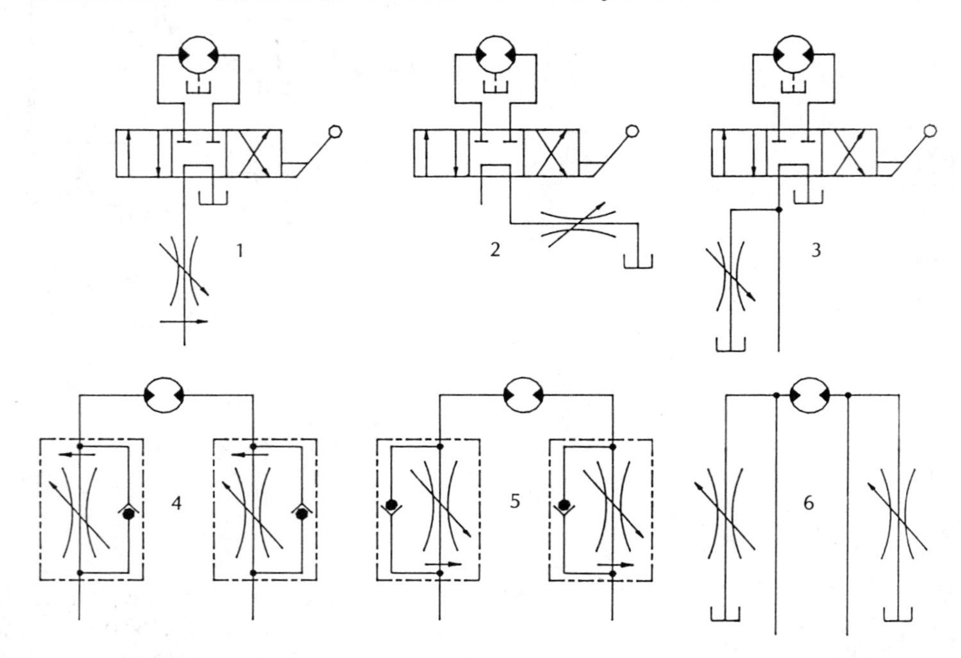

Assignment: Verbally describe each of the flow-control rotary circuits.

1. ____________________

2. ____________________

3. ____________________

4. ____________________

5. ____________________

6. ____________________

Like flow-controlled linear circuits, flow-controlled rotary circuits will produce varying internal pressures. In each case, the pressure differential will be determined by the method of flow control and the fluid velocity within the circuit. However, calculation of pressure drop in the

flow-controlled rotary circuit is greatly simplified, since opposing areas are equal within the motor design. Yet, the general parameters established for linear flow control still apply. *The pressure drop across the flow control should range between 49 and 400 psi {338 and 2,758 kPa} to eliminate turbulent flow.* This internal operational pressure, in combination with the pressure necessary to overcome the external torque load on the motor, will determine operating pressure in both meter-in and meter-out circuitry. The pressure-relief valve setting for a flow-controlled rotary circuit containing no other components (i.e., brake valves) will be established by these calculations (see Figure 9-14).

FIGURE 9-14

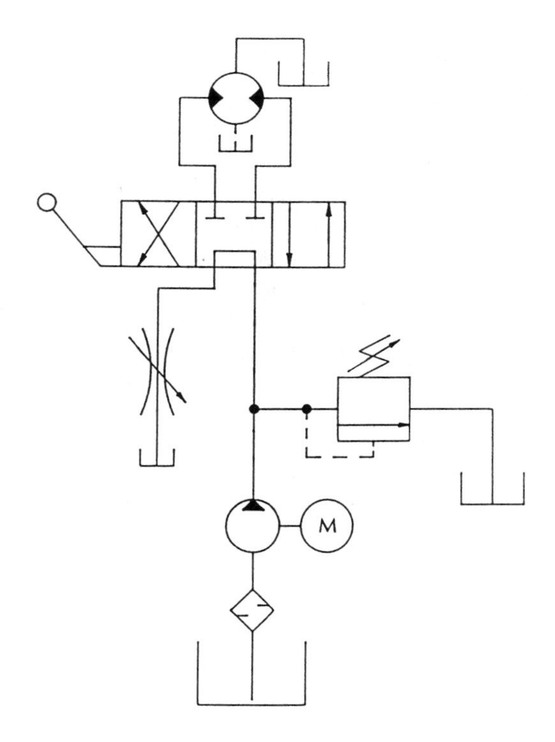

This meter-out rotary circuit is operating a motor having a 3.25-in.^{3}/rev. {*53.3-cm^3/rev.*} displacement. The load on this motor will develop 750 in. lbs. {*412.5 N cm*} of torque.

Problem 9-9

What pressure will be necessary to overcome the torque and starting loads on the motor?

P (psi) {*kPa*} = __________

Problem 9-10 What are the minimum and maximum settings for the pressure-relief valve to ensure that turbulent flow does not occur?

Maximum P (psi) {*kPa*} = ________

Minimum P (psi) {*kPa*} = ________

As would be expected, *pressure-reducing-type pressure-compensated flow controls are typically used in meter-out circuits*, while *pressure-relieving compensation is common in meter-in*. Also, temperature compensation may be incorporated.

[] [] [] 9.7 PRACTICAL DESIGN PROBLEM

Parts Suppliers Incorporated operates a large warehouse to supply its subsidiaries with bulk quantities of metal fasteners. Various fasteners are packaged in boxes and stored on pallets containing thirty-six boxes each. The pallets are retrieved from storage with a forklift and transported to a constantly operating conveyor line for transmission to the central shipping and receiving dock.

Currently, the conveyor is driven by a large electric motor that only operates in one direction. This works very well for shipping purposes. However, it has been suggested that the same line could be used to resupply stock, if only the line could be driven in both directions. Then, remote **electrical controls** could be used to stop the line at any location for removal of supplies.

The drive roller for the conveyor system has a 15-in. {*38.1-cm*} diameter drum, and the maximum load imposed on this roller has been calculated to be 500 lbs. {*2.2 kN*} (see Figure 9-15). You have been asked to design and specify all components for a hydraulic drive unit to power the conveyor in both rotation directions. It has been suggested that the drive speed of the belt be variable within a range of 110 to 200 in./min. {*279.4 to 508 cm/min.*} in both directions. Complete this assignment by specifying all components according to the schedule provided and producing a schematic for your circuit design. Of course, some of these components may not be needed in your design, while others may need to be added. *Feel free to make such additions or deletions and provide specifications accordingly.*

FIGURE 9-15

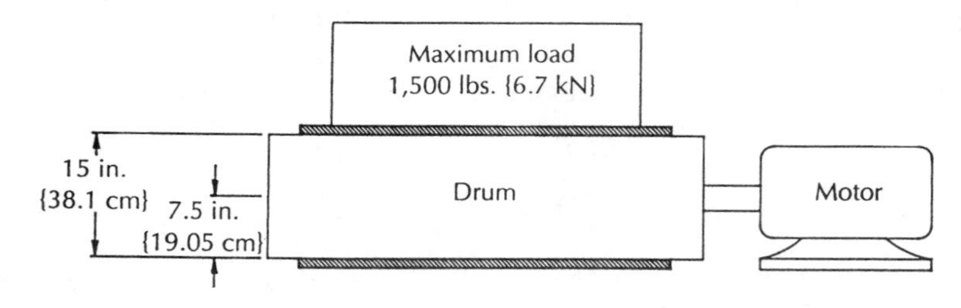

Figure 9-15 relates to the proposed direct-drive hydraulic motor and current drum design for the conveyor system.

Problem 9-11 What is the maximum torque load that will be imposed on the hydraulic motor?

T (in. lbs.) {*N m*} = __________

Problem 9-12 How far will the belt advance in one revolution of the drum?

Distance (in.) {*cm*} = __________

A Hydraulic Motor

Operating Speed (max.)	Torque Load	Type	Design	Volumetric Efficiency
_____ rpm	_____ in. lbs. {*N m*}	_______	_______	_____ %

Displacement	Required Delivery	OPERATING PRESSURE (at Motor) Full Delivery	OPERATING PRESSURE (at Relief Valve) Minimum Delivery	Starting Pressure (at Motor)
_____ in.^{3}/rev. {*cm^3/sec.*}	_____ gpm {*lpm*}	_____ psi {*kPa*}	_____ psi {*kPa*}	_____ psi {*kPa*}

Torque Rating	Mounting	Port Size
_____ in. lbs./100 psi {*N m/1,000 kPa*}	_________	_____ in. {*mm*}

B Pump

Type	Design	Delivery	Drive Speed	Volumetric Efficiency
______	______	_____ gpm {*lpm*}	______ rpm	______ %

Displacement at V.E.	PORT SIZE Inlet	PORT SIZE Outlet
______ in.^{3}/rev {*cm^3/rev*}	_______ in. {*mm*}	_______ in. {*mm*}

C Flow Control Valve

Method	Type	Motor Speed at Flow Control (Lowest Setting)	Delivery through Valve at Flow Control Setting
________	________	________ rpm	______ gal./min. {*l/min.*}

Allowed Pressure Drop	Pressure Compensation Design	Delivery Range	Port Size
______ psi {*kPa*}	__________	0 to _____ gpm {*lpm*}	______ in. {*mm*}

D Reservoir

CAPACITY	
Minimum	Maximum
______ gal. {*l*}	______ gal. {*l*}

E Filtration

Type	Element	*RATING* Element	*RATING* Flow	Port Size
__________	__________	________ mesh	________ gpm {*lpm*}	______ in. {*mm*}

Description	Mounting
__________	__________

F Master System Directional Control

Positions	Ways	Element	Activation	Port Size
__________	__________	__________	__________	______ in. {*mm*}

G Pressure Relief Valve

Type	Element	Setting	Mounting	Port Size
________	________	______ psi {*kPa*}	________	______ in. {*mm*}

H Electric Motor

Speed	Voltage	Phase	Power at V.E.
______ rpm	________	________	______ hp {*W*}

I Conductors

Inlet Type	Inlet Description	Working Type	Working Description	*INLETS (I.D. DECIMAL)* Minimum	*INLETS (I.D. DECIMAL)* Maximum
______	________	________	________	______ in. {*cm*}	______ in. {*cm*}

	WORKINGS (I.D. DECIMAL) Minimum	*WORKINGS (I.D. DECIMAL)* Maximum	*LINES—NOMINAL* Inlets	*LINES—NOMINAL* Workings
Pump	______ in. {*cm*}	______ in. {*cm*}	______ in. {*mm*	______ in. {*mm*}
Motor	______ in. {*cm*}	______ in. {*cm*}		
Relief	______ in. {*cm*}	______ in. {*cm*}		

J Draw an accurate schematic of the proposed bidirectional rotary circuit with flow control.

K The following are variations for the Practical Design Problem (9.7):

Bidirectional Rotary Circuit with Flow Control: Drum, Belting, and Belt Speed Considerations

	A	B	C	D	E
Drum Diameter	20 in.	18 in.	12 in.	50 *cm*	45 *cm*
Belt Load	800 lbs.	450 lbs.	900 lbs.	3 *kN*	1.8 *kN*
Belt Velocity, Min.	120 in./min.	90 in./min.	125 in./min.	400 *cm/min.*	350 *cm/min.*
Belt Velocity, Max.	210 in./min.	160 in./min.	210 in./min.	750 *cm/min.*	550 *cm/min.*

9.8 SUGGESTED ACTIVITIES

1. Disassemble a vane-type motor in your laboratory. Describe either verbally or with sketches the methods employed to pre-extend the vanes and any flow-directing devices incorporated to allow bidirectional rotation.
2. Fabricate a mock-up of the bidirectional rotary circuit with meter-in, meter-out, and bleed-off flow control. Operate the circuit and observe rotational speed variations.

9.9 REVIEW QUESTIONS

1. What is torque, and how is it calculated?
2. Which factors determine the torque in an operational hydraulic motor?
3. Which factors determine rotational speed in an operational hydraulic motor?
4. Which factors determine power (horsepower or wattage) necessary to drive a rotary hydraulic circuit in terms of hydraulic motor design and operation?
5. Describe torque development in an external gear-type motor.
6. Describe torque development in an axial piston-type motor.
7. What special arrangements must be made to the basic vane-type pump to make it operate as a vane-type motor?
8. Discuss hydraulic vane pre-extension in a vane-type hydraulic motor.
9. Describe torque development in an unbalanced vane-type motor.
10. Describe the effects that open-, closed-, and tandem-centered three-position, four-way directional controls have on rotary circuit operation.
11. Discuss the effects that brake valves have on rotary circuit operation in terms of application and operation.
12. Discuss the rationale for the use of only one flow control to modulate rotational speed in both rotation directions in a rotary circuit.

10

Auxiliary Fluid and Power Supply

The power supply, various valving, actuators, and conductors constitute the vital functioning elements of a hydraulic circuit. However, other components that provide auxiliary fluid supply, condition the fluid, or enhance circuit operation may be included. These include **accumulators** and **intensifiers**.

10.1 ACCUMULATORS

Accumulators function as auxiliary power supplies for operating circuitry. These components are simply containers that store a supplemental supply of fluid under pressure. The **dead-weight accumulator**, although seldom used in industrial circuitry, works well to explain the potential of accumulators. The dead-weight accumulator is similar to the ram-type cylinder in appearance. This vertically mounted accumulator has a heavy weight imposed on its rod. When the circuit is operating, the cylinder will fill with fluid when system pressure becomes great enough to overcome this load in consideration of the cross-sectional area of the accumulator's piston (see Figure 10-1).

One application of the accumulator is to replenish fluid to a cylinder that must hold a load steadily for an extended period of time. Often, the power supply for this type of circuit will be shut down; and the closed cylinder ports on the center position of the directional control are used to maintain pressure and position. However, some slippage and leakage will

FIGURE 10-1

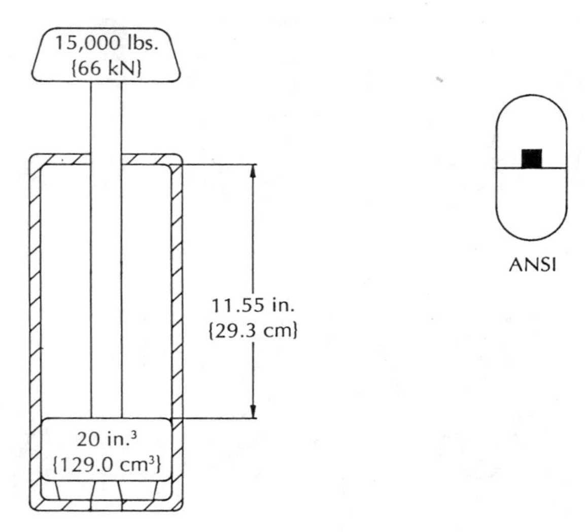

This accumulator has a 20-in.^{2} {*327.7-cm*^{2}} cross-sectional area piston and a potential stroke of 11.55 in. {*29.3 cm*}. It is loaded with a 15,000-lb. {*66-kN*} weight.

Problem 10-1

Calculate the charging pressure and supply potential for the accumulator.

a. P (psi) {*kPa*} = __________

b. Supply (in.^{3}) {*cm*^{3}} = __________

occur around the directional control's spool. This results in a loss of pressure and dislocation of the load. The slow leakage may be resupplied by the accumulator at its charging pressure (see Figure 10-2).

The primary purpose of the circuit described in Figure 10-2 is to maintain pressure regardless of circuit malfunction. The supply of fluid is fairly inconsequential in this design. However, *replenishing circuits may also be used as safety devices that reposition an actuator during malfunction.* In this case, the potential supply from the accumulator is critical, while pressure is a secondary consideration. *The capacity of the accumulator is generally sized to totally displace the cylinder in either or both stroke directions or to cause one complete rotation of a motor or an accumulation of actuators* (see Figure 10-3).

The **replenishing circuit** illustrated in Figure 10-3 can be made automatic by adjusting the design of the directional control used to discharge the accumulator. A normally open, two-position, one-way directional control that is pilot activated from the main circuit branch is used in this design (see Figure 10-4).

FIGURE 10-2

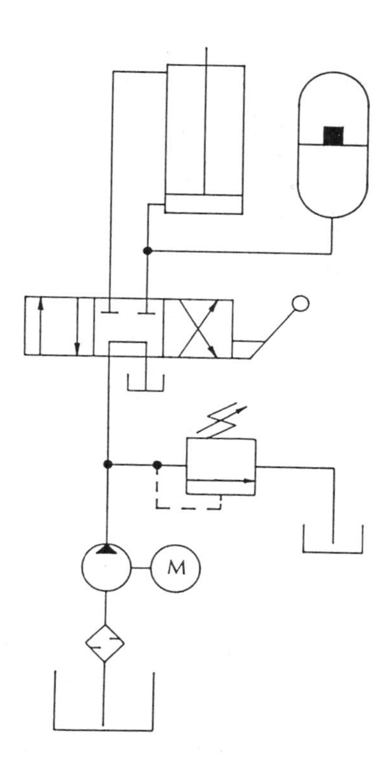

This replenishing circuit charges the accumulator when the cylinder extends. During (or after) extension, the charged accumulator will resupply any leakage from the cylinder when the directional control is shifted to the center position. This will occur even if the power supply is shut down.

In the replenishing circuit illustrated in Figure 10-4, the accumulator may charge when DCV-1 is in any position. When the circuit is energized, DCV-2 is normally open until a specified pressure develops in the circuit. At that point, the pilot pressure will overcome the biased spring and close DCV-2. DCV-2 will remain closed because pressure will always be evident in the power supply during operation. Should the pump fail and pressure drop in the circuit, DCV-2 will open and discharge the accumulator. DCV-1 may then be shifted into either position to complete the cylinder cycle.

Accumulator circuitry may also be used to control surging and pulsation in the hydraulic circuit. As noted in Chapter 4, supply from a pump is erratic; and even with phased discharge of various pumping chambers, some pulsation will be evident. However, the charged accumulator, located in the power supply, will discharge a portion of its capacity at a constant rate. This will tend to smooth the irregular pulses created by the pump. *The accumulator will also act as a shock absorber for hydraulic circuitry.* When rapid loading or shifting of fluid flow occurs in a hydraulic circuit, shock waves are created in the fluid stream as a result of inertial forces. Even the dead-weight accumulator will act to absorb some of these shock waves, although other types of accumulators perform this function more efficiently.

FIGURE 10-3

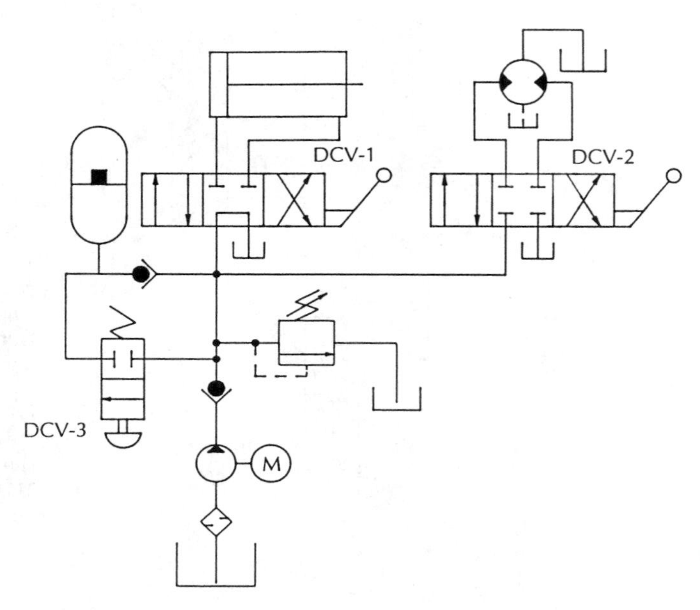

This replenishing circuit charges the accumulator to a level determined by the setting on the pressure-relief valve. If the hydraulic circuit fails, DCV-3 can be shifted to supply enough fluid to alternately recycle the cylinder and the motor.

The circuit described above uses a cylinder that has a 4-in. {*10.2-cm*} diameter piston, a 1-in. {*2.5-cm*} diameter rod, and a 10-in. {*25.4-cm*} stroke opposed by a 9,000-lb. {*40.0-kN*} load in both extension and retraction. The hydraulic motor is sized so that it will overcome a 1,500-in. lb. {*825.0-N cm*} torque load using the maximum pressure required in the linear circuit.

Problem 10-2

Describe the pressure setting and supply necessary in the accumulator to completely recycle the circuit.

P (psi) {*kPa*} = __________

Supply (in.^{3}) {*cm^3*} = __________

10.1.1 Spring-Type Accumulators

A less cumbersome accumulator design uses a spring-loaded piston fitted into a cylinder. Similar to a spring-loaded ram in appearance, this accumulator relies on system pressure acting against a spring-loaded piston for charging. When the spring compresses, increasingly higher forces will be exerted upon the piston until the spring is completely compressed. The

FIGURE 10-4

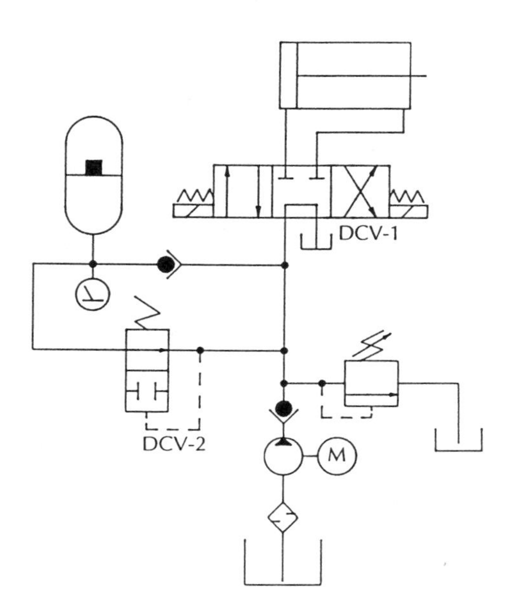

relationship between the compression distance and the force exerted by the helical spring is linear. For example, a spring that is fully extended **(D1)** exerts no force **(F1)**. The spring, when fully compressed, may be displaced to a foreshortened length **(D3)** and exerts a greater force **(F3)**. The force exerted **(F2)** at any increment of compression length **(D2)** can be interpolated using the following formula:

$$F2\ (\text{lbs.})\ \{kN\} = [D1\ (\text{in.})\ \{cm\} - D2\ (\text{in.})\ \{cm\}] * \frac{F3\ (\text{lbs.})\ \{kN\} - F1\ (\text{lbs.})\ \{kN\}}{D1\ (\text{in.})\ \{cm\} - D3\ (\text{in.})\ \{cm\}}$$

Problem 10-3

A fully extended spring has a length of 5 in. {*12.7 cm*}. When compressed to a length of 3 in. {*7.6 cm*}, it exerts 500 lbs. {*2.2 kN*} of force. What force will be exerted when the spring is compressed to a length of 4.5 in. {*11.4 cm*}?

F2 (lbs.) {*kN*} = __________

Of course, the force value for an uncompressed spring will be 0. However, the formula will still be correct; and it may be used to calculate the effect of the spring constant between any two distances of compression.

In the **spring-type accumulator**, the force will be exerted over the area of the piston and will define the accumulator's charge. Thus, a relationship exists between the resultant spring pressure and the potential supply from the accumulator (see Figure 10-5).

FIGURE 10-5

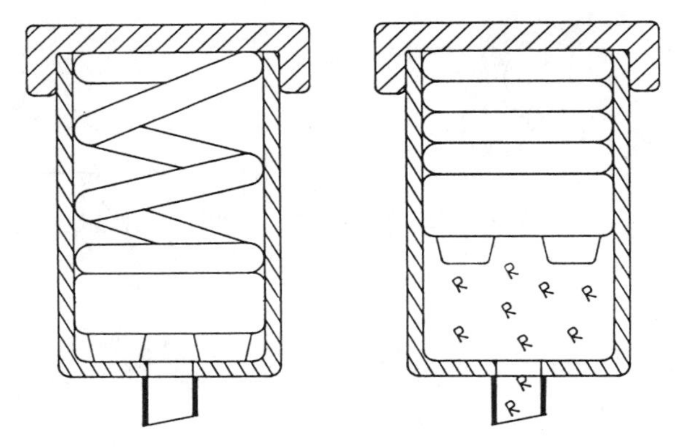

This spring-type accumulator has a 4-in. {*10.2-cm*} diameter piston. When it is assembled, the length of the spring is 5 in. {*12.7 cm*} and the partially compressed spring exerts 1,256 lbs. {*5.6 kN*} of force.

When fully compressed, the spring length is 2 in. {*5.1 cm*} and the force exerted by the spring equals 7,536 lbs. {*33.2 kN*}.

Problem 10-4

What charging pressure will be available when the accumulator has a potential supply of 18.84 in.^{3} {*308.7 cm^3*}?

P (psi) {*kPa*} = __________

Spring-type accumulators have very rapid reaction and require little physical space to supply high pressure. Unlike the dead-load type, a spring-type accumulator loses pressure as it discharges its supply. For this reason, it is typically used in circuitry to maintain pressure rather than to resupply actuators. As a result, *spring-type accumulators are normally small and have low displacement*. Furthermore, variability potential is low because the accumulator must be disassembled and a different spring inserted for this to occur.

10.1.2 Gas-Charged Accumulators

The highest degree of variability potential for accumulators exists in the gas-charged types. In the **gas-charged accumulator**, a container pressurized with dry nitrogen gas acts to impose a load on the hydraulic system. Seldom will air be used to pressurize a gas-type accumulator, and oxygen should never be used because of the potential for combustion under high-pressure loads. Although the device may operate with the gas in direct contact with the hydraulic fluid **(nonseparated types)**, a barrier between the gas and the oil is preferred **(separated types)**. This separator divides the accumulator into two chambers. Figure 10-6 illustrates three designs for gas-charged accumulators.

FIGURE 10-6

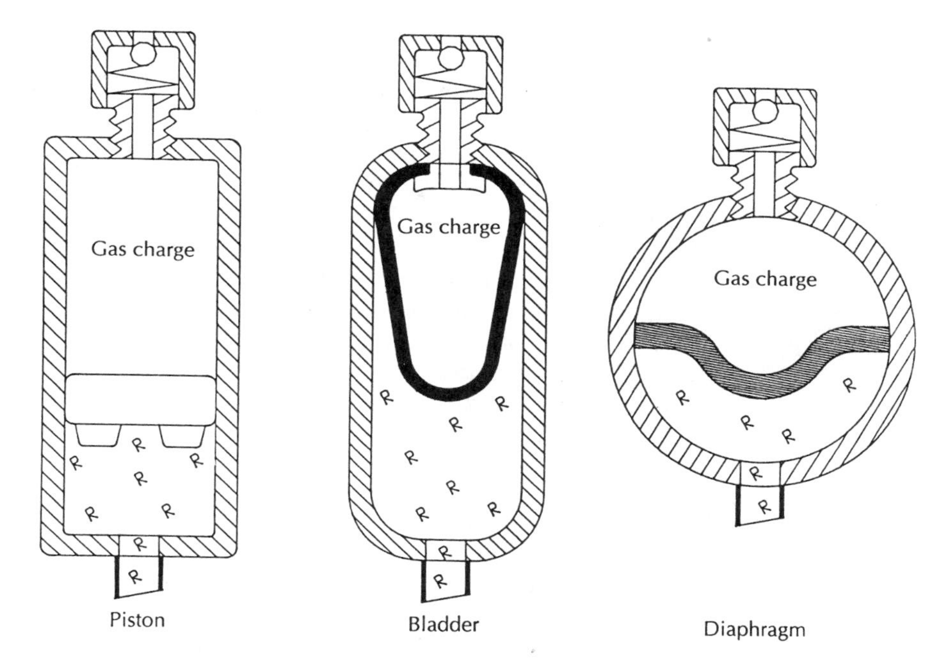

10.1.2.1 Gas-accumulator operation The gas chamber is **precharged** with variable gas pressure loads. As hydraulic pressure becomes greater than the gas pressure, the oil chamber will increase in volume as the gas chamber decreases. This will compress the gas and result in even higher pressure developing in the gas chamber in accordance with the gas law established by British scientist Robert Boyle. **Boyle's Law** states that *"the volume of an enclosed gas varies inversely with its absolute pressure, if the temperature remains constant":*

$$V1\ (\text{in.}^3)\ \{cm^3\} * P1\ (\text{psi}_a)\ \{kPa_a\} = V2\ (\text{in.}^3)\ \{cm^3\} * P2\ (\text{psi}_a)\ \{kPa_a\}$$

Note that Boyle's Law deals with absolute pressure (psi_a or Kpa_a) or the reading from a pressure gauge plus atmospheric pressure. Although atmospheric pressure will vary according to weather conditions and elevation, for the sake of this presentation, it will be standardized to be 14.7 psi {*101 kPa*}.

When the gas-charged accumulator discharges, the loaded pressure will drop. Thus, this accumulator must be overcharged to allow sufficient supply before pressure drops below a value needed to operate the circuitry. *The important design calculation is to specify the volumetric size of the accumulator that is necessary to achieve the needed supply at pressure.* Thus, three pressure-equivalent volumes are used to specify the accumulator.

A potential supply **(S)** of hydraulic fluid will be required to cycle the circuit as determined by the size of the circuit's actuator or actuators. The accumulator is precharged to a specified pressure **(P1)**, typically one-third of the maximum pressure required by any actuator to perform its function. At this point, the total volume **(V1)** of the accumulator will be filled with gas. The circuit will then be allowed to rise to a load pressure **(P2)**. Load pressure is typically established at 400 psi {*2,758 kPa*} more than the pressure necessary to perform actuator functions and is set by the pressure-relief valve. This will assure that turbulent flow does not occur during the pressure drop when discharged. At this point, the gas will be compressed to its load volume **(V2)**. Finally, the isolated accumulator can activate its supply and the circuit will drop to its discharge pressure **(P3)**, which, of course, is the maximum pressure required by any actuator. At discharge pressure, the volume of gas **(V3)** in the accumulator will rise once more. *The discharge volume of gas will be equal to the load volume of gas plus the specified supply*:

$$\mathbf{V3 = V2 + S}$$

To specify the size **(V1)** for the accumulator, two independent Boyle's Law equations are related:

$$\mathbf{P1 * V1 = P2 * V2}, \text{ and}$$

$$\mathbf{P2 * V2 = P3 * V3}$$

Since **V3 = V2 + S**, the second equation can be solved when using the required supply:

$$\mathbf{P2}\ (\text{psi}_a)\ \{kPa_a\} * \mathbf{V2}\ (\text{in.}^3)\ \{cm^3\} = \mathbf{P3}\ (\text{psi}_a)\ \{kPa_a\} * [\mathbf{V2}\ (\text{in.}^3)\ \{cm^3\} + \mathbf{S}\ (\text{in.}^3)\ \{cm^3\}]$$

Problem 10-5

A gas-charged accumulator circuit has an allowable pressure drop from 1,150 (P2) to 750 (P3) psi_g {*7,929 to 5,171* kPa_g} while supplying (S) 462 in.^{3} {*7,571* cm^3} of fluid. What gas volumes will be contained in the accumulator during load charge (V2) and discharge (V3)?

V2 (in.^{3}) {cm^3} = ______________

V3 (in.^{3}) {cm^3} = ______________

The charging volume **V1**, which equates to the accumulator size, can then be calculated using the charging pressure **P1** and the derived values for **V2** and **P2**:

$$\mathbf{V1}\ (\text{in.}^3)\ \{cm^3\} = \frac{\mathbf{P2}\ (\text{psi}_a)\ \{kPa_a\} * \mathbf{V2}\ (\text{in.}^3)\ \{cm^3\}}{\mathbf{P1}\ (\text{psi}_a)\ \{kPa_a\}}$$

Problem 10-6

In the previously described accumulator circuit, the charging pressure (P1) is 250 psi_g {*1,724 kPa_g*}. What size accumulator (V1) is needed to complete the design as specified?

V1 (in.^{3}) {*cm^3*} = __________

10.1.2.2 Reservoir accommodations for accumulators Circuitry operating with accumulators must have larger reservoirs to hold the potential discharged hydraulic fluid. Incorrectly, it may be assumed that the depletion of fluid from the reservoir to accommodate load volume can be adjusted by simply pouring more fluid into the reservoir after loading. The potential may exist to overflow the reservoir, then, when the accumulator discharges. Rather, the normal design procedure for sizing the reservoir based on delivery should be performed; then, an additional capacity should be made to hold the highest volume of fluid that may be held in the accumulator. The largest supply (volume) of fluid will be contained in the accumulator at load pressure. The volume of fluid (not gas) at that time may be calculated as the difference between precharge volume **(V1)** and load volume **(V2)**:

$$S_{acc\ max}\ (\text{in.}^3)\ \{cm^3\} = \mathbf{V1}\ (\text{in.}^3)\ \{cm^3\} - \mathbf{V2}\ (\text{in.}^3)\ \{cm^3\}$$

Normal procedure would then be to determine the necessary reservoir capacity by dividing the delivery by ½ and ⅓ and adding the derived accumulator supply to each of those values.

10.1.2.3 Characteristics of gas-charged accumulators The specifications for gas-charged accumulators, as described, are accurate only if the accumulator discharges at a slow rate. Otherwise, turbulence and the resulting heat generated will reduce the efficiency of the accumulator. In accumulator applications operating under extremely high pressure drops, exponential adjustments must be made for changes in temperature that occur. However, at pressure drops less than 400 psi {*2,758 kPa*}, the change in temperature is relatively low enough to be considered negligible. Inconsequentially, the change in pressure from changes in temperature in accumulator operation will tend to result in lower pressure and volume requirements; and the formulation provided in the previous section will accommodate most circuitry.

The spherical shape of the **diaphragm design** allows the most capacity for space required in gas-charged accumulators. However, the **piston design** normally creates the fewest maintenance problems (see Figure 10-7). The **bladder design** allows for greater supply potential over a given pressure drop due to the expansion potential of the rubber bag sepa-

FIGURE 10-7

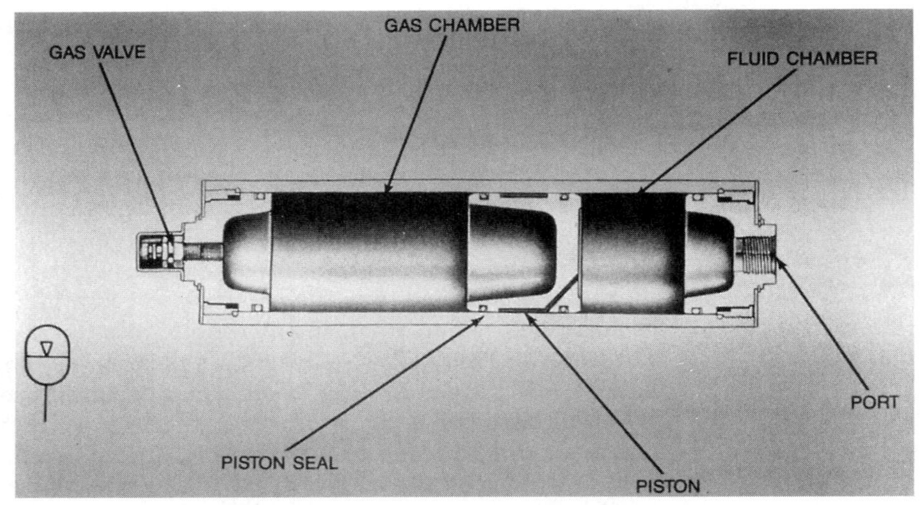

Piston-type gas-charged accumulator. (*Courtesy Vickers, Inc.*)

FIGURE 10-8

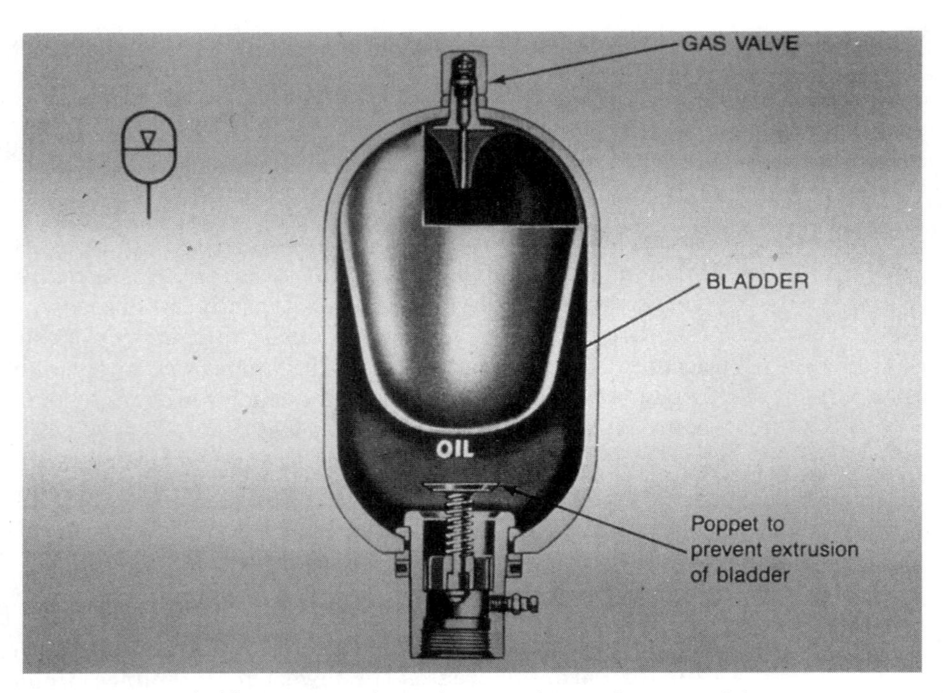

Bladder-type gas-charged accumulator. (*Courtesy Vickers, Inc.*)

rator. Notice the spring-loaded valve in the bladder design in Figure 10-8. It is included to keep the expanded rubber bag from extruding into the port opening and potentially causing damage to the bag and rupture in extreme cases.

Gas-charged accumulators are especially suited to replenishing applications and are adaptable to high supply and pressure requirements. They also are superior to other accumulators for shock absorption and pulsation control because of the cushioning effect created by the compressed gas. In general, the gas-charged type is the best all-around type of accumulator. As with all accumulator circuits, precaution should be taken to discharge the gas-charged accumulator before disassembling the circuit. Typically, a **shunt to tank** from the accumulator containing a two-position, one-way, lever-activated directional control is used to perform the discharge (see Figure 10-9).

10.2 INTENSIFIERS

Occasionally, a hydraulic circuit operation will require an extremely high pressure only during a very short distance at the end of a cylinder stroke. Obviously, it would be advantageous to isolate a large portion of the circuit from this intensified pressure. This would lessen the bulk of other components and reduce wasted power that would be needed in other portions of the cycle to accommodate this high pressure.

The **intensifier**, or **booster**, is a hydraulic device that multiplies a high-volume, lower-input pressure into a high-pressure, low-displacement

FIGURE 10-9

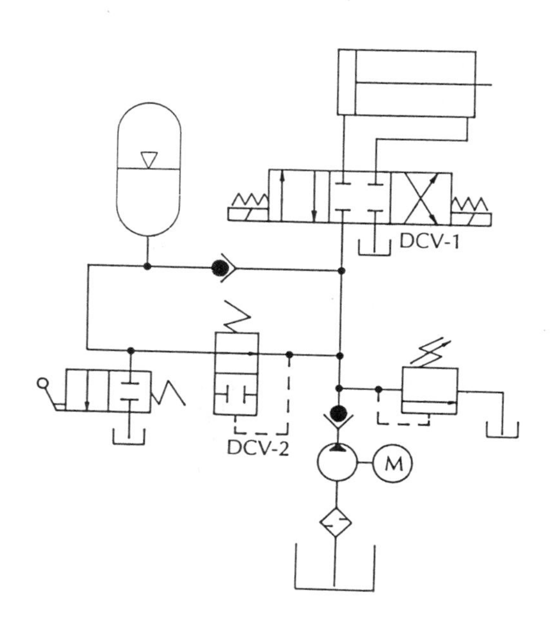

output. The intensifier is similar in design to a cylinder. However, it has an enclosed rod that acts to supply hydraulic fluid to an actuator. *Intensifiers may be either single or double acting.* Figure 10-10 illustrates the design and operation of the single acting intensifier.

FIGURE 10-10

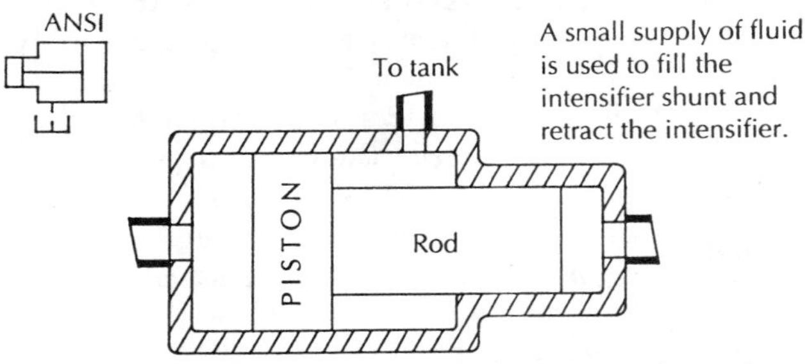

A single-acting intensifier has a 4-in. {*10.2-cm*} diameter piston, a 1-in. {*2.5-cm*} diameter rod, and a 6-in. {*15.2-cm*} stroke. It is used in a circuit having an operating pressure of 800 psi {*5,516 kPa*}.

Problem 10-7 What pressure will be evident at the intensifier outlet?

P (psi) {*kPa*} = __________

Problem 10-8 What supply will be provided by a full stroke of the intensifier?

S (in.^{3}) {*cm^3*} = __________

10.2.1 Intensifier Circuitry

The intensifier primarily operates to provide a high-pressure charge. To allow the intensifier circuit to operate advantageously, *other components must be incorporated to isolate the intensifier*. Figure 10-11 illustrates an intensifier circuit that allows full pump supply to advance a cylinder to a position, then supply intensified pressure at the end of the stroke.

The intensifier circuit described in Figure 10-11 uses a **two-position, two-way, plunger-activated directional control** to selectively isolate and direct flow to either the intensifier piston end or to the cylinder. To understand the operation of the intensifier circuit, first *consider the action when the output cylinder is being retracted.*

FIGURE 10-11

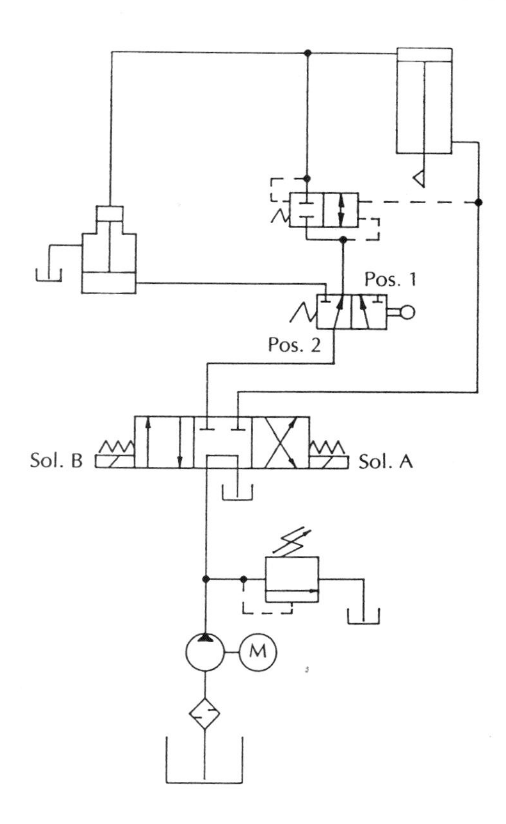

When **solenoid A** is energized, supply is directed to the head end of the cylinder. At this time, the **"selector"** directional control is being held in **Position 1** by the cylinder cam activation of the plunger. During initial retraction, the exhaust from the base end of the cylinder is used to charge the rod end and retract the intensifier, since the piston end of the intensifier is exhausted through the selector directional control and the three-position, four-way directional control. After initial cylinder retraction, the cam will clear the plunger on the selector directional control and shift it into its normal **Position 2**. The cylinder will then retract as a normal linear circuit with exhaust from the base end of the cylinder being passed through the selector and three-position, four-way directional controls.

The intensifier action occurs during extension. When **solenoid B** is energized, supply is directed to the selector directional control. In its normal position, supply is then passed through this directional control to extend the cylinder as a normal linear circuit. During this portion of the extension stroke, the intensifier supply port is blocked at the selector directional control in **Position 2**. When the cylinder extends to a specified location, the cam on the cylinder rod shifts the selector directional control to **Position 1**, closing the "normal" supply to the cylinder. However, at this time, supply is directed to the piston end of the intensifier; and high-pressure output is evident at the rod end. *This high pressure is directed to the base end of the cylinder during the end of extension.*

Notice that the majority of the circuit is further isolated from this intensified pressure by the **pilot override check valve**. It separates the rod-end output of the intensifier from the rest of the circuit. During cylinder retraction, the pressure on the head end of the cylinder will shift the pilot override check valve through pilot activation and allow free exhaust flow through it. Exhaust flow will then be directed through the other directional controls and returned to tank.

10.2.2 Designing Intensifier Circuitry

Design and specification procedures for most intensifier circuitry are similar to techniques used for other circuitry. However, since the intensifier shunt operates at extremely high pressure, special consideration must be made to allow this shunt to operate correctly during both "normal" and intensification phases. Normally, this will require oversizing of conductor wall thickness and specification of components that contact the intensifier shunt to ensure that they are capable of withstanding high pressure.

Since the primary function of the intensifier is to provide high pressure, *in many circuits, the intensifier supply is inconsequential*. In these cases, the stroke of the cylinder is specified by linear movement requirements. Activation of the intensifier shunt occurs at the end of the stroke, and no allowance is made for cylinder movement resulting from intensifier supply. However, the specified **burst strength** of the conductors in the shunt must be great enough to withstand intensified pressure. The remaining portions of the circuit will not be exposed to this high pressure and may be designed and specified as normal.

In other intensifier circuits, not only will high pressure be necessary but a minor displacement of the actuator must also be accommodated. In these cases, the design parameter becomes the necessary cylinder supply to afford cylinder movement under intensified pressure.

10.2.2.1 Pressure requirements for intensifiers A 4-in. {*10.2 cm*} diameter cylinder would have 12.56 sq. in. {*81.7 cm*^{2}} of area operating on the output load. If the cylinder load during this portion of the stroke is 30,000 lbs. {*133.3 kN*}, the intensifier must develop 2,389 psi {*16,472 kPa*} in the intensified shunt (outlet side of the intensifier). If the intensifier has a 1-in. {*2.5-cm*} diameter rod, the force component for the intensified supply may be calculated as follows:

$$F \text{ (lbs.) } \{N\} = P \text{ (psi) } \{Pa\} * A \text{ (in.}^2\text{) } \{m^2\}$$

$$F \text{ (lbs.) } \{N\} = 2{,}389 \text{ (psi) } \{16{,}472{,}000\ Pa\} * 0.785 \text{ (in.}^2\text{) } \{0.00049\ m^2\}$$

$$F = 1{,}875 \text{ (lbs.) } \{8{,}071\ N \text{ or } 8.1\ kN\}$$

If the intensifier has a 2-in. {*5.1-cm*} diameter piston, the power-supply pressure (inlet side of the intensifier) may be calculated as follows:

$$P \text{ (psi) } \{Pa\} = \frac{F \text{ (lbs.) } \{N\}}{A \text{ (in.}^2\text{) } \{m^2\}}$$

$$P \text{ (psi) } \{Pa\} = \frac{1{,}473 \text{ (lbs.) } \{8{,}071\ N\}}{3.14 \text{ (in.}^2\text{) } \{0.00204\ m^2\}}$$

$$P = 469 \text{ (psi) } \{395{,}637\ Pa \text{ or } 3{,}956\ kPa\}$$

Since the intensified pressure is not evident in the power supply, this unintensified value becomes the highest operating pressure in the circuit (not withstanding retraction pressure and other pressures necessary to operate other actuators or valving). Therefore, it *should be* the parameter for determining the relief valve setting or the discharge pressure for an accumulator.

10.2.2.2 Delivery requirements for intensifiers The operating intensifier may impose an additional delivery requirement on the circuit when operating from a cycle rate parameter. In general, the requirements imposed by actuator loads will determine the sizes for the piston and rod in the intensifier. In turn, the intensifier rod area, in combination with supply requirements determined by the actuator (cylinder) during intensified stroke, will determine the necessary minimum stroke for the intensifier:

$$St_{int} \text{ (in.) } \{cm\} = \frac{S_{cyl} \text{ (in.}^3\text{) } \{cm^3\}}{A_{int\ rod} \text{ (in.}^2\text{) } \{cm^2\}}$$

For example, if the cylinder is demanding 12.56 in.^{3} {*206 cm*^{3}} of supply and the intensifier has a 1-in. {*2.5-cm*} diameter rod, the minimum stroke for the intensifier would be as follows:

$$St_{int} \text{ (in.) } \{cm\} = \frac{12.56 \text{ (in.}^3\text{) } \{206\ cm^3\}}{3.14 \text{ (in.}^2\text{) } \{20.3\ cm^2\}}$$

$$St_{int} = 4 \text{ (in.) } \{10.2\ cm\}$$

Once the minimum stroke for the intensifier is determined, the necessary supply for the intensifier can be determined in consideration of the intensifier's piston area:

$$S_{int\ ext}\ (in.^3)\ \{cm^3\} = A_{int\ pis}\ (in.^2)\ \{cm^2\} * St_{int}\ (in.)\ \{cm\}$$

If the intensifier calculated to have a minimum 4-in. {*10.2-cm*} stroke has a 3-in. {*7.6-cm*} diameter piston, the necessary extension supply per cycle would be as follows:

$$S_{int\ ext}\ (in.^3)\ \{cm^3\} = 7.07\ (in.^2)\ \{45.4\ cm^2\} * 4\ (in.)\ \{10.2\ cm\}$$

$$S_{int\ ext} = 28.28\ (in.^3/cyc)\ \{464.1\ cm^3/cyc\}$$

Finally, the determined intensifier extension supply in consideration of the operational rate established for the circuit will determine the additional supply rate demanded by the intensifier:

$$S.R._{int}\ (in.^3/min.)\ \{cm^3/min.\} = S_{int\ ext}\ (in.^3/cyc)\ \{cm^3/cyc\} * C.R.\ (cyc/min.)$$

If the intensifier described in this section is to operate in a circuit at a rate of 6 cycles/min., the additional supply rate imposed upon the circuit would be as follows:

$$S.R._{int}\ (in.^3/min.)\ \{cm^3/min.\} = 28.28\ (in.^3/cyc)\ \{464.1\ cm^3/cyc\} * 6\ (cyc/min.)$$

$$S.R._{int} = 169.7\ (in.^3/min.)\ \{2{,}784.6\ cm^3/min.\}$$

This supply rate demanded by the intensifier, in combination with the supply rate (or rates) demanded by the other operating actuators, will establish the supply rate and pump delivery required for the circuit. Notice that no supply or delivery demand is placed on the circuit to operate the intensifier during retraction if the circuit is designed as in Figure 10-11. The closed-loop design will automatically retract the intensifier with exhaust from the base end of the cylinder during initial retraction of the cylinder.

10.2.2.3 Conductor size determination Typically, three ports exist on the single-acting intensifier. They are the piston-end port, the rod-end port, and a pocket-drain port. The pocket-drain port is normally sized by the manufacturer in consideration of the size of the intensifier. It is a low-volume, low-pressure return to tank. The piston-end port is operating within an unintensified environment and under normal pump delivery. Therefore, it may be sized in compliance with other working lines in the circuit. However, the physical size of the piston in the intensifier in consideration of the pump delivery will determine the velocity at which the intensifier will extend:

$$VEL_{int\ ext}\ (in./min.)\ \{cm/min.\} = \frac{DEL\ (gpm)\ \{lpm\} * 231\ (in.^3/gal.)\ \{1,000\ cm^3/l\}}{A_{int\ pis}\ (in.^2)\ \{cm^2\}}$$

If an intensifier having a 3-in. {*7.6-cm*} diameter piston is operating under a delivery of 2 gpm {*7.6 lpm*}, the extension velocity would be as follows:

$$VEL_{int\ ext}\ (in./min.)\ \{cm/min.\} = \frac{2\ (gpm)\ \{7.6\ lpm\} * 231\ (in.^3/gal.)\ \{1,000\ cm^3/l\}}{7.07\ (in.^2)\ \{45.4\ cm^2\}}$$

$$VEL_{int\ ext}\ (in./min.)\ \{cm/min.\} = \frac{462\ (in.^3/min.)\ \{7,600\ cm^3/min.\}}{7.07\ (in.^2)\ \{45.4\ cm^2\}}$$

$$VEL_{int\ ext} = 65.3\ (in./min.)\ \{167.8\ cm/min.\}$$

In turn, the supply rate at which fluid is exiting the intensifier is determined by the intensifier's extension velocity and the size of the intensifier rod:

$$S.R.\ (in.^3/min.)\ \{cm^3/min.\} = VEL\ (in./min.)\ \{cm/min.\} * A_{int\ rod}\ (in.^2)\ \{cm^2\}$$

If the previously described intensifier having a 1-in. {*2.5-cm*} diameter rod were operating at the derived velocity, the resulting supply rate would be as follows:

$$S.R.\ (in.^3/min.)\ \{cm^3/min.\} = 65.3\ (in./min.)\ \{167.8\ cm/min.\} * 0.7854\ (in.^2)\ \{4.9cm^2\}$$

$$S.R. = 51.3\ (in.^3/min.)\ \{822.2\ cm^3/min.\}$$

The supply rate for fluid leaving the rod end of the intensifier may be specified as a delivery by dividing by 231 in.^{3}/gal. {*1,000 cm*^{3}*/l*}, expressing the result in gpm {*lpm*}. Subsequently, the size of conductors in the intensified shunt and the size of the port on the rod end of the intensifier may be determined using the derived delivery and formulation for determining working lines.

10.2.2.4 Other intensifier determinants and requirements The time necessary to perform a cycle of the intensifier will require calculations based on intensifier extension velocity and stroke. In general, it is best to calculate this duration and disregard that portion of the stroke of the cylinder during intensified extension for determination of cycle time attributed to cylinder extension. This will result in the stroke of the cylinder being described as only the length operating under unintensified conditions during extension, and that same length of stroke will be used to determine requirements for supply and time of cylinder extension.

The pressure that exists within the intensified shunt of the circuit (during intensified operation) is inconsequential to determination of the pressure-relief valve setting and power requirements. However, it should be noted that the system pressure-relief valve will also set the limit for the pressure that exists within this shunt (if it is the highest operating pressure) through the multiplying factor of the intensifier. Although the intensifier shunt conductors are typically smaller in size than other working lines within the system (owing to lower delivery requirement), they also must withstand higher pressures. This will result in a necessity for heavier wall thickness in piping in the shunt. As a general rule, Schedule 40 pipe is appropriate up to 1,000 psi {*6,895 kPa*}, Schedule 80 up to 2,500 psi {*17,237 kPa*}, and Schedule 160 up to 5,000 psi {*34,474 kPa*}. For more specific information of pressure ratings for pipe, hose, and tube, see Chapter 11.

[] [] [] 10.3 PRACTICAL DESIGN PROBLEM

The Knockout Punch Incorporated punches the holes in heavy industrial stainless steel sinks. The punch presses currently being used are driven by electrical motors working through levers and arms to produce the punching movement.

The company has become familiar with a design that would change its presses to hydraulic operation (see Figure 10-12). It is the company's understanding that this design is faster, more flexible, and safer. However, the design has no specifications for components or evaluation of its operational cycle. Therefore, you have been hired to specify all components and report on the circuit's operation.

The press head attached to the cylinder weighs 1,800 lbs. {*7.8 kN*}, and it will require 28,500 lbs. {*126.7 kN*} of force to shear the holes in the metal through intensified extension during the end of the cylinder stroke. For clearance, loading, and setup purposes, the die head must move a distance of 8 in. {*20.3 cm*}. Furthermore, *the punch dies should extend an extra distance of 1 in. after shearing to reject the punched metal.* It is anticipated that a full cycle of the punch should take no more than 5 seconds. The system is also equipped with a gas-charged *accumulator and controls that allow completion of one full cycle* in the event of power-supply failure.

Complete your assignment by reviewing the schematic, specifying all components according to the format provided, and recording your report of the circuit's operation.

FIGURE 10-12

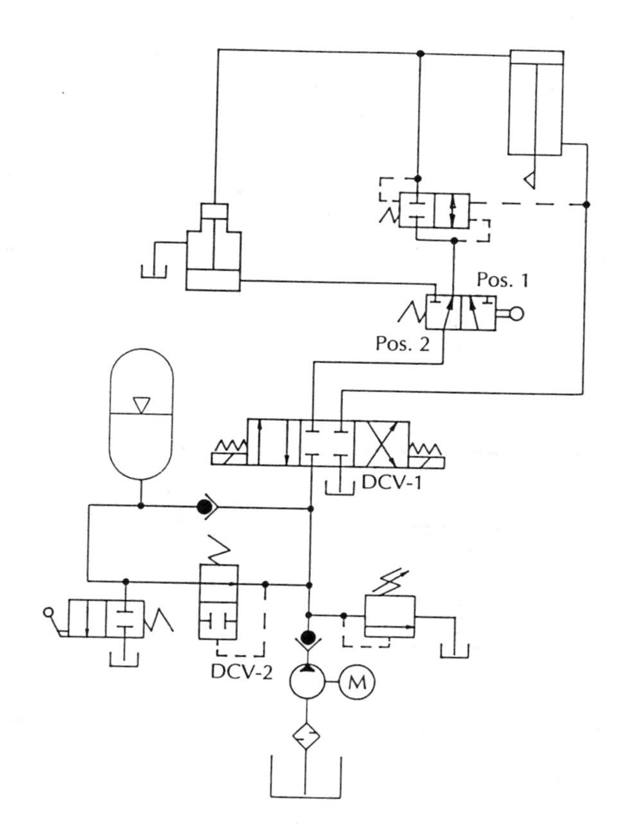

A Single End Rod, Double-Acting Cylinder

PISTON		ROD			
Diameter	Area	Diameter	Area	Annulus Area	Stroke
_____ in. {*cm*}	_____ in.^{2} {*cm*^{2}}	_____ in. {*cm*}	_____ in.^{2} {*cm*^{2}}	_____ in.^{2} {*cm*^{2}}	8 in. {*20.3 cm*} [normal advance] 1 in. {*2.5 cm*} [intensified]

FORCE		PRESSURE		SUPPLY	
Extension	Retraction	Extension	Retraction	Extension	Retraction
28,500 lbs. {*126.7 kN*} [intensified] 1,800 lbs. {*7.8 kN*} [normal]	1,800 lbs. {*7.8 kN*}	_____ psi{*kPa*} _____ psi{*kPa*}	_____ psi {*kPa*}	_____ in.^{3} {*cm*^{3}} [normal advance]	_____ in.^{3} {*cm*^{3}} [full return]

A Single End Rod, Double-Acting Cylinder (*continued*)

Total Supply	Rate	Supply Rate	VELOCITY Extension	VELOCITY Retraction
_____ in.^{3} {*cm*^{3}}	12 cyc/min.	_____ in.^{3}/min. {*cm*^{3}*/min.*}	_____ in./min. {*cm/min*}	_____ in./min. {*cm/min.*}

TIME Extension	TIME Retraction	Port Size	Body Style	Mounting
_______ sec.	_______ sec.	_______ in. {*mm*}	___________	___________

B Intensifier

Necessary Cylinder Displacement	INTENSIFIER PISTON Diameter	INTENSIFIER PISTON Area	INTENSIFIER ROD Diameter	INTENSIFIER ROD Area
________ in.^{3} {*cm*^{3}}	________ in. {*cm*}	_________ in.^{2} {*cm*^{2}}	________ in. {*cm*}	________ in.^{2} {*cm*^{2}}

FORCE Cylinder Piston	FORCE Intensifier Rod	INTENSIFIER PRESSURE Piston	INTENSIFIER PRESSURE Rod	Stroke
28,500 lbs. {*126.7 kN*}	_______ lbs. {*kN* }	________ psi {*kPa*}	________ psi {*kPa*}	________ in. {*cm*}

SUPPLY INPUT FROM PUMP Extension	SUPPLY INPUT FROM PUMP Retraction	Rate	Supply Rate
________ in.^{3} {*cm*^{3}}	0 in.^{3} {cm^3}	12 cyc/min.	_______ in.^{3}/min. {*cm*^{3}*/min*}

Velocity Extension	Time Extension	PORT SIZE Piston End	PORT SIZE Rod End	Mounting
_______ in. /min. {*cm/min.*}	_______ sec.	_______ in. {*mm*}	_______ in. {*mm*}	____________

C Pump

Type	Design	Delivery	Drive Speed	Volumetric Effic.
______	______	______ gpm {*lpm*}	______ rpm	______ %

Displacement at V.E.	PORT SIZE Inlet	PORT SIZE Outlet
______ in.^{3}/rev. {*cm^3/rev.*}	______ in. {*mm*}	______ in. {*mm*}

D Gas-Charged Accumulator

Necessary Supply (S)	Load Volume (V2)	Discharge Volume (V3)	Precharge Pressure (P1)	Load Pressure (P2)
______ in.^{3} {*cm^3*}	______ in.^{3} {*cm^3*}	______ in.^{3} {*cm^3*}	______ psi_a {kPa_a}	______ psi_a {kPa_a}

Discharge Pressure (P3)	Precharge Volume (V1)	Size	Element Type	Port Size
______ psi_a {kPa_a}	______ in.^{3} {*cm^3*}	______ gal. {*l*}	______	______ in.{*mm*}

E Reservoir

CAPACITY Minimum	CAPACITY Maximum
______ gal. {*l*}	______ gal. {*l*}

F Filtration

Type	Element	RATING Element	RATING Flow	Port Size
______	______	______ mesh	______ gpm {*lpm*}	______ in. {*mm*}

F Filtration (*continued*)

Description	Mounting
________	________

G Pressure-Relief Valve

Type	Element	Setting	Mounting	Port Size
________	________	_____ psi {*kPa*}	________	_____ in. {*mm*}

H Electric Motor

Speed	Voltage	Phase	Power at V.E.
_____ rpm	________	________	_____ hp {*W*}

I Master System Directional Control

Positions	Ways	Element	Activation	Port Size
________	________	________	________	_____ in. {*mm*}

J Intensifier Selector Directional Control

Positions	Ways	Element	Activation	Port Size
________	________	________	________	_____ in. {*mm*}

K Pilot Override Check Valve

Positions	Ways	Element	Activation	Port Size
________	________	________	________	_____ in. {*mm*}

L Accumulator Shunt Directional Control—1

Positions	Ways	Element	Activation	Port Size
________	________	________	________	______ in. {*mm*}

M Accumulator Shunt Directional Control—2

Positions	Ways	Element	Activation	Port Size
________	________	________	________	______ in. {*mm*}

N Check Valves

Positions	Ways	Element	Activation	Port Size
________	________	________	________	______ in. {*mm*}

O Inlet and Working Conductors

Inlet Type	Inlet Description	*INLETS (I.D. DECIMAL)*		Working Type	Working Description
		Minimum	Maximum		
______	________	______ in. {*cm*}	______ in.{*cm*}	______	________

WORKINGS (I.D. DECIMAL)		*LINES—NOMINAL*	
Minimum	Maximum	Inlets	Workings
______ in. {*cm*}	______ in. {*cm*}	______ in. {*mm*}	______ in. {*mm*}

P Intensifier Rod End Conductor

Type	Description	Resolved Delivery from Rod End
________	________	______ gpm {*lpm*}

P Intensifier Rod End Conductor (*continued*)

WORKINGS (I.D. DECIMAL)		*LINES—NOMINAL*	
Minimum	Maximum	Inlets	Workings
________ in. {*cm*}	________ in. {*cm*}	________ in. {*mm*}	________ in. {*mm*}

Q Report any changes made in the circuit or specifications that would alter the schematic that was provided.

R The following are variations for the Practical Design Problem (10.3):

Component: Double-Acting, Single End Rod Cylinder

	A	B	C	D
Stroke				
Normal Advance	10 in.	12 in.	24 *cm*	20 *cm*
Intensified Advance	1.5 in.	2 in.	3 *cm*	5 *cm*
Extension Force				
Normal	1,600 lbs.	1,950 lbs.	8.5 *kN*	9.3 *kN*
Intensified	25,000 lbs.	30,000 lbs.	130 *kN*	125 *kN*
Retraction Force	1,600 lbs.	1,950 lbs.	8.5 *kN*	9.3 *kN*
Rate	5 cyc/min.	4 cyc/min.	6 *cyc/min.*	3 *cyc/min.*

10.4 SUGGESTED ACTIVITIES

1. Being careful to ensure that the components are completely discharged before examination, disassemble a spring-type and gas-charged accumulator. Explain the construction and operation of the components verbally, with a sketch, or both.
2. Disassemble a single-acting intensifier and explain its construction and operation verbally, with a sketch, or both.
3. Draw an accurate schematic of a linear circuit having terminal intensification of the cylinder's stroke and auxiliary power accumulator supply. Label all components in the circuit on the schematic.
4. Using a computer spreadsheet program or a programming language, write a program that will determine the necessary size (volume) of an accumulator based on input of supply required by a circuit and discharge pressure.

10.5 REVIEW QUESTIONS

1. Describe how accumulators are used in hydraulic circuits.
2. Name the three basic types of accumulators.
3. Describe the design, operation, characteristics, and uses of dead-weight accumulators.
4. What precaution should be taken before servicing any accumulator used in an operating circuit?
5. What is the purpose of the check valve located after the pump outlet used in an accumulator circuit?
6. Describe the design, operation, characteristics, and uses of spring-type accumulators.
7. Name the three common designs for gas-charged accumulators.
8. Why is dry nitrogen gas normally used to precharge a gas-type accumulator rather than oxygen or air?
9. What is precharge pressure on a gas-type accumulator, how is its pressure value determined, and how does precharge pressure relate to the size (volume) of the accumulator?
10. What is load pressure in a gas-charged accumulator, and how is it set by the operating circuit?
11. At what level should discharge pressure be established for a hydraulic circuit using an accumulator for replenishing?
12. What is the relationship between load volume and discharge volume in a gas-charged accumulator used for replenishing?
13. What effect will accumulator operation have on sizing the reservoir when designing a hydraulic circuit?
14. Describe the design, operation, characteristics, and uses of a single-acting intensifier.

15. What determines the pressure evident on the intensifier rod end when used in a linear circuit?
16. What determines the necessary minimum stroke for an intensifier?
17. Why is the rod-end port of a single-acting intensifier sized differently than other working lines in the circuit?

11

Fluid Distribution

It may appear that all that is necessary to achieve an optimally operating hydraulic circuit is to define parameters and specify components, then link them together with any conductor. If this were true, fluid plumbing would be made from drinking straws that slip-fit into components. Furthermore, the drinking straw would have very thin walls to save on production costs; and there would be no concerns for system pressure or compatibility between the fluid and the conductor. Obviously, this could not be further from the truth, even though some hydraulic circuits are operating with nearly this level of disregard for the lifelines of hydraulic circuitry—the **fluid conductors**.

11.1 TYPES OF FLUID CONDUCTORS

Three distinct types of fluid conductors are prevalent in hydraulic circuitry (see Figure 11-1):

1. Piping
2. Tubing
3. Hosing

Commonly, hydraulic circuits will be found that incorporate only one type of fluid conductor. However, the different types have unique characteristics; and a thorough analysis should be made to *match fluid conductors to conditions that exist within the circuit.*

FIGURE 11-1

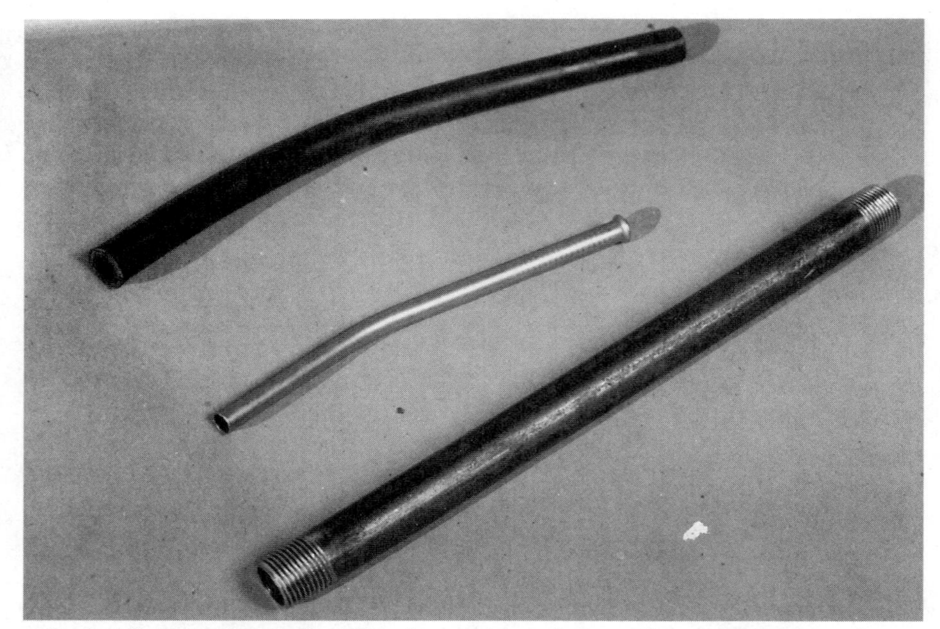

Typical fluid conductors, including hose, tube, and pipe. (*Photograph by the author.*)

11.1.1 Piping

Rigid fluid conductors are collectively known as **pipe**. It should not be inferred that other fluid conductors may not be rigid but that the highest degree of rigidity in circuitry plumbing is available with pipe. Typically, pipe is made from relatively heavy-walled mild steel (American Iron and Steel Institute [AISI] and SAE 1010). Occasionally, alloy steel pipe, such as AISI 4130, may be used to make pipe. Galvanized pipe should not be used in hydraulic applications.

11.1.1.1 Pipe sizing The high tensile strength of mild steel, as compared to materials from which other fluid conductors are made, results in higher **burst strength**. The maximum pressure **(Max. P.)** that a fluid conductor can withstand is a function of tensile rating **(T.S.)**, the outside diameter **(O.D.)**, and the wall thickness **(W.T.)** of the conductor. It may be calculated using a variation of **Barlow's Formula**:

$$\text{Max. P. (psi)}\ \{kPa\} = \frac{\text{W.T. (in.)}\ \{mm\} * [2 * \text{T.S. (psi)}\ \{kPa\}]}{\text{O.D. (in.)}\ \{mm\}}$$

Problem 11-1

A hydraulic circuit is operating using 0.540-in. {*12-mm*} outside diameter pipe having 0.088-in. {*2-mm*} wall thickness. The pipe has been made from AISI 1010 steel having a tensile rating of 55,000 psi {*379,212 kPa*}. What is the maximum pressure that may be accommodated in the circuit without pipe failure?

Max. P. (psi) {*kPa*} = ________

*Industrial standards establish a safety factor of 8:1 in circuitry operating at less than 1,000 psi {*6,895 kPa*}*. Therefore, a conductor having a maximum pressure rating of 8,000 psi {*55,158 kPa*} should be used in circuitry requiring no more than 1,000 psi {*6,895 kPa*}. *This safety factor is reduced to 6:1 for circuitry operating between 1,000 and 2,500 psi {*7,895 *and* 17,500 kPa*} and is further reduced to 4:1 for circuitry operating at more than 2,500 psi {*17,500 kPa*}*. In general, these pressure ranges and safety factors relate to established schedules for nominal pipe sizes. Three schedules are commonly used in hydraulic circuitry. In increasing order of burst strength, these are Schedules 40, 80, and 160. Table 11-1 describes various schedules of standard piping and their characteristics.

The procedure for selection of pipe size (inside diameter) to prevent turbulence in a hydraulic circuit has been well documented throughout this text. That process should be considered the primary step in specifying piping. After determination of the inside diameter for the piping and calculation of the maximum operating system pressure have been made, a suitable nominal size and schedule may be selected. These specifications, along with a determination of the assembly methods to be employed, should adequately describe the conductor.

11.1.1.2 Pipe connections The connection of sections of piping to one another and to other components typically incorporates the use of tapered screw threads. Unlike pipe threads used in household plumbing, *a special-design thread form that eliminates spiral clearance should be used in hydraulic plumbing*. Figure 11-2 compares the National Pipe thread form to the **dry-seal or National Pipe Thread Fluid (NPTF) thread form**.

External threads are normally chased onto the outside of the pipe, while internal threads are tapped into fittings or components. *Fittings are used to change direction, branch, adjust sizes of, or terminate pipe*. Some common pipe fittings are illustrated in Figures 11-3 and 11-4. Like the pipe itself, pipe fittings are typically made from mild steel; however, malleable cast iron may be used for low-pressure environments such as pump inlet lines, drains, and returns.

While tapered threads are the more commonly used method for connecting pipe, they also represent what may be the poorest method. Occa-

TABLE 11-1 Dimensions of Welded and Seamless Wrought Steel Pipe

U.S. Customary by Nominal Size					Metric by Nominal Size				
Fractional (Nominal)	Outside Diameter	Schedule	Wall Thickness	Inside Diameter	Nominal	Outside Diameter	Schedule	Wall Thickness	Inside Diameter
1/8 in.	0.405 in.	40	0.068 in.	0.269 in.	6 mm	1.03 cm	STD	1.73 mm	0.68 cm
1/8 in.	0.405 in.	80	0.095 in.	0.215 in.	6 mm	1.03 cm	XS	2.41 mm	0.54 cm
1/4 in.	0.540 in.	40	0.088 in.	0.364 in.	8 mm	1.37 cm	STD	2.24 mm	0.92 cm
1/4 in.	0.540 in.	80	0.119 in.	0.302 in.	8 mm	1.37 cm	XS	3.02 mm	0.76 cm
3/8 in.	0.675 in.	40	0.091 in.	0.493 in.	11 mm	1.71 cm	STD	2.31 mm	1.25 cm
3/8 in.	0.675 in.	80	0.126 in.	0.423 in.	11 mm	1.71 cm	XS	3.20 mm	1.07 cm
1/2 in.	0.840 in.	40	0.109 in.	0.622 in.	14 mm	2.13 cm	STD	2.77 mm	1.58 cm
1/2 in.	0.840 in.	80	0.147 in.	0.546 in.	14 mm	2.13 cm	XS	3.73 mm	1.38 cm
1/2 in.	0.840 in.	160	0.188 in.	0.464 in.	14 mm	2.13 cm		4.78 mm	1.17 cm
3/4 in.	1.050 in.	40	0.113 in.	0.824 in.	20 mm	2.67 cm	STD	2.87 mm	2.10 cm
3/4 in.	1.050 in.	80	0.154 in.	0.742 in.	20 mm	2.67 cm	XS	3.91 mm	1.89 cm
3/4 in.	1.050 in.	160	0.219 in.	0.612 in.	20 mm	2.67 cm		5.56 mm	1.56 cm
1 in.	1.315 in.	40	0.133 in.	1.049 in.	24 mm	3.34 cm	STD	3.38 mm	2.66 cm
1 in.	1.315 in.	80	0.179 in.	0.957 in.	24 mm	3.34 cm	XS	4.55 mm	2.43 cm
1 in.	1.315 in.	160	0.250 in.	0.815 in.	24 mm	3.34 cm		6.35 mm	2.07 cm
1 1/4 in.	1.660 in.	40	0.140 in.	1.380 in.	33 mm	4.22 cm	STD	3.56 mm	3.51 cm
1 1/4 in.	1.660 in.	80	0.191 in.	1.278 in.	33 mm	4.22 cm	XS	4.85 mm	3.25 cm
1 1/4 in.	1.660 in.	160	0.250 in.	1.160 in.	33 mm	4.22 cm		6.53 mm	2.91 cm
1 1/2 in.	1.900 in.	40	0.145 in.	1.610 in.	39 mm	4.83 cm	STD	3.68 mm	4.09 cm
1 1/2 in.	1.900 in.	80	0.200 in.	1.500 in.	39 mm	4.83 cm	XS	5.08 mm	3.81 cm
1 1/2 in.	1.900 in.	160	0.281 in.	1.338 in.	39 mm	4.83 cm		7.14 mm	3.40 cm
2 in.	2.375 in.	40	0.154 in.	2.067 in.	49 mm	6.03 cm	STD	3.91 mm	5.25 cm
2 in.	2.375 in.	80	0.218 in.	1.939 in.	49 mm	6.03 cm	XS	5.54 mm	4.92 cm
2 in.	2.375 in.	160	0.344 in.	1.687 in.	49 mm	6.03 cm		8.74 mm	4.28 cm
2 1/2 in.	2.875 in.	40	0.203 in.	2.469 in.	59 mm	7.30 cm	STD	5.16 mm	6.27 cm
2 1/2 in.	2.875 in.	80	0.276 in.	2.323 in.	59 mm	7.30 cm	XS	7.01 mm	5.90 cm
2 1/2 in.	2.875 in.	160	0.375 in.	2.125 in.	59 mm	7.30 cm		9.53 mm	5.39 cm
3 in.	3.500 in.	40	0.216 in.	3.068 in.	75 mm	8.89 cm	STD	5.49 mm	7.79 cm
3 in.	3.500 in.	80	0.300 in.	2.900 in.	75 mm	8.89 cm	XS	7.62 mm	7.37 cm
3 in.	3.500 in.	160	0.438 in.	2.624 in.	75 mm	8.89 cm		11.13 mm	6.67 cm

sionally, components or conductors must be removed from circuitry for inspection, repair, or replacement. When tapered pipe threads are reassembled, they must be advanced further into mating components to achieve sufficient compression to reseal. Unfortunately, pipe runs from one location to another in a straight line. When the threaded fasteners are advanced further than normal, the distance between these two locations will need to be foreshortened. Typically, hydraulic components are rigidly secured and something "has to give." Most often, it is the threads. Needless to say, stripped threads do not seal very well. Furthermore, the po-

FIGURE 11-2

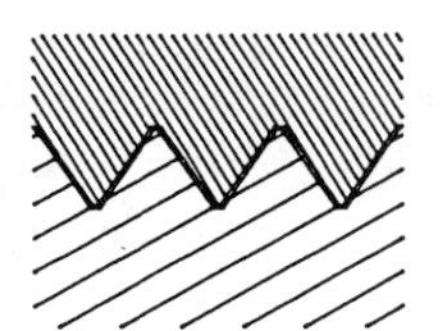

The dry-seal thread form incorporates sharp "V" crests and flat roots, allowing for no spiral clearance.

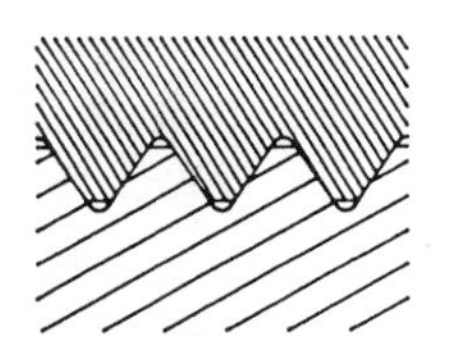

The standard thread form incorporates flat crests and rounded roots. The flanks engage first; this allows for spiral clearance and potential leakage.

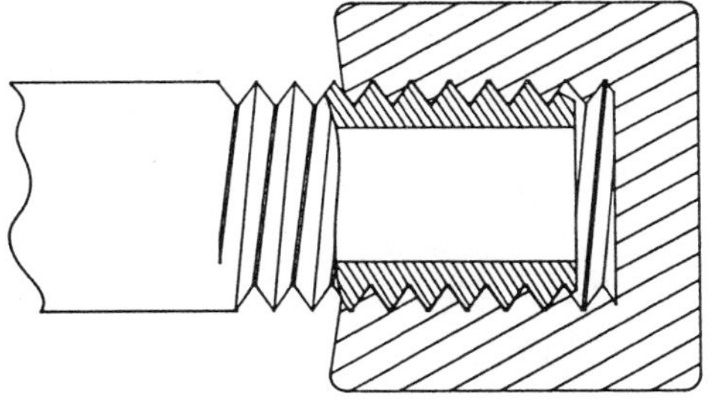

Dry-seal threads may be assembled with no need for pipe "dope" or teflon tape as a sealing device.

FIGURE 11-3

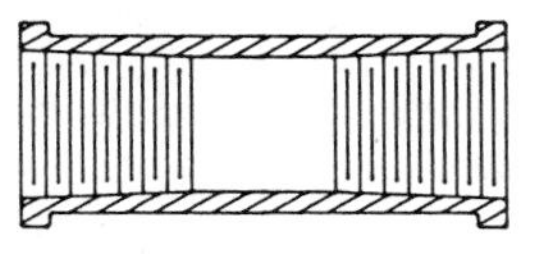

Coupling

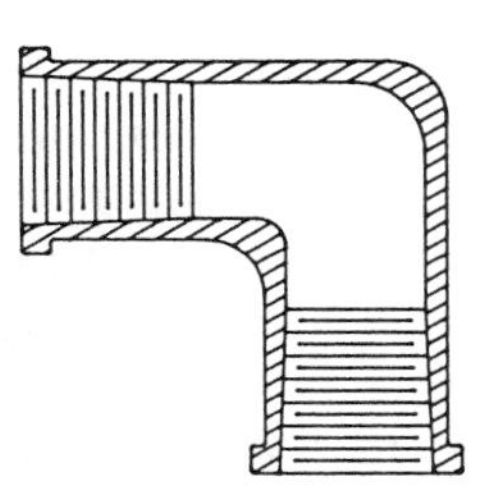

Elbow

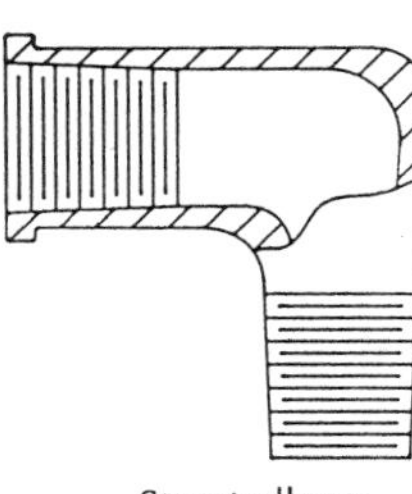

Street elbow

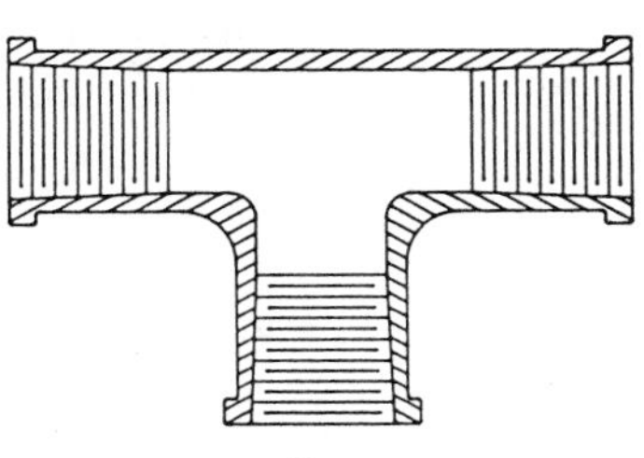

Tee

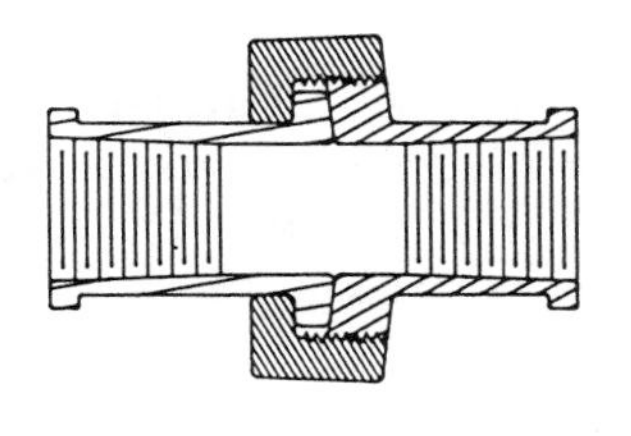

Union

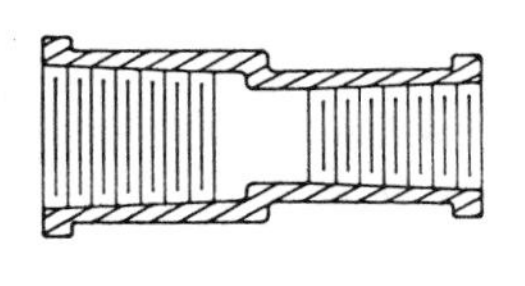

Reducing bushing

FIGURE 11-4

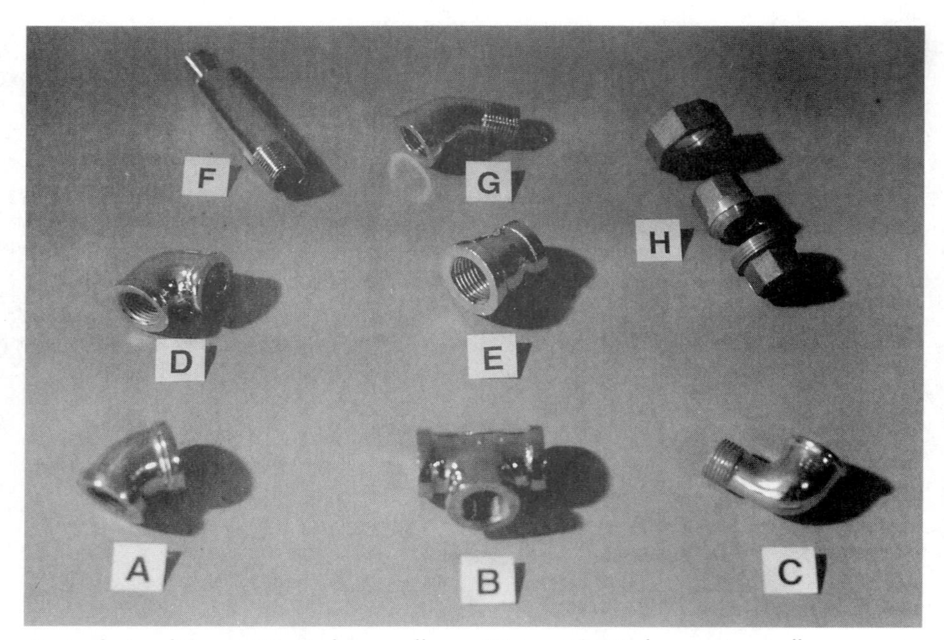

Typical pipe fittings: **(A)** 45-degree elbow, **(B)** tee, **(C)** 90-degree street elbow, **(D)** 90-degree elbow, **(E)** reducing coupling, **(F)** nipple, **(G)** 45-degree street elbow, **(H)** union (disassembled). (*Photograph by the author.*)

tential for fluid contamination with metallic chips increases each time tapered pipe threads are reassembled.

Most of the deficiencies associated with tapered pipe threads can be overcome by using **straight threads**. Even when reassembled, these devices typically maintain a constant distance and do not contaminate the fluid stream with metallic chips. However, since the straight thread has minimal interference with mating components, they do not seal. Therefore, auxiliary seals (typically O-rings) need to be incorporated into the fastener package to prohibit leakage. A typical straight-thread fastener package is illustrated by the flange connector in Figure 11-5.

Even with the advantages offered with straight threads, problems are still inherent with threaded fasteners in general. Primarily, threaded fasteners weaken the pipe by reducing its wall thickness at a very critical point. Surges and shock from rapid changes of fluid flow direction at high pressure are greatest at the connection between components and conductors. Even if the interface between mating thread forms were perfect (which they are not), the extended portion of the thread beyond the interface would leave an unprotected area. Pressure surges that may be as great as 50% over operating pressure may burst these thinned areas of threaded connectors through fatigue caused by cyclic loading.

FIGURE 11-5

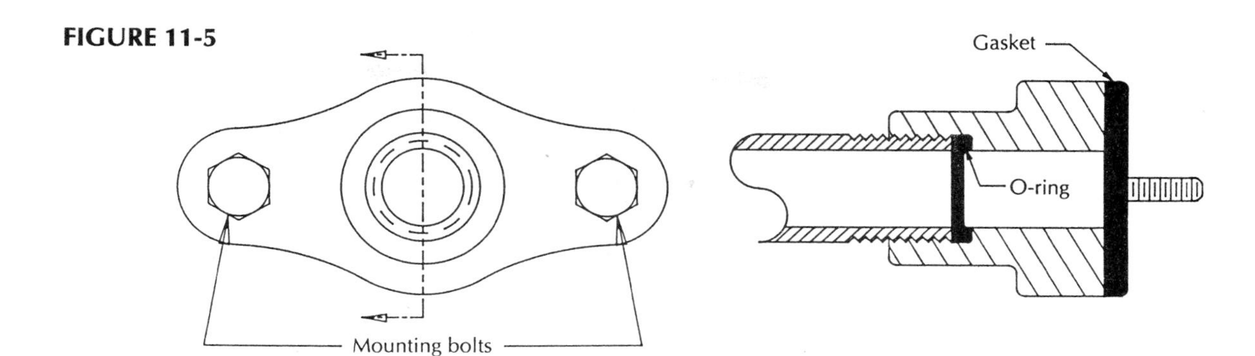

The best sealing mechanism and the ultimate burst strength in a pipe connection is realized with **weld joints** and welded assembly is common in pipe having a nominal diameter greater than 1¼ in. {*30 mm*}. However, care must be taken to ensure that the heat generated through welding does not affect mating parts or damage seals. If the connection must be made directly to a hydraulic component, it should be thoroughly disassembled before welding. Also, the assembly should be thoroughly cleaned after welding to remove any scaling or oxidation that may infiltrate the fluid stream. The preferred method for welding the assembly is to weld the pipe to a flange, then bolt the flange to the component using a gasket seal at the interface (see Figure 11-6).

The welded assembly has disadvantages, too. The most obvious is the equipment and skill necessary to make a good weld joint. Furthermore, the welded assembly may be impractical on pipes smaller than 1 in. in diameter. Disassembly of weld-joint circuitry is nearly impossible without destroying conductors or components. Therefore, circuitry that primarily relies on weld-joint assembly typically incorporates threaded fasteners also for the sake of disassembly.

FIGURE 11-6

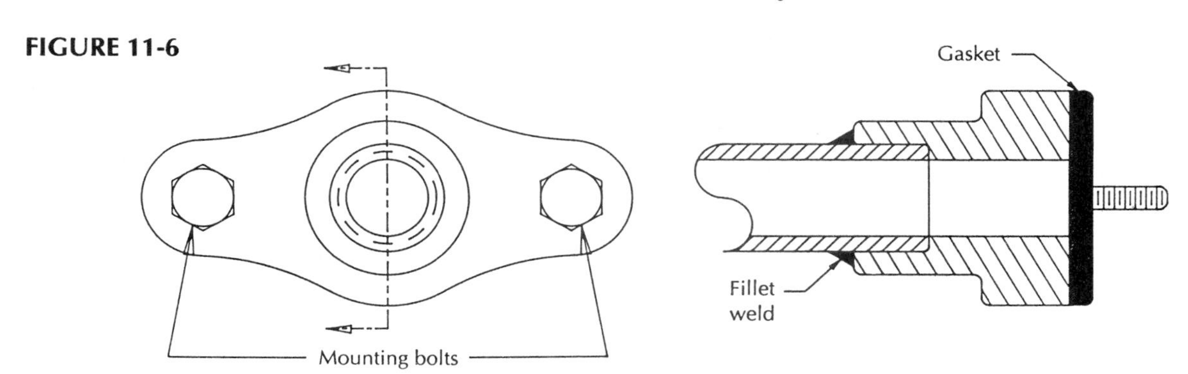

11.1.2 Tubing

Tubing is similar in mechanical characteristics to piping. However, the thin-walled construction, in combination with special alloy constituency, affords a degree of flexibility in tubing. This characteristic allows tubing to be bent to cause a major change in the direction of fluid flow. Still, care must be taken to assure that a gradual bend is made so that folding or collapsing does not occur. *Typically, the radius of the bend should be no less than six times the inside diameter of the tube.*

The most common metallic tubing is made from a copper alloy and is widely used as household water supply line. Copper tubing should not be used in hydraulic circuitry because of chemical reactions between the tubing and various oils and a tendency to work harder and burst through fatigue under high pressure. Other metals such as annealed plain carbon and stainless steels, as well as aluminum, are typically used to produce hydraulic tubing.

11.1.2.1 Sizing tube Since tubing is made from a homogeneous material, specifications for tube follow the same general rules as sizing for pipe. However, tubing typically has thinner walls; and *the nominal size for tubing refers to the outside diameter rather than an approximate inside diameter*. Furthermore, a wide variety of schedules do not exist for tubing. Therefore, tubing should only be used in relatively low-pressure circuitry. Its maximum operating pressure may be calculated using Barlow's Formula and a safety factor of 6:1.

Problem 11-2

A hydraulic circuit is operating under a system pressure of 950 psi {*6,550 kPa*}. The circuit is designed to use 5/8-in. {*16-mm*} type 18-8 stainless steel tubing having a tensile strength of 85,000 psi {*586,054 kPa*}. What is the minimum wall thickness for this tubing if a 6:1 safety factor is necessary?

a. Safety Factor Pressure (psi) {*kPa*} = ________

b. Tubing W.T. (in.) {*mm*} = ________

11.1.2.2 Tubing specifications Tubing may be extruded into one continuous conductor or may be made by roll forming of sheet metal sealed by a continuous weld. Obviously, the former has less potential failure at a given pressure. Beyond this construction factor, the ability of tubing to withstand pressure is affected by the metal from which it is manufactured. Ten-point carbon steel is the material used most often for hydraulic tubing. The fully annealed AISI 1010 steel has a tensile rating of 55,000 psi {*379,212 kPa*}. Although this steel tube bends easily, it may be susceptible to oxidation and other corrosive actions. Stainless steel alloys of the 300 series, containing nickel and chromium, resist oxidation and corrosion. Fully annealed, AISI 304 (Type 18-8) stainless steel, having a tensile rating of 85,000 psi {*586,054 kPa*}, will produce tubing with higher burst strength for any nominal size than plain carbon tubing. However, it is expensive and generally more difficult to bend. Tubing may also be manufactured from aluminum alloys. Two alloys, 5052 (tensile strength of 38,000 psi {*262,000 kPa*}) and 6061 (tensile strength of 45,000 psi {*310,264 kPa*}), are typically used to manufacture aluminum tubing. Easily bendable aluminum tubing has good corrosion resistance and is becoming increasingly popular in use for hydraulic circuitry.

11.1.2.3 Sizes of tubing *Hydraulic tubing may be specified by fractional equivalent of outside diameter in the U.S. customary designations or by millimeter equivalents that relate to outside diameter sizes in the metric specification.* In general, a nominal-size tubing will be available with a limited variety of wall thicknesses. These varieties may be common to many as the types "K", "L," and "M" copper tubing. Table 11-2 summarizes sizes and wall thicknesses for hydraulic tubing.

TABLE 11-2 Tube Specifications

U.S. Customary by Outside Diameter				Metric by Outside Diameter			
Fractional (Nominal)	Decimal	Wall Thickness	Inside Diameter	Nominal	Decimal	Wall Thickness	Inside Diameter
$^1/_8$ in.	0.125 in.	0.035 in.	0.055 in.	4 mm	0.4 cm	0.5 mm	0.3 cm
$^3/_{16}$ in.	0.1875 in.	0.035 in.	0.118 in.	6 mm	0.6 cm	1.0 mm	0.4 cm
		0.035 in.	0.180 in.			1.5 mm	0.3 cm
$^1/_4$ in.	0.250 in.	0.049 in.	0.152 in.			1.0 mm	0.6 cm
		0.065 in.	0.120 in.	8 mm	0.8 cm	1.5 mm	0.5 cm
		0.035 in.	0.243 in.			2.0 mm	0.4 cm
$^5/_{16}$ in.	0.3125 in.	0.049 in.	0.215 in.			1.0 mm	0.8 cm
		0.065 in.	0.183 in.	10 mm	1.0 cm	1.5 mm	0.7 mm
		0.035 in.	0.305 in.			2.0 mm	0.6 mm
$^3/_8$ in.	0.375 in.	0.049 in.	0.277 in.			1.0 mm	1.0 cm
		0.065 in.	0.245 in.	12 mm	1.2 cm	1.5 mm	0.9 cm
		0.035 in.	0.430 in.			2.0 mm	0.8 cm
$^1/_2$ in.	0.500 in˙	0.049 in.	0.402 in.	14 mm	1.4 cm	2.0 mm	1.0 cm
		0.065 in.	0.370 in.	15 mm	1.5 cm	1.5 mm	1.2 cm
		0.095 in.	0.310 in.			2.0 mm	1.1 cm

TABLE 11-2 *(continued)*

U.S. Customary by Outside Diameter			
Fractional (Nominal)	**Decimal**	**Wall Thickness**	**Inside Diameter**
⁵⁄₈ in.	0.625 in.	0.035 in.	0.555 in.
		0.049 in.	0.527 in.
		0.065 in.	0.495 in.
		0.095 in.	0.435 in.
¾ in.	0.750 in.	0.049 in.	0.652 in.
		0.065 in.	0.620 in.
		0.109 in.	0.532 in.
⁷⁄₈ in.	0.875 in.	0.049 in.	0.777 in.
		0.065 in.	0.745 in.
		0.109 in.	0.657 in.
1 in.	1.000 in.	0.049 in.	0.902 in.
		0.065 in.	0.870 in.
		0.120 in.	0.760 in.
1¼ in.	1.250 in.	0.065 in.	1.120 in.
		0.095 in.	1.060 in.
		0.120 in.	1.010 in.
1½ in.	1.500 in.	0.065 in.	1.370 in.
		0.095 in.	1.310 in.
		0.134 in.	1.232 in.
1¾ in.	1.750 in.	0.065 in.	1.620 in.
		0.095 in.	1.560 in.
		0.134 in.	1.482 in.
2 in.	2.000 in.	0.065 in.	1.870 in.
		0.095 in.	1.810 in.
		0.134 in.	1.732 in.

Metric by Outside Diameter			
Nominal	**Decimal**	**Wall Thickness**	**Inside Diameter**
16 mm	1.6 cm	2.0 mm	1.2 cm
		3.0 mm	1.0 cm
18 mm	1.8 cm	1.5 mm	1.5 cm
20 mm	2.0 cm	2.0 mm	1.6 cm
		2.5 mm	1.5 cm
		3.0 mm	1.4 cm
22 mm	2.2 cm	1.0 mm	2.0 cm
		1.5 mm	1.9 cm
		2.0 mm	1.8 cm
25 mm	2.5 cm	3.0 mm	1.9 cm
		4.0 mm	1.7 cm
28 mm	2.8 cm	2.0 mm	2.4 cm
		2.5 mm	2.3 cm
30 mm	3.0 cm	3.0 mm	2.4 cm
		4.0 mm	2.2 cm
35 mm	3.5 cm	2.0 mm	3.1 cm
		3.0 mm	2.9 cm
38 mm	3.8 cm	4.0 mm	3.0 cm
		5.0 mm	2.8 cm
42 mm	4.2 cm	2.0 mm	3.8 cm
		3.0 mm	3.6 cm

11.1.2.4 Tube fittings Because of the thin-walled nature of tubing, integral threads are not chased onto it. With the ease of bending inherent in tubing, the need for elbows to change direction may be eliminated. However, fittings are still necessary for branching purposes and to attach tubing to hydraulic components.

Fittings that slip over the outside of the conductor may be brazed, silver soldered; or, in heavy-wall sections, welded. In other cases, **compres-**

FIGURE 11-7

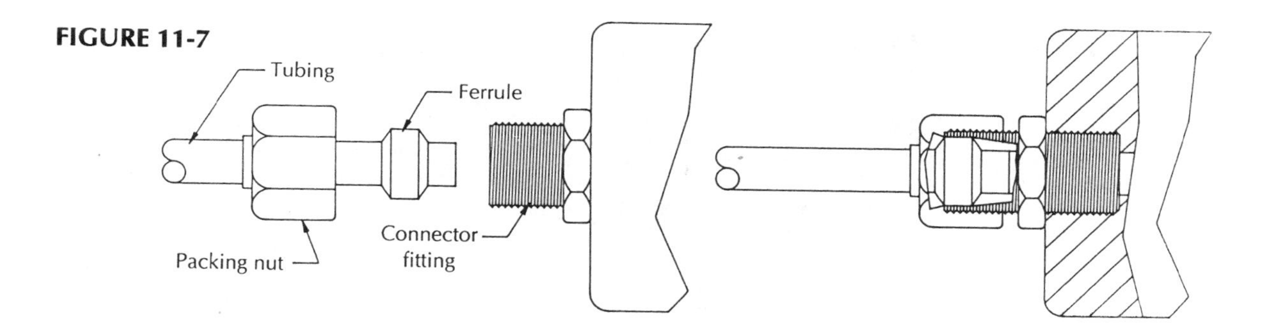

sion fittings may be used. The compression fitting incorporates a threaded packing nut in combination with a ferrule that slides over the tube. The packing nut holds the tube in location, while the compressed ferrule creates the fluid seal. The compression fitting allows for easy assembly and disassembly of tubing (see Figure 11-7).

The **flared fitting** is probably the best and most often used connector for tubing. Taking advantage of the inherent malleability of the tube, a tapered mandrel may be forced into the end of a clamped tube to form a conically shaped bell. When mated with a fitting having a matched conical seat, a seal may be created by drawing the flares together with a packing nut. Two flare angles are commonly used for hydraulic tubing: 37 and 45 degrees. In either case, the flare fitting produces a relatively inexpensive tube fitting that may be easily disassembled and reassembled (see Figures 11-8 and 11-9). A modern variation on the flare fitting includes an O-ring boss on the flare to further reduce leakage and facilitate disassembly–assembly cycles.

11.1.3 Hosing

The lowest degree of rigidity for hydraulic conductors is evident in **hosing**. Although metallic reinforcement may be used in hoses, the basic construction is of plastic or rubber. This accounts for the flexibility and lightness of

FIGURE 11-8

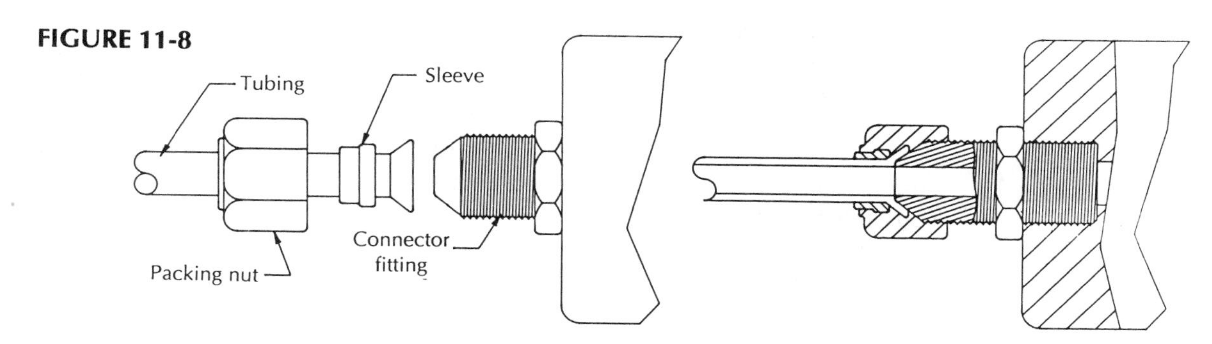

FIGURE 11-9

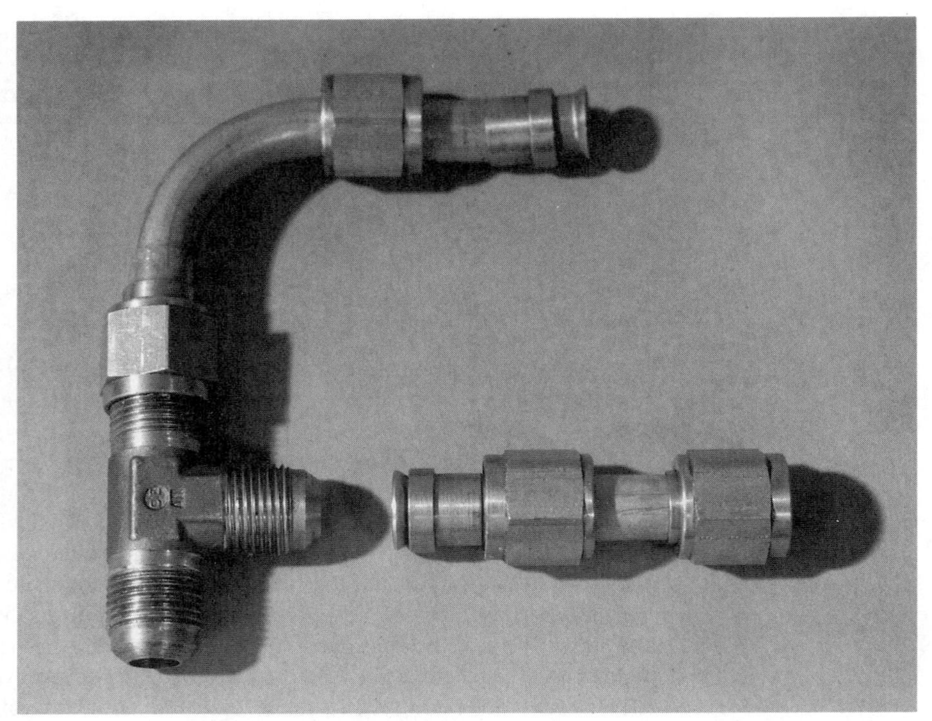

Flare fitting packages are commonly used with tubing. (*Photograph by the author.*)

hoses. Some simple hoses are extruded from crystalline thermoplastics such as nylon or polyethylene. More commonly, in hydraulic applications, the crystalline material is used to provide fluid compatibility; and layers of materials are added to the outside of this inner tube as reinforcement.

11.1.3.1 Hose reinforcement **Reinforcement** is wound around the core of the hosing to provide added strength to the hosing package. Not only will this reinforcement increase the burst strength of the hose, but it will also accommodate compression and expansion surges in the operating circuit. Typically, *the greatest degree of burst strength rating may be achieved with a braided wire winding*, while *ultimate surge protection is provided by spiral winding*. Surges in the braided weave hose package often wear the overlapped reinforcement bundles, causing fraying and possible fracture. For this reason, hosing packages are available that include multiple layers of spiral and braided reinforcement separated by extruded plastic or rubber layers that reduce contact wear (see Figure 11-10).

FIGURE 11-10

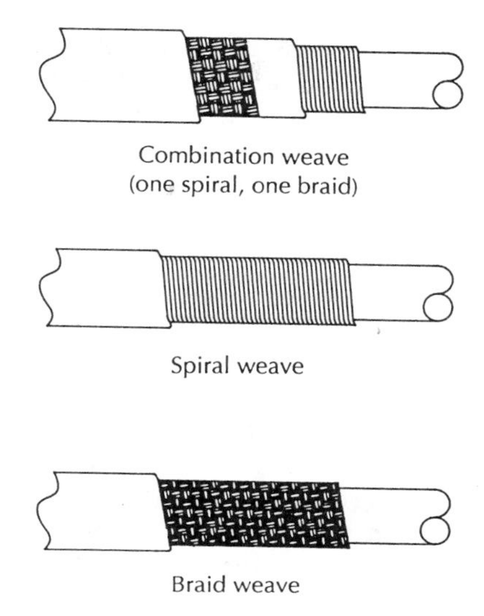

Combination weave
(one spiral, one braid)

Spiral weave

Braid weave

Reinforcements have traditionally been made from fine steel wire bundles of two or more filaments. These bundles are then woven into double plats as they traverse the hose length. Although steel wire remains the most popular reinforcement material, it has lately been combined with synthetic and natural fiber layers of nylon, rayon, or cotton. Exotic synthetic fibers such as aramids and carbon graphite are currently available. The major advantage of the natural and synthetic fibers is weight reduction. The hose package is typically finished with a skin cover extruded over the outermost reinforcement. Although the primary function of the inner core was to resist chemical attack and elevated temperature of the fluid, *the primary functions of the outer skin are to withstand environmental attack and provide resistance to physical abuse*. Synthetic rubbers and elastomers are typically used as outer skins on working and drain lines. Impregnated textile covers may be used on inlet lines.

11.1.3.2 Hose sizing *Hoses, somewhat like pipes, are nominally sized by inside diameters*. The SAE's 100R series designations are typically used in the industry. Although compliance to its standards are not mandatory, most hose manufacturers, at the least, relate their designations to the SAE J517 standards. Hoses generally range from 3/16 in. {*5* mm} to 3 in. {*76 mm*} in diameter. Hoses (using the U.S. customary units) may be described by a number system that specifies incremental 1/16-in. increases in diameter or by fractional equivalent. Thus, a **–6** hose would have a 0.375-in. inside diameter; and a **–8** hose would have a 0.5-in. inside diameter. This dash designation system holds true throughout the SAE's 100R standards except for the 100R5 series, which generally increases the dash rating numbers. For example, a $^{3}/_{16}$-in. nominal hose in the 100R5 standard is designated as a –4, while a $^{1}/_{2}$-in. nominal hose is designated –10. This inconsistency is primarily to accommodate some $^{1}/_{32}$-in. incrementally sized hoses in this series. Throughout their range of specification, metric sizes for hoses remain as inside diameter in millimeters. Currently, hose manufacturers in the United States are including both dash/fractional and metric sizes on their **laylines** (printing on the outside of the hose, like those shown in Figure 11-11).

Unlike other fluid conductors, hose packages are not made from homogeneous materials. Therefore, the theoretical formulation used to describe pressure ratings in piping and tubing is not applicable in hosing. Rather, hose manufacturers, after extensive testing, will determine a rated burst strength for a hose package.

The **proof pressure** (rating pressure) for each hose package will be specified at 50% of **minimum burst pressure**. However, *hoses should be used at an* ***operating pressure*** *of no more than 25% of minimum burst pressure or one-half of the manufacturer's proof pressure*. With a variety

FIGURE 11-11

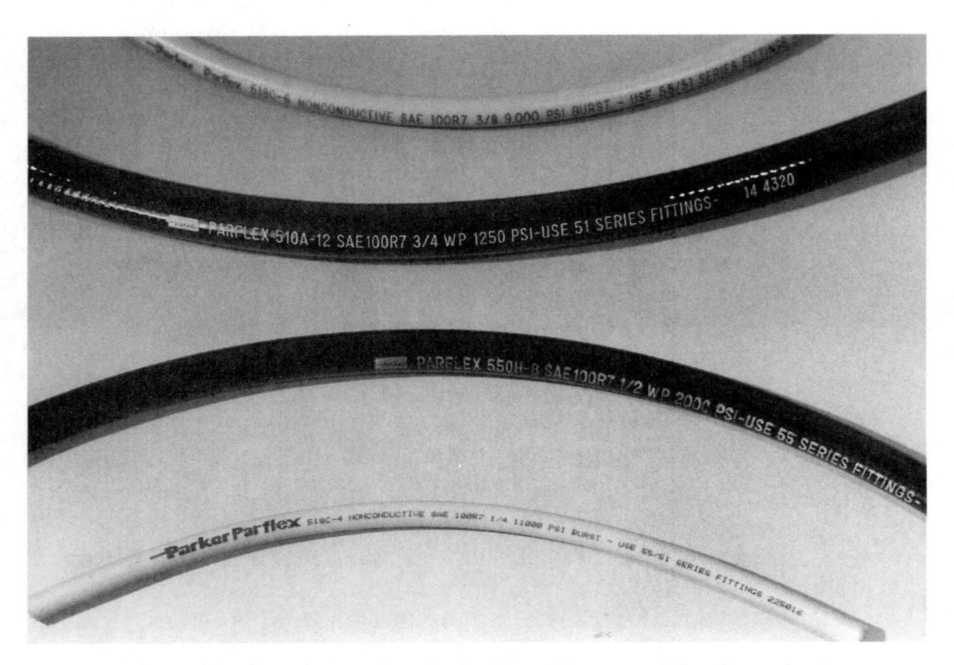

Typical layline designation for hosing. (*Photograph by the author.*)

of hosing materials and multiple layers of reinforcement, a wide range of permissible operating pressures will be available for each nominal hose size. Examination of the SAE's J517 and 100R hose packages and their descriptions, constructions, and minimum burst pressure ratings are summarized in Table 11-3 as U.S. Customary Standards and in Table 11-4 as Metric Standards.

11.1.3.3 Hose fittings The connectors for assembling hoses to components and to other hoses are similar to connectors for piping and tubing. In many cases, piping and tubing connectors are used with hosing; however, an **adaptor** that bridges the interface must be made to the end of the hose. These adaptors typically rely upon a metallic screw thread that shears into the inner tube or outer cover of the hose. The threaded mechanism prevents linear movement of the adaptor, and the self-threading mechanism biting into the rubber or plastic hose creates a seal that prevents spiral clearance around the thread–cover interface.

Three assembly methods are typically used to attach adaptors. The **ferrule locking** mechanism uses a threaded shank into the core of the tubing and a slip-fit ring (ferrule) over the outer cover. When the ferrule is forced over the assembled threaded mechanism, it compresses the hose and tightens the internal seal. Most ferrule locking mechanisms also crimp or swage the ferrule to further secure the outer ring. The **skive-type** adaptor uses both internal and external self-threading rings and shanks. This method requires the removal (skiving) of the outer hose cover. The external threading mechanism (inside the ring) then occurs with the reinforcement layer. A newer method for assembling hose adaptors eliminates the need for skiving. The **no-skive** adaptor uses both external and internal self-threading mechanisms. However, hoses with special thin-walled covers (designated as the "T" types in 100R specifications) allow direct attachment of the ring without skiving. The thread form on the ring is designed to penetrate the outer hose cover and bites into the reinforcement. This eliminates the need for an accurate skive length and the fraying of the reinforcement often associated with skive-type adaptors. Furthermore, the self-sealing nature of the no-skive ring inhibits the passage of air or other corrosives that may attack the reinforcement.

The adaptor shank not only secures and seals the internal passage of fluid but also contains the threaded connection for attachment to other components. This connection may be either internal or external threading devices and have pipe or straight threads. The connector may be solidly attached to the shank, or it may be free to rotate **(swivel connector)**. Figure 11-12 illustrates three assembly methods for hose adaptors and variations of connections (see also Figure 11-13).

TABLE 11-3 SAE 100R U.S. Customary Series Hose Specifications

Type		Tube Material	Reinforcement	Braided Layers	Spiral Layers	Cover Material	Skive or No-Skive	3/16 (-3) [0.188]	¼ (-4) [0.250]	5/16 (-5) [0.312]	3/8 (-6) [0.375]	½ (-8) [0.500]	5/8 (-10) [0.625]	¾ (-12) [0.750]	7/8 (-14) [0.875]	1 (-16) [1.000]	1-1/8 (-18) [1.125]	1¼ (-20) [1.250]	1½ (-24) [1.500]	2 (-32) [2.000]	2½ (-40) [2.500]	3 (-48) [3.000]	3½ (-56) [3.500]	4 (-64) [4.000]
		Descriptions						Minimum Burst Pressure (in 1000 Psi) by Inside Diameter—Nominal (Dash) [Decimal] in Inches																
100R1	A	R	S	1		OWR	S	12	11	10	9	8	6	5	4.5	4		2.5	2	1.5				
	AT	R	S	1		OWR	NS	12	11	10	9	8	6	5	4.5	4		2.5	2	1.5				
100R2	A	R	S	2		OWR	S	20	20	17	16	14	11	9	8	8		6.5	5	4.5	4			
	B	R	S	1	2	OWR	S	20	20	17	16	14	11	9	8	8		6.5	5	4.5	4			
	AT	R	S	2		OWR	NS	20	20	17	16	14	11	9	8	8		6.5	5	4.5	4			
	BT	R	S	1	2	OWR	NS	20	20	17	16	14	11	9	8	8		6.5	5	4.5	4			
100R3		R	Y	2		OWR	S	6	5	4.8	4.5	4	3.5	3	2.25			1.5						
100R4		R	T/M	1T	1M	OWR	S							1.2		1		0.8	0.6	0.4	2.5	2.25	1.8	1.4
100R5*		R	T/S	2T/1S		IT	S	12	12	9		7	6		3.2		2.5							
100R6		R	T	1		OWR	S	2	1.6	1.6	1.6	1.6	1.4											
100R7**		TP	SYN	**	**	HWT	**	12	11	10	9	8	6	5		4								
100R8**		TP	SYN	**	**	HWT	**	20	20		16	14	11	9		8								
100R9	A	R	S		4	OWR	S				18	16		12		12		10	8	8				
	AT	R	S		4	OWR	NS				18	16		12		12		10	8	8				
100R10	A	R	HS		4	OWR	S	20	17.5		15	12.5		10		8		6	5	5				
	AT	R	HS		4	OWR	NS	20	17.5		15	12.5		10		8		6	5	5				
100R11		R	HS		6	OWR	S	50	45		40	30		25		20		14	12	12				
100R12		R	HS		4	OWR	S				16	16		16		16		12	10	10				

*100R5 series sizes are not valid for dash size specifications.

**100R7 and 100R8 series hoses have reinforcement classified as being enough synthetic fiber to sufficiently reach pressure standards.

Key: R, oil-resistant synthetic rubber; TP, thermoplastic; S, steel wire; Y, textile yarn; SYN, synthetic fiber; HS, heavy steel wire; OWR, oil and weather resistant rubber; IT, oil- and mildew-resistant synthetic rubber impregnated; HWT, hydraulic fluid and weather-resistant thermoplastic; S, skive; NS, no-skive.

TABLE 11-4 SAE 100R Metric Series Hose Specifications

Descriptions							Minimum Burst Pressure (in 1000 Kpa) by Inside Diameter—Millimeter Nominal {Centimeter}																
Type	Tube Material	Reinforcement	Braided Layers	Spiral Layers	Cover Material	Skive or No-Skive	4.8 {0.5}	6.4 {0.7}	7.9 {0.8}	9.5 {1.0}	12.7 {1.3}	15.9 {1.6}	19.0 {1.9}	22.2 {2.2}	25.4 {2.5}	28.6 {2.9}	31.8 {3.2}	38.1 {3.8}	50.8 {5.1}	63.5 {6.4}	76.2 {7.6}	88.9 {8.9}	101.6 {10.2}
100R1 A	R	S	1		OWR	S	82.7	75.8	68.9	62	55.2	41.4	34.5	31	27.6		17.2	13.8	10.3				
AT	R	S	1		OWR	NS	82.7	75.8	68.9	62	55.2	41.4	34.5	31	27.6		17.2	13.8	10.3				
100R2 A	R	S	2		OWR	S	137.9	137.9	117.2	110.3	96.5	75.8	62	55.2	55.2		44.8	34.5	31	27.6			
B	R	S	1	2	OWR	S	137.9	137.9	117.2	110.3	96.5	75.8	62	55.2	55.2		44.8	34.5	31	27.6			
AT	R	S	2		OWR	NS	137.9	137.9	117.2	110.3	96.5	75.8	62	55.2	55.2		44.8	34.5	31	27.6			
BT	R	S	1	2	OWR	NS	137.9	137.9	117.2	110.3	96.5	75.8	62	55.2	55.2		44.8	34.5	31	27.6			
100R3	R	Y	2		OWR	S	41.4	34.5	33.1	31	27.6	24.1	20.7	15.5			10.3						
100R4	R	T/M	1T	1M	OWR	S							8.3		6.9		5.5	4.1	2.8	1.7	1.5	1.2	1
100R5	R	T/S	2T/1S		IT	S	82.7	82.7	62		48.3	41.4		22.1		17.2							
100R6	R	T	1		OWR	S	13.8	11	11	11	11	9.7											
100R7*	TP	SYN	*	*	HWT	*	82.7	75.8	68.9	62	55.2	41.4	34.5		27.6								
100R8*	TP	SYN	*	*	HWT	*	137.9	137.9		110.3	96.5	75.8	62		55.2								
100R9 A	R	S		4	OWR	S				124.1	110.3		82.7		82.7		68.9	55.2	55.2				
AT	R	S		4	OWR	NS				124.1	110.3		82.7		82.7		68.9	55.2	55.2				
100R10 A	R	HS		4	OWR	S	275.8	241.3		206.8	172.4		137.9		110.3		82.7	68.9	68.9				
AT	R	HS		4	OWR	NS	275.8	241.3		206.8	172.4		137.9		110.3		82.7	68.9	68.9				
100R11	R	HS		6	OWR	S	344.7	310.3		275.8	206.8		172.4		137.9		96.5	82.7	82.7				
100R12	R	HS		4	OWR	S				110.3	110.3		110.3		110.3		82.7	69	69				

*100R7 and 100R8 series hoses have reinforcement classified as being enough synthetic fiber to sufficiently reach pressure standards.

Key: R, oil-resistant synthetic rubber; TP, thermoplastic; S, steel wire; Y, textile yarn; SYN, synthetic fiber; HS, heavy steel wire; OWR, oil and weather resistant rubber; IT, oil- and mildew-resistant synthetic rubber impregnated; HWT, hydraulic fluid and weather-resistant thermoplastic; S, skive; NS, no-skive.

FIGURE 11-12

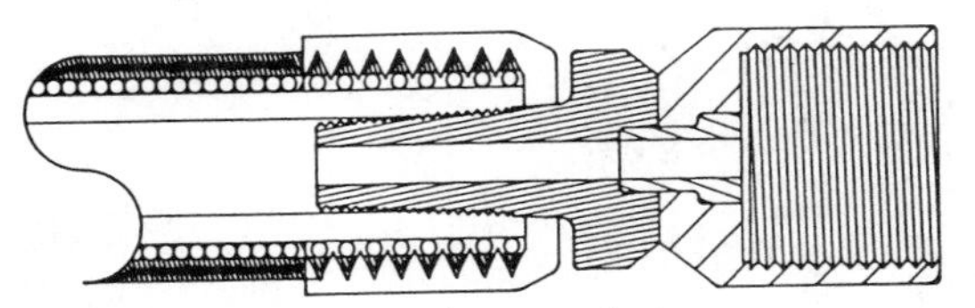

No-skive adaptor, swivel-type, straight thread, female connector

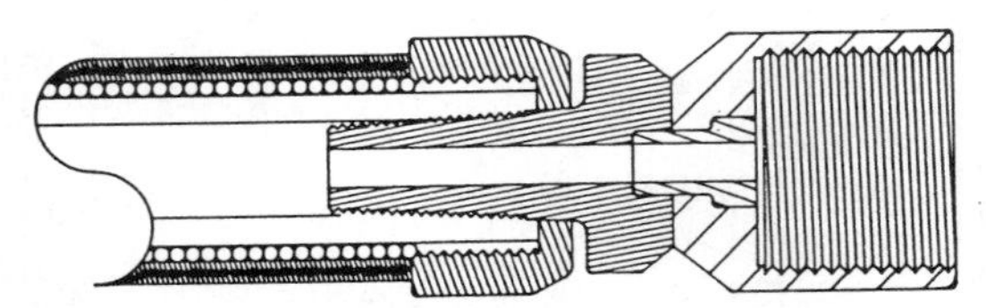

Skive adaptor, swivel-type, straight thread, female connector

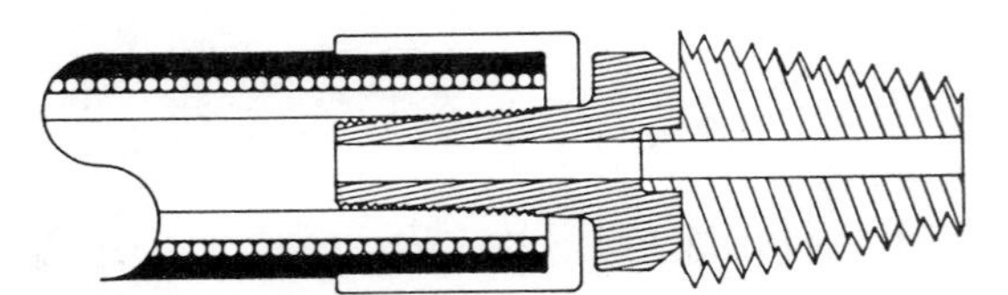

Ferrule adaptor, nonswivel-type, tapered thread, male connector

11.1.3.4 Hose disconnects One of the great advantages of hosing lies in its ability to be easily assembled and disassembled. This characteristic has been even further advanced by the development of small-package, integral check valve connectors called **quick disconnects**. The quick disconnect may be attached to the hose, the hydraulic component, or both. In either case, it commonly contains a check valve that normally excludes the outlet passage of fluid. In some cases, the quick disconnect may have a check in only one side or no check included. A matched pair of female **(socket)** and male **(plug)** connectors is necessary to allow an override of this normally closed condition. When assembled, the socket slides over the plug and is secured into position by a locking ring (on the socket) and ball/seat locators. When secured, internal sealing devices exclude external fluid leakage. In the assembled condition, the check elements physi-

FIGURE 11-13

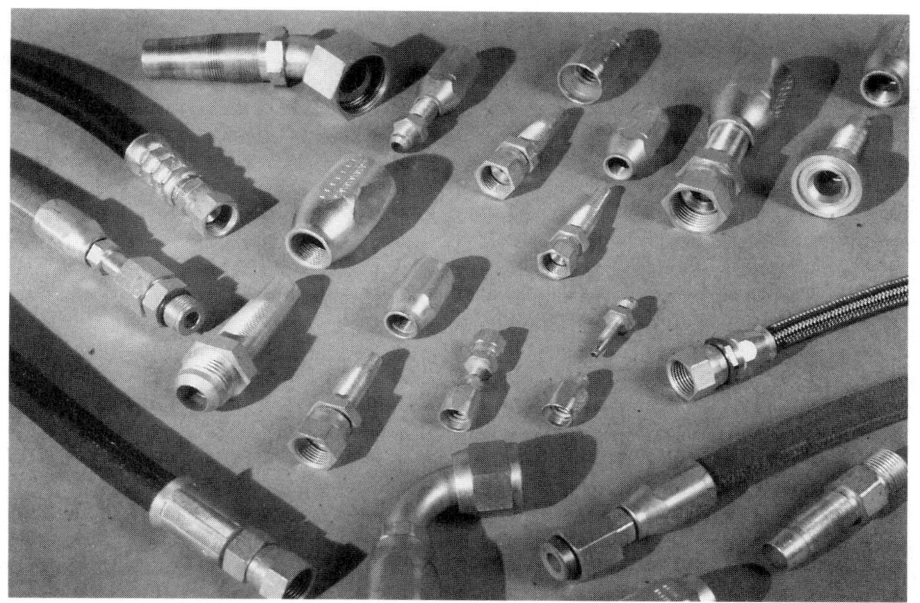

Various assembly methods and hose adaptors. (*Photograph by the author.*)

cally interfere with one another and overcome the springs holding the checks normally closed. This allows free passage of fluid through the quick disconnect only when the socket and plug are mated (see Figures 11-14 and 11-15).

11.2 SELECTION OF FLUID CONDUCTORS

The selection of which type of fluid conductor to use in a hydraulic circuit is typically based on pressure requirements, necessary support for hydraulic components, and service requirements. In any case, the primary consideration is to select a conductor of sufficient size (internal diameter) to allow streamlined fluid flow. Since all types of conductors are not available in an infinite variety of internal diameters, some conductors are eliminated from consideration because of capacity requirements. Concurrently, the selected size conductor must be able to withstand the maximum pressure produced by the circuit. Beyond these considerations, various factors interact to preclude or include any given conductor into a hydraulic circuit.

Fluid compatibility with the conductor is an important consideration. Nearly all hydraulic conductors are compatible with neat petroleum fluids. However, hosing may be attacked by various synthetic fluids used for high-temperature applications. Obviously, the use of hosing in high-temperature

FIGURE 11-14

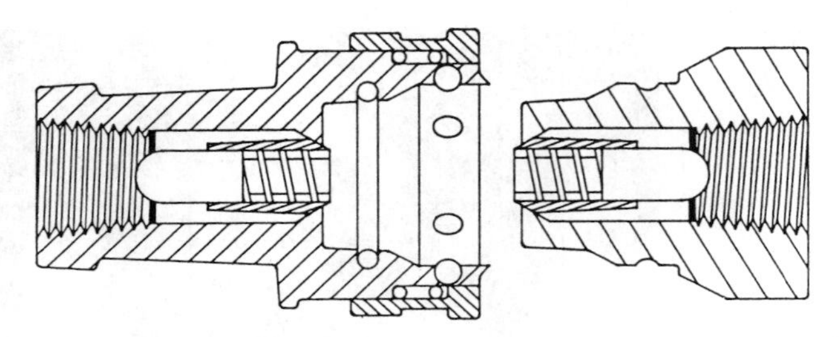

When disassembled, the poppet-type check valves in the center of the disconnects block supply flow.

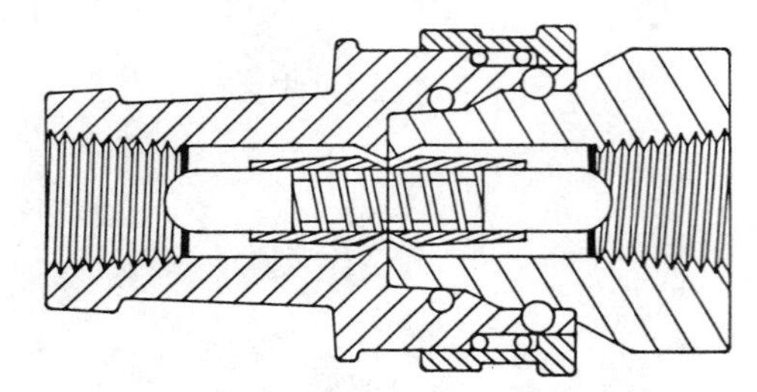

When assembled, the poppet checks interfere with one another and are forced off their seats. This allows free fluid flow through the disconnect.

applications is, in itself, questionable. However, water-based and water-emulsion fluids promote oxidation in piping and tubing. If water-type fluids are to be used, either hosing or stainless steel or aluminum tubing may be the preferred conductor.

Piping will provide the maximum support for hydraulic components. This factor alone should not preclude the use of hosing or tubing because hydraulic components can (and, in most cases, should) be rigidly attached to other supporting members and not rely on the conductor for support. However, the rigid mounting of components presents another problem for conductors. When high-pressure circuitry is rapidly shifted, the ensuing shock waves will induce heavy shock loads on conductors. The ability of piping to withstand this shock is unparalleled. However, the flexibility of tubing and hosing allows these conductors to absorb shock waves by changing shape. Therefore, unexpected bends will be incorporated into straight sections of tubing to accommodate shape change. Also, close attention should be made to adhere to required minimum bend radii (as specified by conductor manufacturers) so that shock does not cause rupture at the weakened bend point.

Since some circuitry requires the movement of components, such as a feedable drill head, certain designs require the conductor to change shape.

FIGURE 11-15

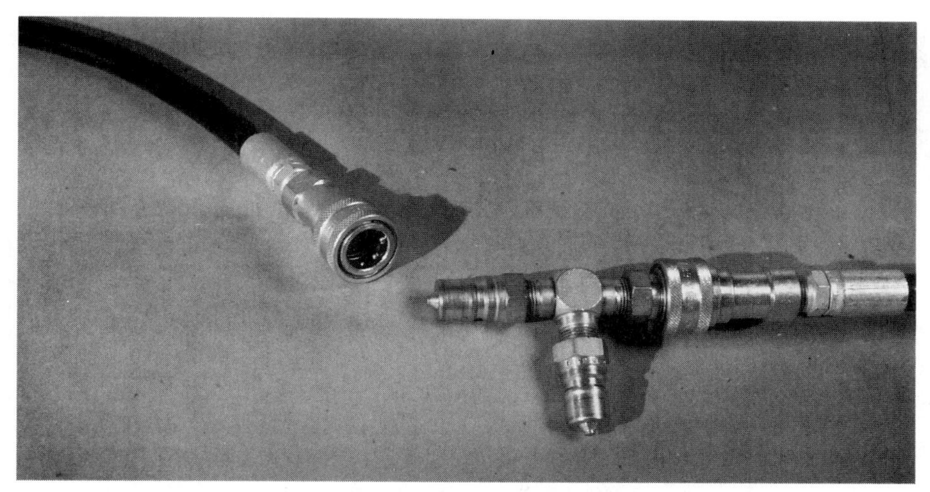

Hose quick-disconnect packages offer a convenient method of making and breaking connections. (*Photograph by the author.*)

Special swivel connectors are available for piping; however, the ability to change shape is greatest in hosing. Also, hosing (through the use of quick disconnects) allows for rapid redesign of circuitry with standard components. The hydraulic power takeoff unit on tractors and other mobile equipment requires this degree of adaptability.

[] [] [] 11.3 PRACTICAL DESIGN PROBLEM

The specification guidelines shown on the following pages describe a hydraulic circuit that is used to clamp and shear heavy sections of plate steel. It is also equipped with a conveyor system that is used to deliver the steel plates to the operator of the shear and retrieve the parts after the shearing operation.

The basic design of the linear side of the circuit is a pressure parameter sequence circuit. The primary operation will be to move a 950-lb. {*4-kN*} clamping frame into position through an 8-in. {*20-cm*} stroke of a 3-in. {*7.5-cm*} diameter single end rod, double-acting cylinder having a 2.5-in. {*6-cm*} diameter rod. This clamping operation will require at least 19,000 lbs. {*84.5 kN*} of force during an additional 1.5-in. {*4-cm*} stroke to activate a toggle action clamp through the use of a single-acting intensifier having a 3-in. {*7.5-cm*} diameter piston and a 1-in. {*2.5-cm*} diameter rod. When the clamping cylinder has reached sufficient force, a pressure parameter sequence will occur and the 750-lb. {*3.3-kN*} shear head will advance through 10 in. {*25 cm*} of stroke and shear the plate with a force of 4,500 lbs. {*20 kN*} on a 4-in. {*10-cm*} diameter single end rod, double-acting cylinder having a 3.5-in. {*9-cm*} diameter rod. After the shearing operation has been completed, the operator will shift a valve and the shear head should retract first and be held rigidly in a retracted position, followed by a pressure parameter sequence retraction of the clamp to release the part. The linear side of the circuit should be designed to operate at a rate of 5 cycles/min.

The operator will also have control of the conveyor system that operates using a hydraulic motor having a displacement of 4 cu. in./rev. {*65.6 cm³/rev.*}. The rotary circuit should be able to supply fluid to produce a variable speed of motor rotation between 60 and 130 rpm and overcome a torque load of 400 in. lbs. {*540 Nm*}. *Other than electrical motor drive and a single reservoir, the systems should be completely independent of one another for maintenance purposes.*

Finally, the linear side of the system should be equipped with auxiliary supply that, when properly designed, should operate both the clamp and shear to complete one full cycle operation (both extension and retraction) and overcome any loads encountered using a gas-charged accumulator.

Your assignment is to study the schematic provided in Figure 11-16 and complete all specifications called for. Notice that you are required to specify all circuit conductors as pipe, tube, and hose.

A Single End Rod, Double-Acting Shear Cylinder

PISTON		*ROD*			
Diameter	Area	Diameter	Area	Annulus Area	Stroke
____ 4 in. {*10 cm*}	____ in.² {*cm²*}	____3.5 in. {*9 cm*}	____ in.² {*cm²*}	____ in.² {*cm²*}	____ 10 in. {*25 cm*}

FORCE		*PRESSURE*		*SUPPLY*	
Extension	Retraction	Extension	Retraction	Extension	Retraction
4,500 lbs. {*20kN*}	750 lbs. {*3.3 kN*}	____ psi {*kPa*}	____ psi {*kPa*}	____ in.³ {*cm³*}	____ in.³ {*cm³*}

			VELOCITY	
Total Supply	Rate	Supply Rate	Extension	Retraction
____ in.³ {*cm³*}	5 cyc/min.	____ in.³/min {*cm³/min*}	____ in./min. {*cm/min.*}	____ in./min. {*cm/min.*}

FIGURE 11-16

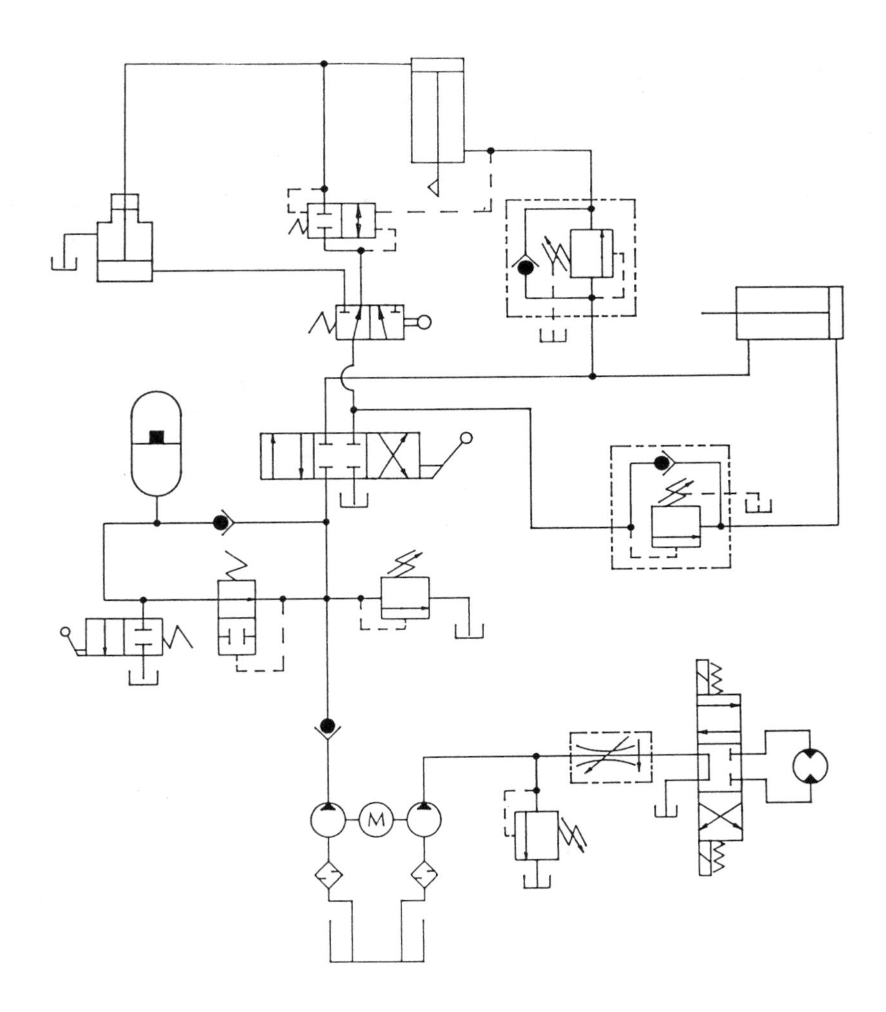

A Single End Rod, Double-Acting Shear Cylinder *(continued)*

Extension Time	Retraction Time	Port Size	Body Style	Mounting
________ sec.	________ sec.	________ in.{*mm*}	________	________

B Single End Rod, Double-Acting Clamp Cylinder

PISTON		*ROD*			
Diameter	Area	Diameter	Area	Annulus Area	Stroke
3 in. {*7.5 cm*}	______ in.^{2} {*cm*^{2}}	2.5 in. {*6 cm*}	______ in.^{2} {*cm*^{2}}	______ in.^{2} {*cm*^{2}}	8 in. {*20 cm*} [normal advance] 1.5 in. {*4 cm*} [intensified]

FORCE		*PRESSURE*		*SUPPLY*	
Extension	Retraction	Extension	Retraction	Extension	Retraction
19,000 lbs. {*84.5 kN*} [intensified] 950 lbs. {*4 kN*} [normal]	950 lbs. {*4 kN*}	____ psi {*kPa*} ____ psi {kPa}	____ psi {*kPa*}	____ in.^{3} {*cm*^{3}} [normal advance]	____ in.^{3} {*cm*^{3}} [full return]

			VELOCITY	
Total Supply	Rate	Supply Rate	Extension	Retraction
____ in.^{3} {*cm*^{3}}	5 cyc/min.	____ in.^{3} /min. {*cm*^{3}*/min.*}	____ in. /min. {*cm/min.*}	____ in. /min. {*cm/min.*}

Extension Time	Retraction Time	Port Size	Body Style	Mounting
________ sec.	________ sec.	________ in. {*mm*}	________	________

C Hydraulic Motor

Operating Speed (max.)	Torque Load	Type	Design	Volumetric Efficiency
130 rpm	400 in. lbs. {*540 N m*}	________	________	________ %

C Hydraulic Motor *(continued)*

Displacement	Required Delivery	OPERATING PRESSURE (at Motor) Full Delivery	OPERATING PRESSURE (at Relief Valve) Minimum Delivery	Starting Pressure (at Motor)
4 in.^{3} /rev. {*65.6 cm^3/rev.*}	_______ gpm {*lpm*}	_______ psi {*kPa*}	_______ psi {*kPa*}	_______ psi {*kPa*}

Torque Rating	Mounting	Port Size
_______ in. lbs./100 psi {*N m/1,000 kPa*}	__________	________ in. {*mm*}

D Intensifier

Necessary Cylinder Displacement	INTENSIFIER PISTON Diameter	INTENSIFIER PISTON Area	INTENSIFIER ROD Diameter	INTENSIFIER ROD Area
_______ in.^{3} {*cm^3*}	3 in. {*7.5 cm*}	_______ in.^{2} {*cm^3*}	1 in. {*2.5 cm*}	_______ in.^{2} {*cm^2*}

FORCE Cylinder Piston	FORCE Intensifier Rod	INTENSIFIER PISTON Piston	INTENSIFIER PISTON Rod	Stroke
19,000 lbs. {*84.5 kN*}	_______ lbs. {*kN*}	_______ psi {*kPa*}	_______ psi {*kPa*}	_______ in. {*cm*}

SUPPLY INPUT FROM PUMP Extension	SUPPLY INPUT FROM PUMP Retraction	Rate	Supply Rate
_______ in.^{3} {*cm^3*}	0 in.^{3} {cm^3}	5 cyc/min.	_______ in.^{3}/min{*cm^3/min*}

Velocity Extension	Time Extension	PORT SIZE Piston End	PORT SIZE Rod End	Mounting
in./min. {*cm/min.*}	________ sec.	________ in. {*mm*}	________ in. {*mm*}	___________

E Pump—Linear Side

Type	Design	Delivery	Drive Speed	Volumetric Efficiency
______	______	______ gpm {*lpm*}	______ rpm	______ %

Displacement at V.E.	PORT SIZE Inlet	PORT SIZE Outlet
______ in.^{3}/rev. {*cm^3/rev.*}	______ in. {*mm*}	______ in. {*mm*}

F Pump—Rotary Side

Type	Design	Delivery	Drive Speed	Volumetric Efficiency
______	______	______ gpm {*lpm*}	______ rpm	______ %

Displacement at V.E.	PORT SIZE Inlet	PORT SIZE Outlet
______ in.^{3}/rev. {*cm^3/rev.*}	______ in. {*mm*}	______ in. {*mm*}

G Gas-Charged Accumulator

Necessary Supply (S)	Load Volume (V2)	Discharge Volume (V3)	Precharge Pressure (P1)	Load Pressure (P2)
______ in.^{3} {*cm^3*}	______ in.^{3} {*cm^3*}	______ in.^{3} {*cm^3*}	______ psi_a {*kPa_a*}	______ psi_a {*kPa_a*}

Discharge Pressure (P3)	Precharge Volume (V1)	Size	Element Type	Port Size
______ psi_a {*kPa_a*}	______ in.^{3} {*cm^3*}	______ gal. {*l*}	______	______ in.{*mm*}

H Reservoir

CAPACITY	
Minimum	Maximum
_______ gal. {*l*}	_______ gal. {*l*}

I Filtration—Linear Side

Type	Element	*RATING* Element	Flow	Port Size
_______	_______	_______ mesh	_______ gpm {*lpm*}	_______ in. {*mm*}

Description	Mounting
_______	_______

J Filtration—Rotary Side

Type	Element	*RATING* Element	Flow	Port Size
_______	_______	_______ mesh	_______ gpm {*lpm*}	_______ in. {*mm*}

Description	Mounting
_______	_______

K Master System Directional Control—Linear Side

Positions	Ways	Element	Activation	Port Size
_______	_______	_______	_______	_______ in. {*mm*}

L Master System Directional Control—Rotary Side

Positions	Ways	Element	Activation	Port Size
______	______	______	______	______ in. {*mm*}

M Intensifier Selector Directional Control

Positions	Ways	Element	Activation	Port Size
______	______	______	______	______ in. {*mm*}

N Pilot Override Check Valve

Positions	Ways	Element	Activation	Port Size
______	______	______	______	______ in. {*mm*}

O Accumulator Shunt Directional Control—1

Positions	Ways	Element	Activation	Port Size
______	______	______	______	______ in. {*mm*}

P Accumulator Shunt Directional Control—2

Positions	Ways	Element	Activation	Port Size
______	______	______	______	______ in. {*mm*}

Q Check Valves

Positions	Ways	Element	Activation	Port Size
______	______	______	______	______ in. {*mm*}

R Flow-Control Valve—Rotary Side

Method	Type	Motor Speed at Flow Control (Lowest Setting)	Delivery through Valve at Flow-Control Setting
________	________	________ rpm	________ gpm {*lpm*}

Allowed Pressure Drop	Pressure-Compensation Design	Delivery Range	Port Size
________ psi {*kPa*}	________	0 to ________ gpm {*lpm*	________ in. {*mm*}

S Sequence Valve—1

Type	Element	Setting	Mounting	Port Size
________	________	________ psi_g {kPa_g}	________	________ in. {*mm*}

T Sequence Valve—2

Type	Element	Setting	Mounting	Port Size
________	________	________ psi_g {kPa_g}	________	________ in. {*mm*}

U Pressure Relief Valve—Linear Side

Type	Element	Setting	Mounting	Port Size
________	________	________ psi_g {kPa_g}	________	________ in. {*mm*}

V Pressure Relief Valve—Rotary Side

Type	Element	Setting	Mounting	Port Size
________	________	________ psi_g {kPa_g}	________	________ in. {*mm*}

W Electric Motor

Speed	Voltage	Phase	Power at V.E.
_______ rpm	___________	___________	_______ hp {*W*}

X Inlet and Working Conductors—Linear Side

Type	Relief Valve Pressure	Safety Factor	Safety Factor Pressure	Pipe Material	Pipe Tensile Rating
Pipe, Tube, and Hose	_____ psi {*kPa*}	____ : 1	_____ psi {*kPa*}	________	_____ psi {*kPa*}

INLETS (I.D. DECIMAL)		*WORKINGS (I.D. DECIMAL)*			
Minimum	Maximum	Minimum	Maximum	Pipe Schedule	Pipe Wall Thickness
______ in. {*cm*}	______ in. {*cm*}	______ in. {*cm*}	______ in. {*cm*}	_________	______ in. {*mm*}

			PIPE—NOMINAL		*TUBE—NOMINAL*	
Tube Material	Tube Tensile Rating	Tube Wall Thickness	Inlets	Workings	Inlets	Workings
_______	____ psi {*kPa*}	____ in. {*mm*}	____ in. {*mm*}	____ in. {*mm*}	____ in. {*mm*}	____ in. {*mm*}

		HOSE—NOMINAL	
Hose Description	Hose Operating Pressure Rating	Inlets	Workings
_________	________ psi {*kPa*}	______ in. or – {*mm*}	______ in. or – {*mm*}

Y Working Conductors—Intensifier Rod End

Type	Intensifier Pressure	Safety Factor	Safety Factor Pressure	Pipe Material	Pipe Tensile Rating
Pipe, Tube, and Hose	_____ psi {*kPa*}	____ : 1	_____ psi {*kPa*}	________	_____ psi {*kPa*}

Y Working Conductors—Intensifier Rod End (*continued*)

Derived Delivery	*WORKINGS (I.D. DECIMAL)* Minimum	Maximum	Pipe Schedule	Pipe Wall Thickness	Pipe Nominal Size
______ gpm {*lpm*}	_______ in. {*cm*}	_______ in. {*cm*}	___________	_______ in. {*mm*}	_______ in. {*mm*}

Tube Material	Tube Tensile Rating	Tube Wall Thickness	Tube Nominal Size	Hose Description	Hose Operating Pressure Rating	Hose Nominal Size
_________	_____ psi {*kPa*}	_____ in. {*mm*}	_____ in. {*mm*}	_________	_____ psi {*kPa*}	_____ in. or – {*mm*}

Z Conductors—Rotary Side

Type	Relief Valve Pressure	Safety Factor	Safety Factor Pressure	Pipe Material	Pipe Tensile Rating
Pipe, Tube, and Hose	______ psi {*kPa*}	_____ : 1	______ psi {*kPa*}	_________	______ psi {*kPa*}

	WORKINGS (I.D. DECIMAL)		*INLETS (I.D. DECIMAL)*	
	Minimum	Maximum	Minimum	Maximum
Pump	_______ in. {*cm*}	_______ in. {*cm*}	_______ in. {*cm*}	_______ in. {*cm*}
Motor	_______ in. {*cm*}	_______ in. {*cm*}		
Relief	_______ in. {*cm*}	_______ in. {*cm*}		

Pipe Schedule	Pipe Wall Thickness	*PIPE—NOMINAL* Inlets	Workings	Tube Material	Tube Tensile Rating	Tube Wall Thickness
________	_____ in. {*mm*}	_____ in. {*mm*}	_____ in. {*mm*}	___________	_____ psi {*kPa*}	_____ in. {*mm*}

TUBE—NOMINAL Inlets	Workings	Hose Description	Hose Operating Pressure Rating	*HOSE—NOMINAL* Inlets	Workings
______ in. {*mm*}	______ in. {*mm*}	___________	______ psi {*kPa*}	______ in. or – {*mm*}	______ in. or – {*mm*}

AA The following are variations for the Practical Design Problem (11.3):

Component: Double-Acting, Single End Rod Shear Cylinder

	A	B	C	D
Piston Diameter	3 in.	4 in.	12 *cm*	10 *cm*
Rod Diameter	2.5 in.	3.5 in.	10 *cm*	9 *cm*
Stroke	12 in.	14 in.	20 *cm*	30 *cm*
Force				
Extenson	3,500 lbs.	4,300 lbs.	30 *kN*	19 *kN*
Retraction	650 lbs.	700 lbs.	5.3 *kN*	3 *kN*
Rate	7 cyc/min.	6 cyc/min.	4 *cyc/min.*	9 *cyc/min.*

Component: Double-Acting, Single End Rod Clamp Cylinder

	A	B	C	D
Piston Diameter	2.5 in.	3.5 in.	10 *cm*	9 *cm*
Rod Diameter	2 in.	3 in.	8 *cm*	7.5 *cm*
Stroke				
Normal Advance	10 in.	12 in.	24 *cm*	20 *cm*
Intensified Advance	1 in.	2 in.	5 *cm*	5 *cm*
Extension Force				
Normal	600 lbs.	1,450 lbs.	6 *kN*	6.5 *kN*
Intensified	15,000 lbs.	17,500 lbs.	92 *kN*	78 *kN*
Retraction Force	600 lbs.	1,450 lbs.	6 *kN*	6.5 *kN*
Rate	7 cyc/min.	6 cyc/min.	4 *cyc/min.*	3 *cyc/min.*

Component: Single-Acting Intensifier

	A	B	C	D
Piston Diameter	4 in.	2.5 in.	8 *cm*	6.5 *cm*
Rod Diameter	1.5 in.	1 in.	2 *cm*	2.5 *cm*

Component: Hydraulic Motor

	A	B	C	D
Operating Speed				
Maximum	150 rpm	125 rpm	100 rpm	90 rpm
Minimum	70 rpm	50 rpm	60 rpm	40 rpm
Torque Load	500 in. lbs.	900 in. lbs.	*600 N m*	*900 N m*
Displacement	6 in.^{3}/rev.	8 in.^{3}/rev.	*7.5 cm^3/rev.*	*11 cm^3/rev.*

11.4 SUGGESTED ACTIVITIES

1. Examine the engine compartment and inspect the undercarriage of an automobile. Identify pipes, tubes, and hoses used in the various assemblies. Note the types of conductors used and how they are connected to automotive components. Why do you think certain types of conductors were used in some applications and other types in other applications?
2. Visit a local hardware store and make a list of the various plumbing fittings they have in stock. Draw sketches of each and label their names beside them.
3. Examine the circuit described in the Practical Design Problem for this chapter. Understanding the actual operation of the circuit (to clamp, shear, and drive a conveyor system), write a brief report on which types of conductors (pipe, tube, or hose) you would use for each line in the circuit.
4. Using outside micrometers and hole gauges or telescoping gauges, measure the inside and outside diameters or various sizes for pipe and tube in your laboratory or shop. Calculate the wall thickness of each and compare these measurements to the standards given in this text.
5. If proper equipment is available, make up some simple pipe, tube, and hose assemblies. You may even wish to test your assemblies on a trainer when finished. Make sure that you do the original testing at low pressure and have burst shields between yourself and the conductors during this activity.

11.5 REVIEW QUESTIONS

1. Name the three types of conductors commonly employed in hydraulic circuitry.
2. What are some advantages and disadvantages in using pipe in a hydraulic circuit?
3. How is pipe nominally sized?
4. Explain the procedure for determining the maximum pressure that may be withstood (burst strength) by pipe and tube used in a hydraulic circuit.
5. Discuss the industrial standards that are used to describe safety factors used for pipe and tube used in operating hydraulic circuits.
6. Describe at least five different pipe fittings that may be used in hydraulic circuits.
7. Discriminate among straight, tapered, and dry-seal thread forms as well as welded assembly in pipe in terms of design of fasteners and uses.
8. What is the main difference between pipe and tube?
9. What are some advantages and disadvantages in using tube in a hydraulic circuit?
10. How is tube nominally sized?
11. Differentiate between compression fittings and flared fittings used in tubing assemblies.
12. What are some advantages and disadvantages in using hose in a hydraulic circuit?
13. How is hose nominally sized?

14. Explain why the same procedures used to determine the pressure ratings in pipe and tube may not be used to determine the pressure ratings for hose.
15. Differentiate between braid and spiral reinforcement used in hose construction in terms of design and use.
16. What are the primary functions of the outer skin in hose construction?
17. How are proof pressure and operating pressure determined for a hydraulic hose?
18. Discuss ferrule locking, skive, and no-skive adaptors used in hose fittings.
19. Explain the design and operation of a hose quick-disconnect package.

12

Power Pneumatics: Principles and Components

Another major area of fluid power involves the use of gas rather than liquid as a medium to transfer pressure and displace loads. Whereas oil was the primary fluid used in hydraulic circuitry, air holds this distinction in **pneumatics**. The question of whether to use air or oil in a fluid power circuit or device is not a simple one to answer. It was once thought that pneumatic circuitry provided cushioning, while hydraulic circuitry should be used to apply or overcome large forces. However, anyone who has tried to manually remove lug nuts on an automobile that were secured with a pneumatic wrench knows that pneumatic tools can apply high forces. Occasionally, the decision to use pneumatics is predicated on the fact that a supply of compressed air is readily available. If there is a choice of which fluid power mechanism to use, a careful evaluation should be made of the parameters under which the circuit is to be operated.

12.1 HYDRAULICS AND PNEUMATICS

Primarily, *air circuitry is useful when rapid response is necessary, while oil circuitry is useful when high-force output is required with smaller actuators*. Beyond this seemingly simple consideration, various factors interact to emphasize the use of air or oil circuitry.

One previously mentioned consideration is availability of fluid. Although this consideration would appear to (and probably does) favor pneumatics, the quality and effect of the available fluid should be evaluated. For example, industrial mining machinery would seem to be adaptable to readily available air supply. However, along with other considerations, the high-volume, high-pressure exhaust air from pneumatic circuitry may react with coal dust and cause explosions. Therefore, mining machinery will use self-contained oil available in hydraulic circuitry. Dirty or dusty environments would appear to be more adaptable to hydraulic circuitry. However, the wetting action of oil on dirt may cause more wear and make the dirt particles more difficult to remove. Therefore, air circuitry is often preferred in dirty or dusty environments.

Ambient heat is another consideration that may preclude use of hydraulics. Although oil is more stable than air when heated, the excellent insulating properties of "dead air" in pneumatic circuitry eliminates much heat transfer. Therefore, pneumatic circuitry may be preferred in high-ambient-heat mechanisms such as die-casting machinery. Inversely, hydraulic circuitry is normally preferred at low ambient temperature. Moisture inherent in free air supply will condense when cooled and possibly frost or freeze at very low temperatures, while oil, heated through operation or with heaters in the reservoir, works well at low ambient temperatures.

Finally, space considerations may be a limiting factor in pneumatics. Air compressors are typically limited in pressure output to a value below 250 psi {*1,724 kPa*}. Therefore, pneumatic actuators will need to be physically larger than hydraulic actuators to overcome a given load. However, since central air supplies are common, an overall reduction in space requirements for pneumatic circuits may compensate for larger actuators. Hydraulic circuits typically have large reservoirs and pumps incorporated into each unit.

12.2 PNEUMATIC POWER SUPPLY

Like hydraulics, pneumatic circuitry relies on an external drive unit to cause air flow. However, unlike hydraulics, the pumping action in pneumatics must accommodate the unique qualities of a gaseous medium.

The principles of pneumatics have been summarized previously in the discussion of the gas-charged accumulator (see Chapter 10). In that presentation, it was shown that air supply will expand to fill any void (the definition of a gas). However, when confined, the air compresses in direct proportion to the value of the load imposed upon it (Boyle's Law).

In the pneumatic circuit, a prepressurized charge of air is supplied from an **air compressor** unit. The unit normally takes free air from the atmosphere through an inlet air filter into a compressor that serves a function similar to a hydraulic pump.

12.2.1 Air Compressors

The basic function of any compressor is to transfer air from its inlet to its outlet. The rate at which this transfer occurs is the compressor's delivery and is rated in standard cubic feet per minute **(scfm)** {*standard cubic meters per minute* ***(scmm)***}. A standard cubic foot {*meter*} of air is defined as one cubic foot {*cubic meter*} of air at one atmosphere (14.7 psi_a at sea level and at 68 degrees Fahrenheit) {*101.4* kPa_a *at sea level and at 20 degrees Celsius*}.

Air compressors are similar in design and operation to hydraulic pumps. As may be expected, gear, vane, and piston compressors are common. Also, hydrodynamic devices such as turbines and propeller designs may be used. Since partial compression of the air will occur in the compressor, it is important that its intake be of sufficient capacity to afford a high quantity of air transfer. Therefore, external and internal gear compressors are not common. Rather, **double-lobe** and **screw compressors** are typical.

The **unbalance vane** design is common, although this compressor will have fewer vanes than similar unbalanced vane hydraulic pumps. The decreased number of vanes allows an enlarged chamber size for the vane compressor inlet. Furthermore, the vanes in these compressors may be made from graphite or reinforcd phenolic to reduce inertial effects and afford lubrication at the vane–reaction ring interface.

Although axial and radial piston compressors may be found, the **crankshaft** variety is more common. Often, the crankshaft compressor will have multiple pistons operating in a "V" design similar to the automobile engine. In this arrangement, the cylinders will be phased from a single-throw journal so that some are in their intake strokes while the others are in compression. With this design, one cylinder may serve as a primary stage that supplies compressed air to the second cylinder for further compression in high-pressure compressors. Single-stage compressors are typically incapable of attaining values above 150 psi {*1,034 kPa*}. Most multistage compressors are of the piston type, and dual-stage designs are capable of attaining 500 psi {*3,447 kPa*}. Three-stage piston compressors may attain 2,500 psi {*17,237 kPa*}, and four-stage designs may reach 5,000 psi {*34,474 kPa*}. Figure 12-1 illustrates some of the common types of air compressors (see also Figure 12-2).

12.2.2 Receiver Tanks

Although those devices that are used to create potential supply in pneumatics are called compressors, by design they are actually pumps. Although, by design, the chambers that exist in each type must decrease in

FIGURE 12-1

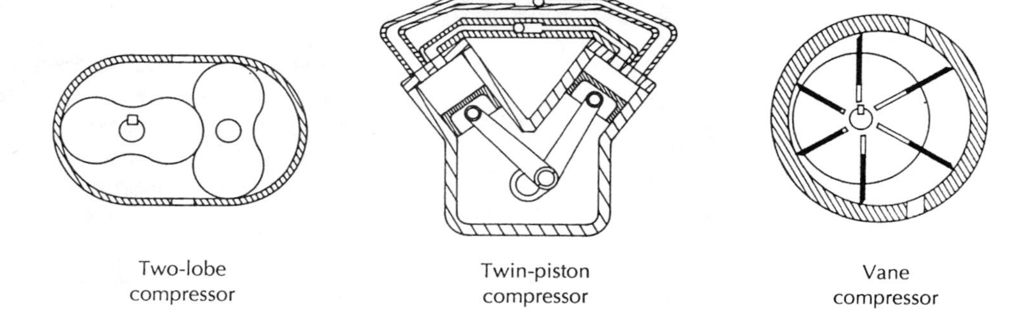

Two-lobe compressor

Twin-piston compressor

Vane compressor

volume from inlet to outlet, very little compression of the gas occurs within the pump. Even though the pump may deliver a volume of air from the outlet at a somewhat elevated pressure, this relatively low volume would seldom be sufficient to power tools requiring fast operation. Furthermore, the somewhat elevated pressure existing from this pump (compressor) delivery would nearly always be resisted by atmospheric pressure, negating its effective force. Rather, the delivery is normally directed to a holding device that consolidates the continuous flow from the compressor to increase effective delivery and intensify the potential pressure.

Air transferred from the compressor is confined in a receiver tank that holds an ever-increasing quantity in a confined space. This action prepressurizes the air for supply purposes. The compression action occurs at a ratio of about 7.8:1 cu. ft./100 psi_a {*11.4:1* m^3 */ 1,000* kPa_a}. Thus, a compressor that delivers 15.6 scfm {*0.44 scmm*} of air would require 1 min. (Time) to prepressurize a 1-cu. ft. {*0.022-*m^3} tank (**CAP**acity) to 200 psi_a {*1,379* kPa_a}:

$$P\ (psi_a) = \frac{[DEL\ (scfm)\ /\ 7.8] * 100\ (psi_a)}{CAP\ (cu.\ ft.)\ /\ T\ (min.)},\ \text{or}$$

$$P\ \{kPa_a\} = \frac{[DEL\ \{scmm\}\ /\ 11.4] * 1{,}000\ \{kPa_a\}}{CAP\ \{m^3\}\ /\ T\ \{min.\}}$$

Problem 12-1

A compressor unit having a receiver tank capacity of 5 cu. ft. {*0.14* m^3} is powered by a compressor that delivers 78 scfm {*2.2 scmm*} of air. What pressure will develop in the tank in 1 min. of operation?

P (psi_a) {kPa_a} = ____________________

FIGURE 12-2

Multistage compressor. (*Courtesy Gast Mfg., Inc.*)

From this example, it would appear that any theoretical pressure could be attained in a compressor unit by simply operating the compressor for an extended period of time. In keeping with previously described studies performed by Robert Boyle, this would be true. For example, a compressor delivering 1 scfm of air into an enclosed receiver tank would cause the pressure within that tank to rise 14.7 psi for each minute of operation. {*A compressor delivering 1 scmm of air would cause the pressure within that tank to rise 101.4 kPa per minute of operation.*} As this compression occurs, the air molecules within the tank are packed increasingly closer together. Friction created through molecular impact causes heat generation in keeping with the studies performed by Gay-Lussac. Gay-Lussac's Law states that *if the volume of a contained gas is held constant, the absolute pressure exerted by the gas is directly proportional to its absolute TEmperature:*

$$\frac{P1}{P2} = \frac{TE1}{TE2}$$

Absolute temperature is defined as degrees **Rankine** and may be calculated by adding the constant 460 to degrees Fahrenheit in the U.S. customary system. {*In the SI, absolute temperature is defined as degrees* ***Kelvin*** *and may be calculated by adding the constant 273 to degrees Celsius.*} Each increment of pressure rise within the closed receiver tank will result in increased temperature.

Problem 12-2

A compressor unit before starting has a receiver tank at atmospheric pressure (14.7 psi_a or 0 psi_g) {*101.4* kPa_a *or 0* kPa_g} and 68 degrees Fahrenheit {*20 degrees Celsius*}. What temperature will be evident in the tank when it is fully charged to 150 psi_g or 164.7 psi_a {*1,034* kPa_g *or 1,135.4* kPa_a}?

TE2 (°F) {°*C*} = ____________________

The reader may very well be astounded by the amount of heat generated through simple compression of air to such a relatively low pressure value. Obviously, this amount of heat would melt down any compressor. However, most of this heat is instantaneously dissipated through the metal components to the atmosphere and by the use of auxiliary coolers or chillers in the compressor unit. To put this operation into perspective, consider the fact that a simple lawn mower engine continuously cools an explosion of nearly 2,400 degrees Fahrenheit {*1,316 degrees Celsius*} to a value of 200 degrees Fahrenheit {*93 degrees Celsius*} in exhaust gases. Of course, the development of high pressure in a compressor does not occur instantaneously; and the resulting temperature may be dissipated over an extended period of time.

Pneumatic systems performing under conditions that allow for dissipation of all heat generated through compression are said to be operating **isothermally**. Those that maintain all the generated heat are said to be operating **adiabatically**. In adiabatic operation, the volume of the gas increases parabolically with increases in temperature. The net effect would be a necessary adjustment to the isothermal form of Boyle's Law:

$$P1 * V1^2 = P2 * V2^2$$

In actuality, pneumatic circuitry typically performs somewhere between the extremes of true isothermal and adiabatic operation. As a result, the amount of heat gained or lost through heating or cooling requires manipulation of both forms of Boyle's Law. The ability of a gas to transfer heat is expressed as its specific heat. The specific heat is expressed as the amount of energy necessary to raise the temperature of 1 lb. {*1 kg*} of material 1 degree Fahrenheit {*1 degree Celsius*}. In the U.S. customary standard, specific heat is expressed in BTU/lb. (British Thermal Units per pound). *In the SI, specific heat is expressed in kJ/kg {kiloJoules per kilogram}*. The specific heat of air has been determined to be 0.2375 BTU/lb. {*0.5519 kJ/kg*}. In adiabatic experimentation performed by Regnault, the change in heat transfer has been calculated to be 0.1689 BTU/lb. {*0.3924 kJ/kg*}. The compression ratio **(C)** for adiabatic pneumatic heat loss or gain may then be expressed as follows:

$$C = \frac{0.2375\ (\text{BTU/lb.})}{0.1689\ (\text{BTU/lb.})} \text{ or } \frac{0.5519\ \{kJ/kg\}}{0.3924\ \{kJ/kg\}}$$

$$C = 1.41$$

Of course, as a ratio, *C* is a dimensionless number. It represents the exponential adjustment necessary for Boyle's Law under pneumatic adiabatic conditions:

$$P1 * V1^{1.4} = P2 * V2^{1.4}$$

The net effect of adiabatic operation is an expansion of gases in the compressor operation. This will decrease the pressure evident and require increased energy to develop a specific pressure in the receiver tank. For example, it requires about one-third more power to compress air to 100 psi_a adiabatically than isothermally. Therefore, air compressors are typically equipped with many devices used to remove the heat generated through compression.

12.2.3 Other Compressor Unit Components

Pneumatic systems have been tested at values as high as 2,000 psi_g {*13,790* kPa_g}, and theoretical studies have occurred to extrapolate higher-pressure characteristics. However, pneumatic circuitry typically operates at a peak of 250 psi_g {*1784* kPa_g}, and more commonly at 150 psi_g {*1,034* kPa_g}. Therefore, the compressor receiver tank is fitted with a **pop-off valve** similar to the relief valve in a hydraulic circuit.

The pop-off valve limits the maximum pressure in a compressor unit. At the same time, a normally closed pressure switch connected to the electrical power supply opens and stops the rotation of the drive motor and compressor. Figures 12-3 and 12-4 illustrate the design and operation of a compressor unit.

FIGURE 12-3

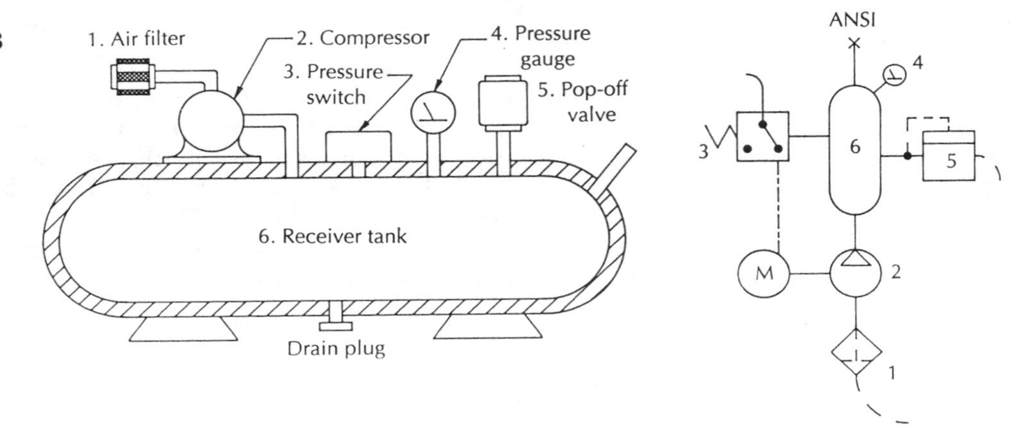

FIGURE 12-4

Typical compressor unit. (*Photograph by the author.*)

12.3 AIR DISTRIBUTION

Unlike the hydraulic circuit designer, the designer of pneumatic circuitry may not even be concerned with fluid supply. Often, it is a foregone conclusion that a supply of pressurized air is readily available within industrial plants and factories. Therefore, pneumatic circuit design often begins at the point of supply air designated at a specific pressure.

12.3.1 Air Conditioning

Air entering and entrained within the pneumatic system is not solely composed of nitrogen, oxygen, carbon dioxide, and trace gases. In nearly all cases, some degree of water vapor will be evident in the air supply. If the water remained in a gaseous state throughout the system, no detrimental effects would occur. However, as pressure drops during operation, the other gaseous molecules in the air separate and may allow condensation (liquefaction) of the entrained water. In essence, this is why clouds composed of water vapor may or may not drop water. When the gaseous water suspended in the high-pressure environment of the cloud reaches a low enough barometric pressure area, the gases separate and allow condensation of the water vapor. In turn, the denser water then separates from the gaseous cloud and falls through gravitational forces. Of course, this is a function of pressure and temperature that define the dew point or the combination of conditions that will cause condensation.

12.3.1.1 Separators Most often, the distribution of air from the compressor unit begins within the receiver tank with a **separator**. The separator removes water vapor from the air stream. Some separators also act as filters, relying on very fine elements that exclude the larger water molecules, while allowing gaseous air flow to pass. Other separators rely on centrifugal action. Within the separator are a series of impeller plates driven by the air stream that flows through them. The impellers direct the high-velocity air to the inner walls of the separator. The impaction of the air against the walls causes a pressure drop that releases the entrained water vapor, and it gathers there. The buildup of water droplets is allowed to escape thorough a drain cock in the lower casing of the separator (see Figure 12-5).

Other water-separating mechanisms may employ special chemical packs called desiccants that absorb moisture into the pores of the contained silica gel. These packages are similar to the packs that may be found in various shipping containers to absorb moisture. Like these shipping packs, desiccant separators in pneumatic systems require periodic replacement or drying to maintain their effectiveness.

12.3.1.2 Intercoolers Many pneumatic power supplies will also contain mechanisms to more effectively remove the heat generated through compression. Most often, compressors will be equipped with thin fins on their casings, resulting in more surface exposure to the atmosphere, to release generated heat by acting as a heat sink. Often, especially in multistage compressors, these fins do not sufficiently remove the generated heat, in consideration of the fact that the air does not remain in the compressor for very long. Supplemental cooling may be necessary.

FIGURE 12-5

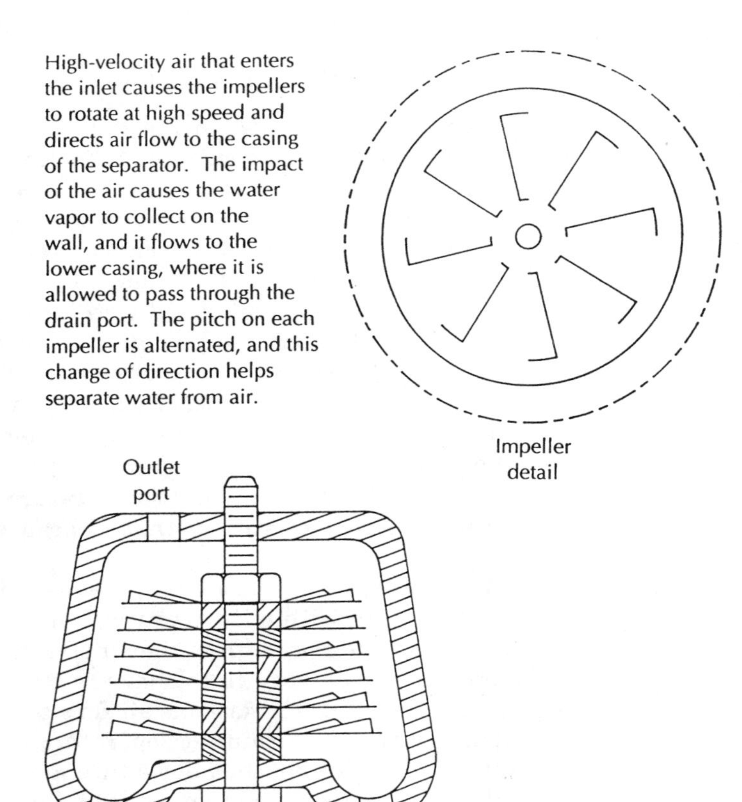

High-velocity air that enters the inlet causes the impellers to rotate at high speed and directs air flow to the casing of the separator. The impact of the air causes the water vapor to collect on the wall, and it flows to the lower casing, where it is allowed to pass through the drain port. The pitch on each impeller is alternated, and this change of direction helps separate water from air.

Intercoolers are special heat-sink mechanisms having a great deal of surface exposure and jacketing to allow water circulation and more effectively cool air supply. This rapid cooling also results in a great deal of water condensation. So, intercooling packages are typically equipped with separators as well.

12.3.2 Pneumatic Supply Control and Conductors

For various reasons, including noise abatement, pneumatic compressors are often remotely located from the eventual point of use. However, one final control element is typically included in the pneumatic power supply before distribution of supply air to those remote locations.

12.3.2.1 Pressure control at the power supply The pressure switch interconnected to the pop-off valve typically has a double trip mechanism (typically called kick-in and kick-out points in the trade). The control will allow you to turn on the compressor anytime receiver tank pressure falls below a given value and then remain on until the receiver tank pressure rises to another predetermined pressure. Thus, some recovery time is always needed if the upper limit is the desired operating pressure. Otherwise, the compressor would be continually operating under supply demand.

Another requirement for many pneumatic power supplies is an **unloading valve**. When cycled off, pressure remains in the outlet line from the compressor (pump) to the receiver tank. This pressure will create forces in the pump that resist rotation and may cause overloading when the compressor kicks in. The unloading valve will temporarily open and exhaust air during initial kick-in to relieve this back pressure.

12.3.2.2 Pneumatic distribution networks After all the conditioning and control afforded by the air compressor is attained, the air supply is normally distributed through piping to various locations. The important consideration in distribution of air supply is to use conductors with extremely smooth interior surfaces and to eliminate directional changes in the pipe when possible. Pressure drop in the distribution system of pneumatic systems is even more critical than pressure drop in hydraulics due to the lower potential pressure in pneumatic systems.

In Schedule 40 pipe, the pressure drop is a function of the diameter and length of the pipe, the supply rate of air flowing through the pipe, and the compression ratio of the entrained air. The value of the pressure drop may then be determined by using a variation of the Harris Formula:

$$\Delta P\ (\text{psi}_g) = \frac{0.1025 * L\ (\text{ft.}) * [\text{S.R. (cu. ft./sec.)}]^2}{[P_{pipe}\ (\text{psi}_a)/P_{atm}\ (\text{psi}_a)] * [\text{I.D. (in.)}]^{5.31}},\ \text{or}$$

$$\Delta P\ (kPa_g) = \frac{420{,}000 * L\ (m) * [\text{S.R.}\ (m^3/sec.)]^2}{[P_{pipe}\ (kPa_a)/P_{atm}\ (kPa_a)] * [\text{I.D.}\ (cm)]^{5.31}}$$

The most apparent challenge to application of this formulation is the decimal exponent calculation for the inside diameter. Table 12-1 contains calculated values for $\text{I.D.}^{5.31}$ for several Schedule 40 pipe sizes.

In addition, fittings (elbows, tees, and valves) impose an additional distance that should be included in calculations for the length (L) of pipe in the distribution system. Table 12-2 shows values of apparent length for some standard fittings used in Schedule 40 pipe. (Note that the "tee" has two specifications—one for the straight through path, and the other for the turned path.)

TABLE 12-1 Calculated Values for Schedule 40 Pipe I.D.[5.31]

U.S. Customary			Metric		
Fractional (Nominal)	**Inside Diameter**	**^5.31**	**Nominal**	**Inside Diameter**	**^5.31**
1/8 in.	0.269 in.	0.0009	6 mm	0.68 cm	0.129
1/4 in.	0.364 in.	0.0047	8 mm	0.92 cm	0.642
3/8 in.	0.493 in.	0.0234	11 mm	1.25 cm	3.27
1/2 in.	0.622 in.	0.0804	14 mm	1.58 cm	11.347
3/4 in.	0.824 in.	0.3557	20 mm	2.10 cm	51.403
1 in.	1.049 in.	1.2892	24 mm	2.66 cm	180.35
1 1/4 in.	1.380 in.	5.5304	33 mm	3.51 cm	786.29
1 1/2 in.	1.610 in.	12.538	39 mm	4.09 cm	1,771.13
2 in.	2.067 in.	47.256	49 mm	5.25 cm	6,668.77
2 1/2 in.	2.469 in.	121.419	59 mm	6.27 cm	17,119.48
3 in.	3.068 in.	348.771	75 mm	7.79 cm	54,208.06

TABLE 12-2 Calculated Values for Schedule 40 Pipe Fittings—Apparent Lengths

	U.S. Customary in Feet						
	3/8 in.	**1/2 in.**	**3/4 in.**	**1 in.**	**1 1/4 in.**	**1 1/2 in.**	**2 in.**
Gate Valve	0.3	0.35	0.44	0.56	0.74	0.86	1.1
Globe Valve	14	18.6	23.1	29.4	38.6	45.2	58
Tee (through)	0.5	0.7	1.1	1.5	1.8	2.2	3.3
Tee (turn)	2.5	3.3	4.2	5.3	7	8.1	10.44
45-Degree Elbow	0.5	0.78	0.97	1.23	1.6	1.9	2.4
90-Degree Elbow	1.4	1.7	2.1	2.6	3.5	4.1	5.2
	Metric in Meters						
	11 mm	***14 mm***	***20 mm***	***24 mm***	***33 mm***	***39 mm***	***49 mm***
Gate Valve	0.091	0.107	0.134	0.171	0.226	0.262	0.335
Globe Valve	4.267	5.669	7.041	8.961	11.765	13.777	17.678
Tee (through)	0.152	0.213	0.335	0.457	0.549	0.671	1.006
Tee (turn)	0.762	1.006	1.28	1.615	2.134	2.469	3.182
45-Degree Elbow	0.152	0.238	0.296	0.375	0.488	0.579	0.732
90-Degree Elbow	0.427	0.518	0.64	0.792	1.067	1.25	1.585

Problem 12-3

A pneumatic distribution system using Schedule 40 3/4-in. {*20-mm*} (nominal) diameter pipe has an overall pipe length of 200 ft. {*61 m*} and includes four 90-degree elbows and two 45-degree elbows. If the receiver tank for this system is operating at 125 psi_g {*861.8 kPa_g*} and a supply rate of 80 scfm {*2.3 scmm*}, what pressure drop will be evident through the system?

ΔP (psi_g) {*kPa_g*} = ____________________

One final consideration in pneumatic distribution is termination of the system. Traditionally, globe valves have been used as shutoffs for air-distribution systems. This choice has typically been made because of the ready availability and low cost of this type of valve, as it is the shutoff typically used in water plumbing systems in houses. For household water supply, this valve may be adequate, since the water system is hydrodynamic and has no great loads to overcome at the outlet. However, pressure drops as high as 75% may be evident in some globe valves used in pneumatic systems. If a combination of flow control and shutoff capability is required, a better choice would be the gate valve. Gate valves may be designed with as little as 25% pressure drop. In modern practice, the air-distribution system is terminated with directional controls that are conditionally fully open or fully closed. There is very little pressure drop through these valves when they are open, and they may be considered as on–off switches for pneumatics. A quick disconnect serves this function, as well as providing a ready means for attachment of other conductors.

12.3.3 Filter, Lubricator, Regulator Units

The bridge between the air-distribution system and the pneumatic circuit is normally the **filter, lubricator, regulator unit**. The **FLR**, as it is commonly known, consists of three separate components, although it is typically packaged as a single unit.

12.3.3.1 Filter The filter portion of the FLR may appear to be unnecessary, since air has been previously filtered before it entered the compressor and at the separator. However, the passage of the airstream through the distribution system often contaminates it with oxidation deposits inherent on the inside of steel piping. Also, condensation on the inside of piping may result in water once again being entrained within the airstream. Therefore, the filter portion of the FLR provides final assurance that clean, dry air will be supplied through the small orifices in pneumatic components. Figure 12-6 illustrates the typical design of an air filter used in an FLR unit.

12.3.3.2 Lubricator The lubricator in the FLR treats the dry air with a small charge of oil. Unlike hydraulic systems, pneumatic systems are not self-lubricating. The lubricator affords this vitally needed lubrication to separate mating parts of pneumatic components with a thin film of oil. This function is so important that line lubricators are often included in complex pneumatic circuitry to supplement the lubricator in the FLR. Lubricators have various designs.

The most common and most easily controlled of these is the venturi type. Similar in operation to the carburetor used in internal-combustion

FIGURE 12-6

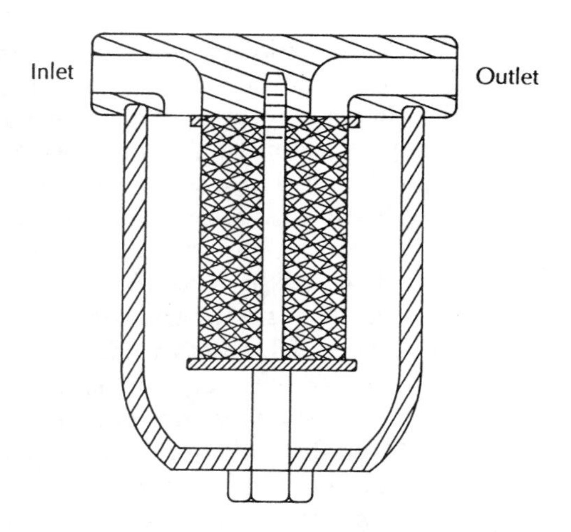

engines, the venturi lubricator relies on pressure differential to force a small amount of oil into the airstream as it passes through the lubricator. This lubricator has a small reservoir of oil contained in a lower chamber casing made of glass or transparent plastic for easy inspection. The space above the oil level in the lower chamber is pressurized by air taken from the inlet side of the lubricator. Beyond the air inlet, there is a restriction (venturi) in the air passageway that increases the velocity and lowers the pressure at this point. A stand tube connects the oil reservoir and the venturi. The higher inlet pressure acting on the reservoir forces oil through the stand tube to the venturi. Here, it mixes with the lower-pressure airstream before being discharged to the larger-diameter outlet passageway. Care should be taken to ensure that sufficient but not excessive lubrication is provided. Therefore, a needle valve is typically connected in the stand tube to adjust oil flow (see Figure 12-7).

12.3.3.3 Regulator Because air circuitry is typically an extension of an air-distribution system, independent pressure control is necessary. Unlike hydraulic circuitry, pneumatic circuitry cannot use a pressure-relief valve at this location. The relief valve, sensing inlet-side pressure for activation, would limit the pressure not only in the desired location but also throughout the air-distribution system. Therefore, pneumatic circuitry uses an outlet-side piloted pressure-reducing valve to limit pressure only to the desired circuit. These special pneumatic pressure-reducing valves are known as **regulators**.

Although the regulator is normally the smallest component in the FLR unit, it is probably the most important. The normally open valve affords variable pressure control through compression and relaxation of a biased

FIGURE 12-7

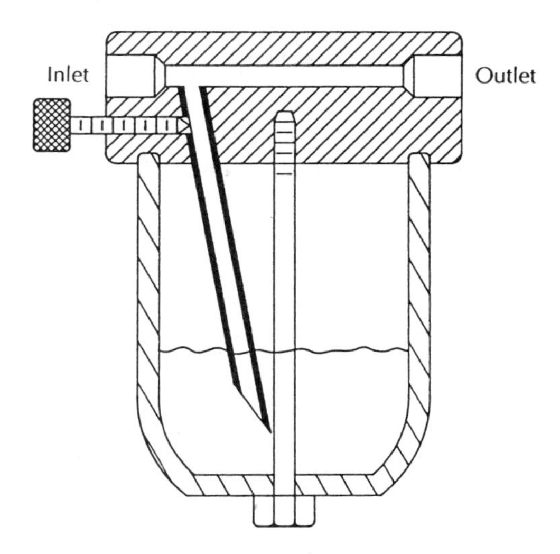

spring acting on the valve's spool. The setting is controlled externally through an adjustment screw. A pressure gauge is conveniently located to this adjustment screw for monitoring and setting of circuit pressure (see Figure 12-8).

FIGURE 12-8

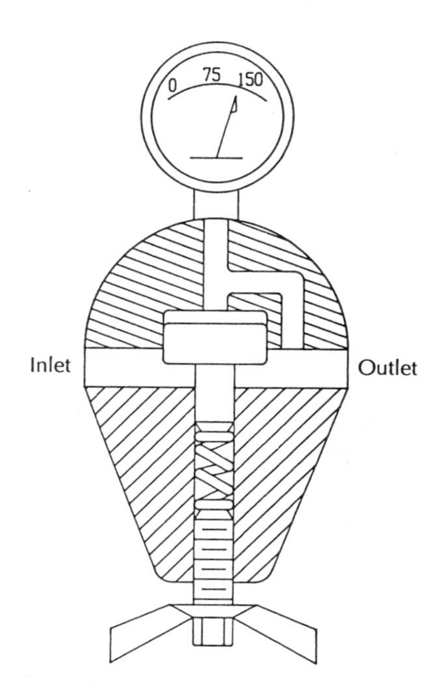

12.3.3.4 The FLR package The three individual components, plumbed together, afford a clean, lubricated, regulated air supply for pneumatic circuitry. The order in which the components are assembled is critical. The filter should be the first component to receive air supply. This affords clean air for passage through the small passageways of the other two. In any case, the filter should be located before the lubricator. If not, the filter, as it removes water, may also strip the air of lubrication. Normally, the regulator bridges the filter and the lubricator, although the order may be filter to lubricator to regulator. The assembled FLR package is illustrated in Figures 12-9 and 12-10.

12.4 THE PNEUMATIC LINEAR CIRCUIT

From the FLR unit, pressurized air may be distributed through a directional control to a cylinder. A surge of compressed air will overcome the load on the cylinder if the supply pressure acting on the cylinder is great enough (Pascal's Law).

12.4.1 Pneumatic Pressure Drop

The previous description of a linear pneumatic circuit differs, but only slightly, from a linear hydraulic circuit. However, the big difference in pneumatic operation begins not in initially overcoming the load, but in sustaining sufficient pressure during the entire cylinder stroke to fully

FIGURE 12-9

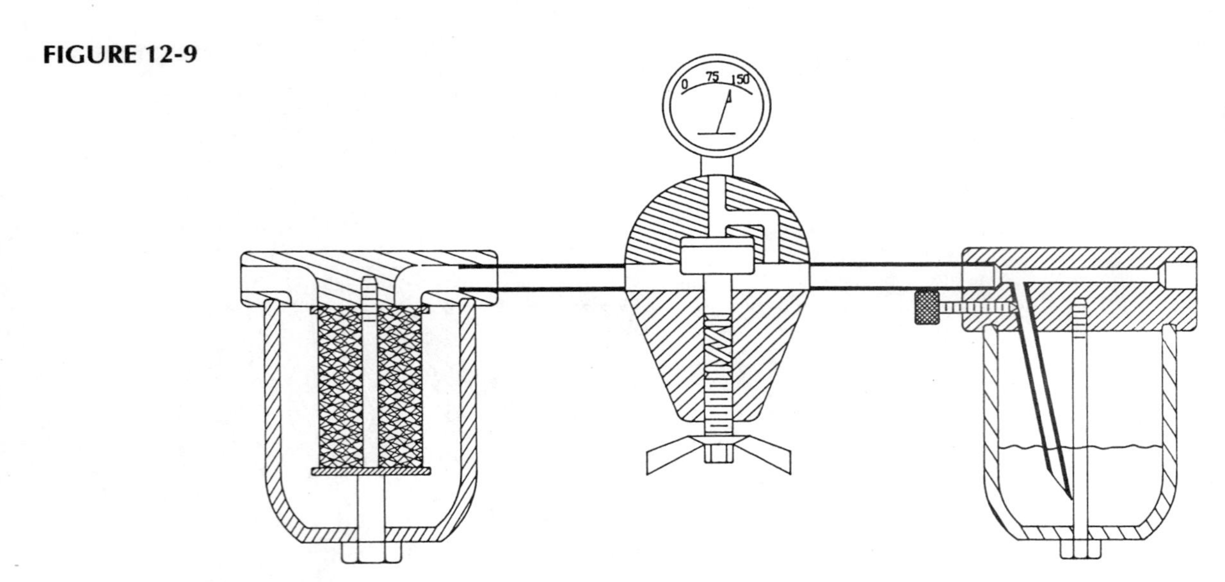

FIGURE 12-10

The FLR unit offers a convenient termination for the air-distribution system as well as providing flexible pressure control and final cleansing and lubrication of the air supply for pneumatic systems. (*Photograph by the author.*)

dislocate the load. Since the volume in the cylinder increases as it moves, an ever-increasing volume is evident throughout the system during cylinder stroke. Accordingly, the pressure throughout the system will drop in keeping with the isothermal expression of Boyle's Law:

$$\frac{P1 * V1}{V2} = P2$$

Problem 12-4

A pneumatic linear circuit is operating isothermally using a 10-in. {*25.4-cm*} diameter ram with a stroke of 20 in. {*50.8 cm*}. This circuit is supplied by a receiver tank having a capacity of 2 cu. ft. {*0.057 m^3*} prepressurized to 200 psi_g {*1,379 kPa_g*}. Assuming that the FLR is set to 200 psi_g {*1,379 kPa_g*} and that temperature does not change, what pressure will be available at the end of the cylinder stroke?

P2 (psi_g) = ______________

Obviously, in nonreplenished pneumatic circuits, such as those used on some braking systems in large trucks, cylinder displacement should be minimized. However, the vast majority of pneumatic circuitry incorporates a compressor supply system as previously described. Therefore, the pneumatic cylinder will move at a relatively constant and instantaneous rate if sufficient time is allowed for the compressor to replenish the receiver tank.

12.4.2 Temperature Effect on Pneumatics

Notice that the previous problem assumed that constant temperature was maintained. This is often not the case and accounts for the reluctance to have pneumatic cylinders hold loads in a constant position over a period of time. Charles's Law states that if the pressure on a confined gas is held constant, the volume of the gas will change in direct proportion to its change in absolute temperature:

$$\frac{V1}{V2} = \frac{TE1}{TE2}$$

As a result, the loss of volume when temperature decreases will cause a vertical load to somewhat retract. The degree to which this contraction occurs is a function of the cross-sectional area of the cylinder's piston:

$$St\ (in.)\ \{cm\} = \frac{V1\ (cu.\ in.)\ \{cm^3\} * TE2\ (^\circ R)\ \{^\circ K\}}{TE1\ (^\circ R)\ \{^\circ K\} * A_{pis}\ (sq.\ in.)\ \{cm^2\}}$$

Problem 12-5

A ram having a 10-in. {*25.4-cm*} diameter piston and a 6-in. {*15.2-cm*} stroke is fully extended vertically supporting a load at 65 degrees Fahrenheit {*15.6 degrees Celsius*}. How much will the ram retract if no further supply is afforded and the temperature drops to 50 degrees Fahrenheit {*10 degrees Celsius*}?

St (in.) {*cm*} = ____________________

Of course, in keeping with Gay-Lussac's Law, if the volume of gas contained within a cylinder is held constant and the temperature changes, the pressure will change in direct proportion:

$$P2\ (psi_a)\ \{kPa_a\} = \frac{TE2\ (^\circ R)\ \{^\circ K\} * P1\ (psi_a)\ \{kPa_a\}}{TE1\ (^\circ R)\ \{^\circ K\}}$$

Problem 12-6

A 10-in. {*25.4-cm*} diameter cylinder is supporting a 10,000-lb. {*44-kN*} load in a fully extended position at 55 degrees Fahrenheit {*12.8 degrees Celsius*}. Assuming that the load cannot move (the volume remains constant) and the temperature rises to 70 degrees Fahrenheit {*21 degrees Celsius*}, what is the new pressure in the cylinder canister?

P2 (psi_g) {kPa_g} = ______________

Seldom do the effects of pressure, volume, and temperature operate in isolation. More often, a complete integration of the gas laws is required to describe a general isothermal gas law:

$$\frac{\text{P1 (psi}_a\text{) } \{kPa_a\} * \text{V1 (cu. in.) } \{cm^3\}}{\text{TE1 (°R) } \{°K\}} = \frac{\text{P2 (psi}_a\text{) } \{kPa_a\} * \text{V2 (cu. in.) } \{cm^3\}}{\text{TE2 (°R) } \{°K\}}$$

Problem 12-7

A 10-in. {*25.4-cm*} diameter piston ram having a 6-in. {*15.2-cm*} fully extended stroke and no external load is operating under a pressure of 75 psi_g {*517.1* kPa_g} and at a temperature of 40 degrees Fahrenheit {*4.4 degrees Celsius*}. If an external load is added to the ram that retracts the piston to an extended distance of 4 in. {*10.2 cm*} and the temperature rises to 50 degrees Fahrenheit {*10 degrees Celsius*}, what is the resulting pressure in the cylinder?

P2 (psi_g) {kPa_g} = ______________

12.4.3 Pneumatic Cylinder Speed

Unlike hydraulic actuation, the rate of movement of a pneumatic actuator is difficult to specify accurately. Since the pneumatic system is prepressurized with air, the rate of movement is affected by the ability of the gas to expand and fill the void in the actuator during movement. Since this void (volume) is constantly changing, along with constantly fluctuating pressure, actuator movement will occur in small bursts. Normally, these incremental bursts of speed are not noticeable in pneumatic actuator movement until a **critical pressure drop** is reached.

The critical pressure drop in a pneumatic circuit is a function of the constituency of the gas used to transmit power. In the case of air, this critical pressure drop is 53%. At or above a critical pressure drop value a pneumatic circuit will operate **isothermally** (at constant temperature), and the random impacts caused by heat gain or loss (**adiabatic** operation) are negligible. Therefore, pneumatic circuitry should be designed to maintain an isothermal condition. As long as isothermal operation is maintained, the decrease in velocity during cylinder movement is proportional to the pressure drop in the cylinder.

An approximation of the theoretical velocity may be determined by calculating the average velocity during the stroke of a pneumatic cylinder under given conditions. The cylinder will move at an initial velocity that is a function of the prepressurized system charge and the pressure needed to overcome the load on the cylinder.

The initial movement of the cylinder is caused by excessive pressure above and beyond that required to displace the load **(L.P.)**. One atmosphere of pressure is always evident and must be included as an external load. The total amount of excess pressure **(E.P.)** is equal to the regulator pressure **(R.P.)** minus a combination of load pressure and atmospheric pressure:

$$\text{E.P. } (\text{psi}_g) = \text{R.P. } (\text{psi}_g) - [\text{L.P. } (\text{psi}_g) + 14.7\ (\text{psi}_g)], \text{ or}$$

$$\text{E.P. } \{kPa_g\} = \text{R.P. } \{kPa_g\} - [\text{L.P. } \{kPa_g\} + 101.4\ \{kPa_g\}]$$

The supply **(S)** potential from the compressor unit may be calculated by expressing the excess pressure in atmospheres (dividing by 14.7 {*101.4*}), then multiplying the result by the compressor DELivery:

$$\text{S (cu. ft./min.)} = \frac{\text{E.P. } (\text{psi}_g)}{14.7\ (\text{psi}_g)} * \text{DEL (cu. ft./min.), or}$$

$$\text{S } \{m^3/min.\} = \frac{\text{E.P. } \{kPa_g\}}{101.4\ \{kPa_g\}} * \text{DEL } \{m^3/min.\}$$

The initial velocity (VEL_{init}) attained by the pneumatic cylinder is dependent on the supply and the appropriate cross-sectional area. Since the supply is rated in cubic feet per minute and the cylinder area is rated in square inches, a conversion is necessary. There are 1,728 cu. in. in 1 cu. ft. {*1,000,000 cm³ in 1 m³*}; therefore:

$$\text{VEL}_{\text{init}} \text{ (in./min.)} = \frac{\text{S (cu. ft./min.)} * 1{,}728 \text{ (cu. in./cu. ft.)}}{\text{A (sq. in.)}}, \text{ or}$$

$$\text{VEL}_{\text{init}} \{cm/min.\} = \frac{\text{S } \{cm^3/min.\} * 1{,}000{,}000\ \{cm^3/m^3\}}{\text{A } \{cm^2\}}$$

Problem 12-8

A single-acting pneumatic cylinder (ram) having a piston diameter of 4 in. {*10.2 cm*} is resisted by a 628-lb. {*2.79-kN*} load. It is operating in a circuit supplied by a compressor unit delivering 2 scfm {*0.057 scmm*}, and the regulator is set to 150 psi_g {*1,034.2* kPa_g}. What is the initial velocity of the cylinder?

VEL_{init} (in./min.) {*cm/min.*} = ________________

As previously described, the pressure in the circuit will drop as the cylinder moves to the end of its stroke. The pressure at the end of the stroke may be calculated by using Boyle's Law. For example, a receiver tank having a volumetric capacity of 2 cu. ft. {*0.057* m^3} at 161.7 psi_a {*1,114.9* kPa_a} actually contains a standard volume (S.V.) of 20 scf {*0.57 scm*} of air:

$$\text{S.V. (scf)} = \frac{161.7\ (\text{psi}_a) - 14.7\ (\text{psi}_a)}{14.7\ (\text{psi}_a)} * 2\ (\text{cu. ft.})$$

$$\text{S.V. (scf)} = \frac{147\ (\text{psi}_a)}{14.7\ (\text{psi}_a)} * 2\ (\text{cu. ft.})$$

$$\text{S.V. (scf)} = 20$$

or

$$\text{S.V. }\{scm\} = \frac{1{,}114.9\ \{kPa_a\} - 101.4\ \{kPa_a\}}{101.4\ \{kPa_a\}} * 0.057\ \{m^3\}$$

$$\text{S.V. }\{scm\} = \frac{1{,}013.5\ \{kPa_a\}}{101.4\ \{kPa_a\}} * 0.057\ \{m^3\}$$

$$\text{S.V. }\{scm\} = 0.57$$

If a cylinder having a 10-sq. in. {*64.5-*cm^2} cross-sectional area and a 17.28-in. {*43.9-cm*} stroke is connected to the supply and the regulator is set to 150 psi_a {*1,034.2* kPa_a}, the final volume will increase by 172.8 cu. in. {*2,831.6* cm^3} or 0.1 cu. ft. {*0.0028* m^3}. The supply pressure will decrease accordingly:

$$\text{P1 } (\text{psi}_a) * \text{V1 (cu. ft.)} = \text{P2 } (\text{psi}_a) * \text{V2 (cu. ft.)}$$

$$\frac{150\ (\text{psi}_a) * 2\ (\text{cu. ft.})}{2.1\ (\text{cu. ft.})} = \text{P2 } (\text{psi}_a)$$

$$142.9 = \text{P2 (psi}_a\text{), or}$$

$$128.2 = \text{P2 (psi}_g\text{)}$$

or

$$\text{P1 } \{kPa_a\} * \text{V1 } \{m^3\} = \text{P2 } \{kPa_a\} * \text{V2 } \{m^3\}$$

$$\frac{1{,}034.2 \ \{kPa_a\} * 0.057 \ \{m^3\}}{0.0598 \ \{m^3\}} = \text{P2 } \{kPa_a\}$$

$$985.8 = \text{P2 } \{kPa_a\}\text{, or}$$

$$884.3 = \text{P2 } \{kPa_g\}$$

Also, the terminal velocity (VEL_{term}) of the cylinder will decrease in proportion to the decrease in pressure. For example, if the initial velocity (VEL_{init}) of the previously described cylinder is 300 in./min. {*762 cm / min.*}, the terminal velocity (VEL_{term}) may be calculated:

$$\frac{VEL_{init} \text{ (in./min.)}}{\text{P1 (psi}_a\text{)}} = \frac{VEL_{term} \text{ (in./min.)}}{\text{P2 (psi}_a\text{)}}$$

$$\frac{300 \text{ (in./min.)} * 142.9 \text{ (psi}_a\text{)}}{150 \text{ (psi}_a\text{)}} = VEL_{term} \text{ (in./min.)}$$

$$258.8 = VEL_{term} \text{ (in./min.)}$$

or

$$\frac{VEL_{init} \ \{cm / min.\}}{\text{P1 } \{kPa_a\}} = \frac{VEL_{term} \ \{cm / min.\}}{\text{P2 } \{kPa_a\}}$$

$$\frac{762 \ \{cm / min.\} * 984.8 \ \{kPa_a\}}{1{,}034.2 \ \{kPa_a\}} = VEL_{term} \ \{cm / min.\}$$

$$725.6 = VEL_{term} \ \{cm / min.\}$$

Assuming that the velocity loss is linear throughout the cylinder stroke, the average velocity (VEL_{avg}) may be calculated using the following formula:

$$VEL_{avg} = \frac{VEL_{init} + VEL_{term}}{2}$$

Using the 10-sq. in. {*64.5-cm²*} cross-sectional area cylinder, the average velocity would be as follows:

$$VEL_{avg}\ (\text{in./min.}) = \frac{300\ (\text{in./min.}) + 285.8\ (\text{in./min.})}{2}$$

$$VEL_{avg}\ (\text{in./min.}) = \frac{585.8\ (\text{in./min.})}{2}$$

$$VEL_{avg}\ (\text{in./min.}) = 292.9$$

or

$$VEL_{avg}\ \{cm/min.\} = \frac{762\ \{cm/min.\} + 725.6\ \{cm/min.\}}{2}$$

$$VEL_{avg}\ \{cm/min.\} = \frac{1{,}487.6\ \{cm/min.\}}{2}$$

$$VEL_{avg}\ \{cm/min.\} = 743.8$$

Using the previously described cylinder, the time (T) for one extension stroke may be calculated:

$$T\ \{(\text{min.})\} = \frac{St\ (\text{in.})\ \{cm\}}{VEL_{avg}\ (\text{in./min.})\ \{cm/min.\}}$$

$$T\ \{(\text{min.})\} = \frac{17.28\ (\text{in.})}{292.9\ (\text{in./min.})} \text{ or } \frac{43.9\ \{cm\}}{743.8\ \{cm/min.\}}$$

$$T\ \{(\text{min.})\} = 0.059, \text{ or } T\ \{(\text{sec.})\} = 3.54$$

Even if we assume no frictional loss, heat loss, or variation in load, determination of the speed of an air cylinder is a complex operation. Variations in temperatures, loads, and atmospheric conditions will greatly affect any theoretical velocity determination. Therefore, air circuitry is not typically used to produce controlled feed rates, although flow-control valves will often be included. Simple needle valve–type flow controls (often referred to as chokes) are extensively used in pneumatic circuitry. Most often, these chokes are included to allow continual "tuning" of velocity through throttling of exhaust air. They may also appear as mechanisms used to retard the rapid response of air signals when used as pilots for shifting and signaling directional controls. In reality, the complex air circuit typically emphasizes the rapid response inherent in pneumatics to order actions by using a network of directional controls.

12.5 ROTARY ACTUATION IN PNEUMATIC CIRCUITRY

Pneumatic circuitry requiring rotary output is becoming increasingly popular with the wide-scale use of portable pneumatic power tools. Two types of rotary actuators are popular: the **rotary motor** and the **oscillator**.

12.5.1 Pnuematic Rotary Motors

Anyone who has watched a professional motor car race or visited a local professional garage is familiar with the operation of a pneumatic rotary motor device used to remove and secure lug nuts during tire changes. These pneumatic tools are typically called air-impact wrenches. Pneumatic rotary motors are also employed to power portable drills, jackhammers, concrete vibrators and saws, or in almost any area where sparks from electical motors could be dangerous.

12.5.1.1 Rotary motor types Theoretically, any device used as a pneumatic compressor could also be employed as a pneumatic rotary motor. Most popularly, pneumatic motors are either vane or piston types.

Vane motors can be produced in very small physical packages, reducing the weight of the tool and resulting in extremely high velocity potential (as high as 30,000 rpm). Typically, vane motors are employed in low-torque applications. However, they commonly produce more power output per unit weight than piston types.

Piston-type rotary motors are usually either of the radial-piston or axial-piston designs. Crankshaft-type motors are not popular because of limitations imposed upon speed by the inertia of the larger masses employed in their construction. The larger variety of piston motors use the radial design. Radial-piston rotary motors are low-speed, high-torque motors that may be fitted with up to six pistons. Axial-piston motors are typically the smaller variety, operating at higher speed and with less potential torque development. Radial piston rotary motors may be rated as highly as 25 hp {*18.6 kW*}, while the axial variety is typically limited to 3 hp {*2.2 kW*}.

12.5.1.2 Rotary motor operation Pneumatic rotary motors are specified and may be purchased based upon their power rating. Through experimentation, manufacturers will rate their pneumatic motors in horsepower {*wattage*} units at a given speed. Determination of system requirements are then possible using formulation:

$$hp = \frac{T\ (ft.\ lbs.) * SP\ (rpm)}{5{,}252},\ or$$

$$kW = \frac{T\ \{N\ m\} * SP\ \{rpm\}}{9{,}550}$$

Problem 12-9

If a manufacturer rates its pneumatic motor at 1.5 hp {*1.1 kW*} at 3,000 rpm, what torque is this motor capable of producing?

T (ft. lbs.) {*N m*} = ________________

The torque capability of the pneumatic rotary motor will reveal its effective displacement at a specified pressure:

$$DIS\ (cu.\ in./rev.) = \frac{T\ (ft.\ lbs.) * 24\pi}{P\ (psi_g)},\ or$$

$$DIS\ \{cm^3/rev.\} = \frac{T\ \{N\ m\} * 2{,}000\pi}{P\ \{kPa_g\}}$$

Problem 12-10

A rotary motor exerting a torque of 2.63 ft. lbs. {*3.5 N m*} is operating in a pneumatic circuit regulated to 125 psi_g {*861.8* kPa_g}. What is the effective displacement for this motor?

DIS (cu. in./rev.) {*cm³/rev.*} = ________________

Knowing the effective displacement of the pneumatic rotary motor, the delivery needed for operation may be determined as follows:

$$DEL\ (scfm) = \frac{DIS\ (cu.\ in./rev.) * SP\ (rpm)}{1{,}728\ (cu.\ in./cu.\ ft.)},\ or$$

$$DEL\ \{scmm\} = \frac{DIS\ \{cm^3/rev.\} * SP\ \{rpm\}}{1{,}000{,}000\ \{cm^3/m^3\}}$$

This necessary delivery is a theoretical construct based on demand at the output end of a rotary pneumatic circuit. As such, it would be adequate for calculations of compressor delivery, if compressor recovery were not a factor. More often, in pneumatic circuitry operating below 125 psi_g

{*861.8* kPa_g} and in intermittent use, this delivery is typically doubled for determination of compressor delivery output. In continuously operating pneumatic circuitry, the motor demand for delivery is typically quadrupled to determine compressor delivery.

Problem 12-11

A pneumatic rotary motor having a 1.75 cu. in./rev. {*26.68* cm^3*/rev.*} displacement is operating at 3,000 rpm. What delivery demand should be made for the circuit compressor if the system is operating intermittently?

DEL (scfm) {*scmm*} = ________________

12.5.2 Pneumatic Oscillators

Pneumatic rotary motors are continuously operating output devices. Occasionally, a discretely operating rotary device is desirable. The pneumatic oscillator is such a device. Some very sophisticated gearing operations may be employed to convert the cyclical nature of a pneumatic cylinder to discrete rotary output. Other oscillators rely on physical stops that limit rotation of a rotary motor.

Determination of system demands from oscillator operation is the same as for any other pneumatic linear circuit or rotary circuit dependent on the osciallator design. If the oscillator is of the limited rotating motor design, delivery demand must be adjusted to reflect the degree of rotation required. For example, an oscillator having 180 degrees of rotation would require a delivery equal to one-half of that required from a rotary motor having the same displacement.

Other than rotary output device designation, the pneumatic rotary or oscillating circuit may be identical to pneumatic linear circuitry. Figure 12-11 shows schematic diagrams for both simple rotary and oscillating circuitry.

In the rotary circuit illustrated, rotation is available in only one direction and will continue as long as the palm button is activated. When the palm button is released, rotation will immediately stop. In the oscillating circuit illustrated, when the lever on the directional control is shifted to one extreme position, partial rotation will occur in one direction. When the lever is shifted into the other extreme position, partial rotation will occur in the other direction. Releasing the lever at any time during partial rotation should hold the oscillator rigidly in any location because the directional control will shift to its center position at that time.

12.6 PNEUMATIC DIRECTIONAL CONTROL

Structurally, pneumatic directional controls are nearly identical to hydraulic directional controls. There are rotary and sliding spool design, with the sliding spool being the more popular, as well as ball and poppet

FIGURE 12-11

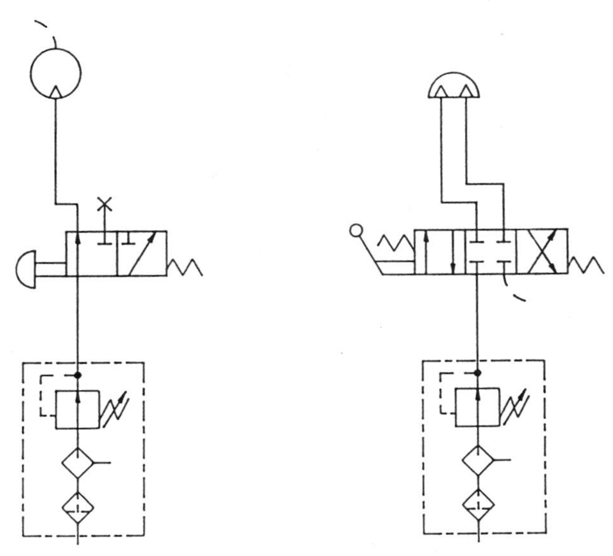

Unidirectional rotary circuit Bidirectional oscillating circuit

element check valves. Master control valves having three ports (cylinder, pressure, and exhaust) are commonly used with ram-type and spring-return cylinders. Four-way (sometimes called five-way) valves are used with double-acting cylinders. The inconsistency in nomenclature as to whether this design is a four- or five-way valve stems from the fact that these air valves typically have five ports. Since exhaust may be directed to the atmosphere, there is no need to consolidate the exhaust ports inside the valve casing. Therefore, the four-way valve will typically have one pressure port and two exhaust and cylinder ports (see Figures 12-12 and 12-13).

Although three-position pneumatic directional controls are available, most are two-position. Among the more commonly used two-position pneumatic controls is the one-way, plunger-activated type, typically known as a **limit switch**.

12.6.1 Flip-Flop Circuitry

One of the more impressively operating pneumatic circuits is the **flip-flop circuit**. The flip-flop circuit is a basic linear circuit using a double end rod, double-acting cylinder or an oscillator. A cam attached to each end rod activates a limit switch at the end of each stroke. When the limit switch opens, a pulse of air is directed to the pilot-activated master directional control. This automatically reverses the cylinder or oscillator direction. Figure 12-14 schematically illustrates the pneumatic flip-flop circuit using cylinder ouput.

FIGURE 12-12

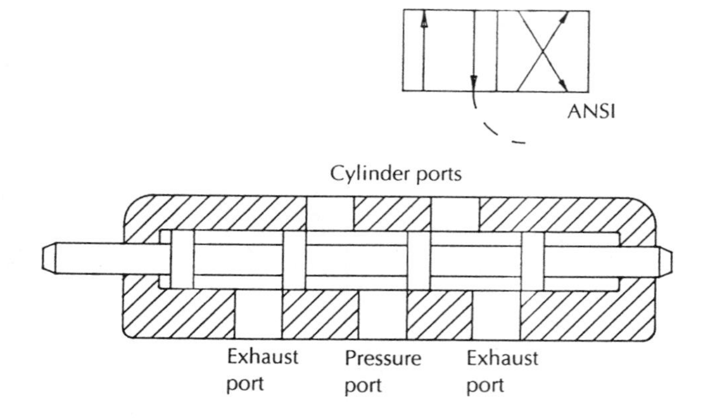

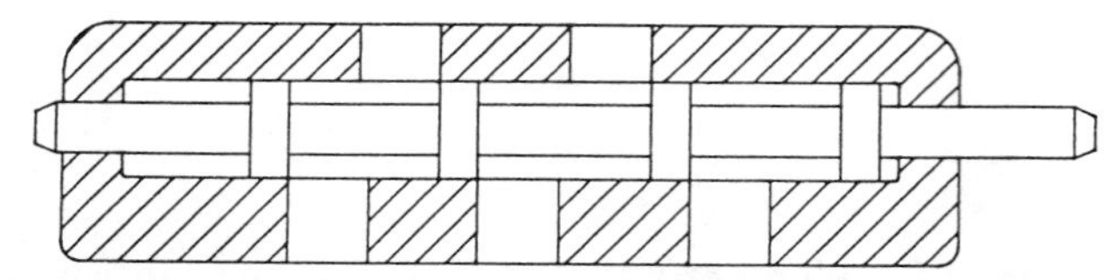

FIGURE 12-13

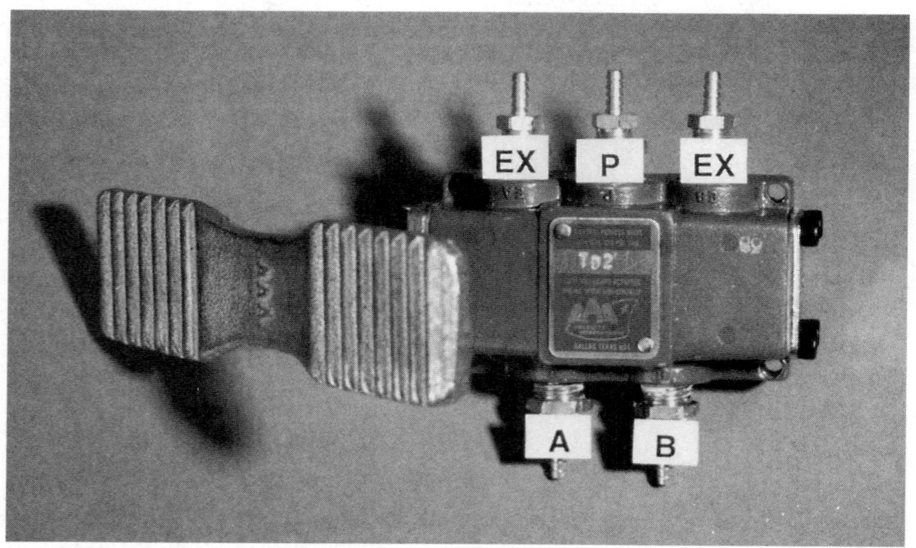

Pneumatic four-way directional control normally has five ports: pressure (P), "A" and "B" ports, and two exhaust ports (EX). (*Photograph by the author.*)

Variability in the operation of the flip-flop circuit is available through alterations within the master directional control. Within the directional control, the pilot activation is caused by a small piston or cylinder con-

FIGURE 12-14

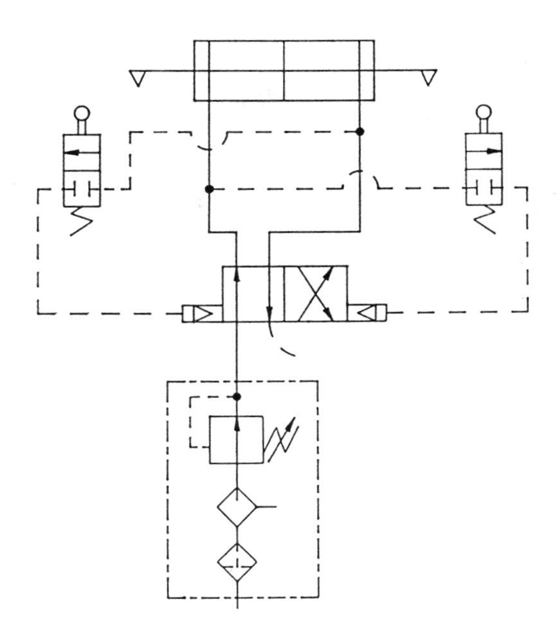

nected to each end of the sliding spool. The stroke of the small cylinder is sized to coincide with full shifting of the master directional control's sliding spool (see Figure 12-15).

From this basic directional control package, many possible arrangements exist. For example, by placing a restriction (flow control) in the pilot port, a delay in shifting can be caused. This arrangement is typically called a **timer**. Although fairly accurate time control may be achieved with certain valves, current technology limits the maximum delay to less than one minute. Also, it should be noted that extremely clean and dry air must be used to afford accurate time control in pneumatics. Therefore, supplemental air supply that is highly filtered and monitored is typically

FIGURE 12-15

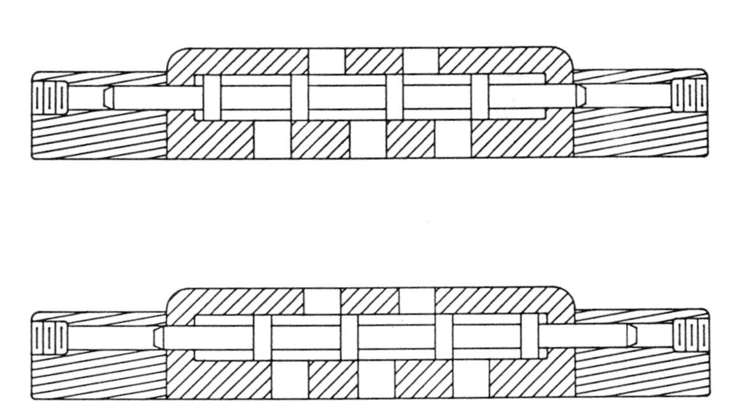

used for timer circuitry. Furthermore, the independent pressure control afforded with an independent pilot supply, in combination with variable restriction in the flow control, allows a broad range of timing possibilities. Figure 12-16 shows the schematic design of a pneumatic sequence circuit with time delay on secondary operations and independent pilot supply.

FIGURE 12-16

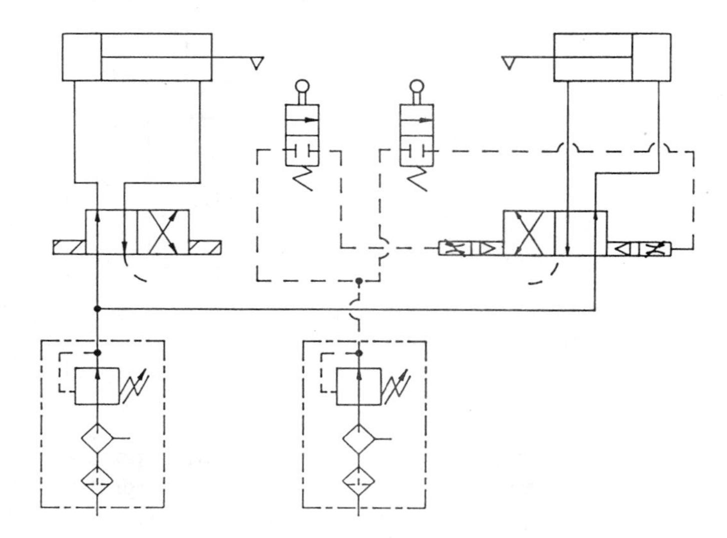

Another variation involves the use of mechanical springs acting on the small shifting cylinders in the directional control. With this arrangement, the valve will not shift until a pilot pressure that overcomes the spring force is evident. So, although pressure parameter sequence is not common in pneumatics, it is possible through this arrangement.

12.6.2 Electrical Activation in Pneumatics

Many limit switches for pneumatics are specialized push button–type electrical switches. In turn, these switches activate solenoids in master directional controls for flip-flop and sequencing purposes. With the sophistication of modern-day electronics, it is possible to activate pneumatic circuitry from pressure, light, heat, noise, and many other sources.

12.7 TRENDS IN PNEUMATICS

There is a major trend in pneumatics toward increased use of digital logic. Through the use of solenoid-activated directional controls interfaced with computers and programmable logic controllers, a wide variety of instan-

FIGURE 12-17

The multistation pneumatic bar code testing cell relies on microprocessor control signals to perform both rotary and linear timed actuation. (*Photograph by the author.*)

taneously reacting circuit design has become available. Combinations of flip-flop, sequence, and timing circuits are easily controlled through the programming of input–output ports of even the simplest microprocessors (see Figure 12-17).

Digital fluid logic (fluidics) has made great advances since the refinement and development of relatively simple switching devices using the Coanda effect. Basically, these devices rely on the nature of a moving fluid stream to bend toward the direction of a force applied to it after passing the point of that force. Consider the water flowing from a faucet. If you insert your finger into the side of the water flow, the water will have a tendency to bend toward your finger after the point of insertion. Controllable pilot pressures exerted 180 degrees apart into an airstream may be used to bend the airflow into one path or another (see Figures 12-18 and 12-19).

Through the use of banks of Coanda valves, logic control systems similar to binary digital computer networks are possible. The marriage of digital logic in pneumatics and electronics with the power and controllability of hydraulics has placed the fluid power industry on the leading edge of developing technology.

FIGURE 12-18

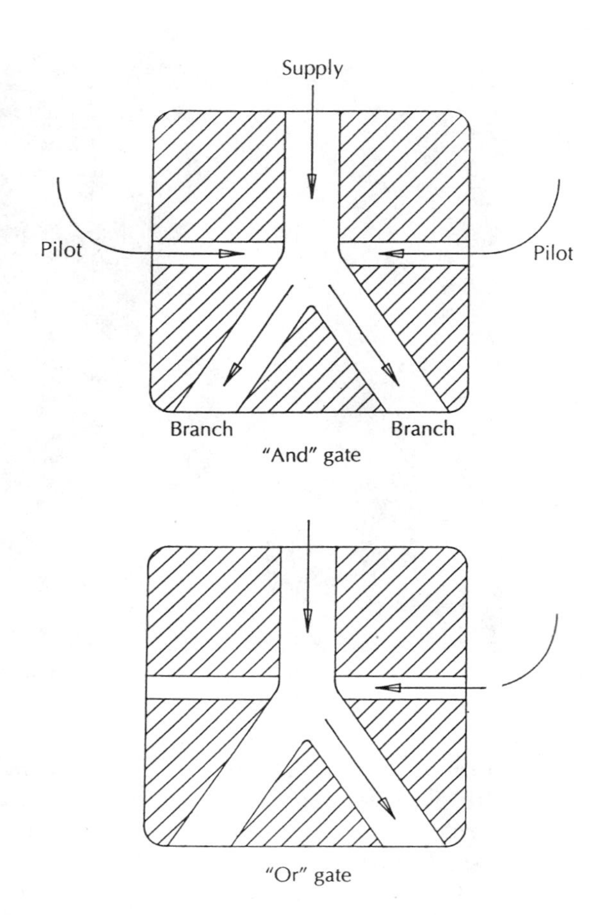

12.8 SUGGESTED ACTIVITIES

1. Visit a local building materials supply store or scan a major supplier's catalog. Make a listing of all the pneumatic tools that are available from those sources. If you have an opportunity, get a demonstration on their uses. Try to determine if the actuator in the device is linear, rotary, or oscillating.
2. Make a small receiver tank from a large-diameter piece of pipe and two end caps. Drill and tap two holes in your receiver tank to apply air fittings that will accept a pressure gauge and an air supply. Calculate mathematically the volume of your receiver tank. Observe the amount of time that it takes the supply to fill your receiver tank to the level that your receiver tank pressure gauge reaches the same setting as the supply pressure gauge. Calculate the number of standard cubic feet {*meters*} or air contained in your tank when fully charged.
3. Set up a mock-up of a pneumatic linear and rotary circuit on a laboratory trainer. Operate the circuitry and observe the action of the actuator, the gauge on the FLR, and time for cycling (extension, retraction, speed of rotation). Summarize your observations in a brief report.

FIGURE 12-19

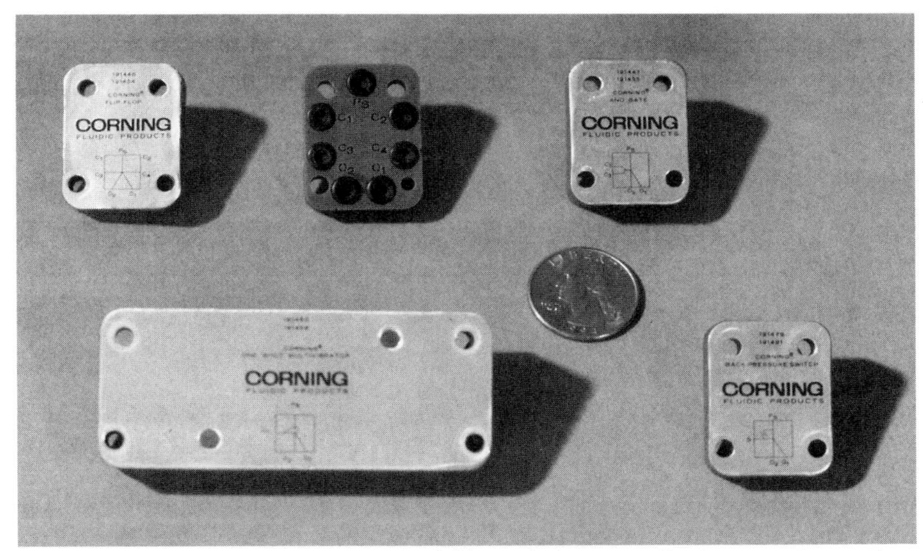

Coanda valves are used to perform digital logic functions in fluidic operations. (*Photograph by the author.*)

4. Set up a mock-up of a pneumatic flip-flop circuit. Start the circuit and record the time for extension and retraction. Let the circuit run awhile, then repeat your timing for extension and retraction. Are they consistent with initial time? If not, what does this tell you about recovery time for the compressor?
5. Dream up an application for a pneumatic circuit (the more complicated, the better!). Describe loads and speed–cycle parameters for your circuit. Draw a schematic that represents a solution for your dream. Specify all components using correct design procedures for your circuit using an electronic spreadsheet program to perform your calculations.

12.9 REVIEW QUESTIONS

1. Describe situations that would be favorable for use of pneumatic power rather than hydraulic power.
2. Discuss designs of air compressors (pumps) typically used in pneumatic power supplies.
3. Describe Gay-Lussac's Law and summarize its effect on the operation of a pneumatic receiver tank.
4. What is a standard cubic foot or *standard cubic meter* of air?
5. What is the function of a pop-off valve on a pneumatic power supply?
6. What is the function of a separator in a pneumatic circuit, and how does a centrifugal separator operate to perform this function?

7. What are unloading valves, and why are they used in pneumatic power supplies?
8. Why do globe valves make poor shutoff devices for pneumatic circuits?
9. Why is pressure drop in a pneumatic distribution system so critical in operation?
10. What are the three components of an FLR unit, and how does each one operate?
11. Discuss the relationships among Boyle's Law, Charles's Law, and Gay-Lussac's Law and their effects on the operation of a pneumatic cylinder.
12. What is the difference between isothermal and adiabatic operation in pneumatics?
13. Describe the procedure used to determine the time required for cycling a pneumatic cylinder under isothermal operation.
14. Describe the procedure for determining load and speed capabilties for pneumatic motors.
15. What is a pneumatic oscillator?
16. Why are four-way pneumatic directional controls sometimes called five-way directional controls?
17. Describe the operation of a pneumatic linear flip-flop circuit.
18. From which parameter (position or pressure) are pneumatic sequence circuits normally designed? Why?
19. What is fluidics, and how are Coanda valves used to perform fluidic functions?

Descriptions and Inside Diameters of Conductors

U.S. Customary by Nominal Size (in Inches)

	Pipe			Tube				
(Nominal)	**Sch. 40**	**Sch. 80**	**Sch. 160**	**K**	**L**	**M**	**N**	**Hose**
1/8 in.	0.269 in.	0.215 in.		0.055 in.				
3/16 in.				0.118 in.				0.188 in.
1/4 in.	0.364 in.	0.302 in.		0.180 in.	0.152 in.	0.120 in.		0.250 in.
5/16 in.				0.243 in.	0.215 in.	0.183 in.		0.312 in.
3/8 in.	0.493 in.	0.423 in.		0.305 in.	0.277 in.	0.245 in.		0.375 in.
1/2 in.	0.622 in.	0.546 in.	0.464 in.	0.430 in.	0.402 in.	0.370 in.	0.310 in.	0.500 in.
5/8 in.				0.555 in.	0.527 in.	0.495 in.	0.435 in.	0.625 in.
3/4 in.	0.824 in.	0.742 in.	0.612 in.	0.652 in.	0.620 in.	0.532 in.		0.750 in.
7/8 in.				0.777 in.	0.745 in.	0.657 in.		0.875 in.
1 in.	1.049 in.	0.957 in.	0.815 in.	0.902 in.	0.870 in.	0.760 in.		1.000 in.
1-1/4 in.	1.380 in.	1.278 in.	1.160 in.	1.120 in.	1.060 in.	1.010 in.		1.250in.
1-1/2 in.	1.610 in.	1.500 in.	1.338 in.	1.370 in.	1.310 in.	1.232 in.		1.500 in.
1-3/4 in.				1.620 in.	1.560 in.	1.482 in.		
2 in.	2.067 in.	1.939 in.	1.687 in.	1.870 in.	1.810 in.	1.732 in.		2.000 in.
2-1/2 in.	2.469 in.	2.323 in.	2.125 in.					2.500 in.
3 in.	3.068 in.	2.900 in.	2.624 in.					3.000 in.

Metric by Nominal Size (in Centimeters)

	Pipe			Tube				
Nominal	**STD**	**XS**	**XXS**	**K**	**L**	**M**	**N**	**Hose**
4 mm				0.30 cm				
5 mm								0.48 cm
6 mm	0.68 cm	0.54 cm		0.40 cm	0.3			
7 mm								0.64 cm
8 mm	0.92 cm	0.76 cm		0.60 cm	0.5	0.4		0.79 cm
10 mm				0.80 cm	0.7	0.6		0.95 cm
11 mm	1.25 cm	1.07 cm						
12 mm				1.00 cm	0.9	0.8		
13 mm								1.27 cm
14 mm	1.58 cm	1.38 cm	1.17 cm		1.00 cm			
15 mm				1.20 cm	1.10 cm			
16 mm					1.20 cm	1.00 cm		1.59 cm
18 mm						1.50 cm		
19 mm								1.90 cm
20 mm	2.10 cm	1.89 cm	1.56 cm		1.60 cm	1.50 cm	1.40 cm	
22 mm				2.00 cm	1.90 cm	1.80 cm		2.22 cm
24 mm	2.66 cm	2.43 cm	2.07 cm					
25 mm						1.90 cm	1.80 cm	2.54 cm
28 mm					2.40 cm	2.30 cm		
29 mm								2.86 cm
30 mm						2.40 cm	2.20 cm	
32 mm								3.18 cm
33 mm	3.51 cm	3.25 cm	2.91 cm					
35 mm				3.10 cm	2.90 cm			
38 mm						3.00 cm	2.80 cm	3.81 cm
39 mm	4.09 cm	3.81 cm	3.40 cm					
42mm				3.80 cm	3.60 cm			
49 mm	5.25 cm	4.92 cm	4.28 cm					
51 mm								5.08 cm
59 mm	6.27 cm	5.90 cm	5.39 cm					
64 mm								6.35 cm
75 mm	7.79 cm	7.37 cm	8.67 cm					
76 mm								7.62 cm
89 mm								8.89 cm
102 mm								10.16 cm

Cylinder Annulus Areas by Piston and Rod Size

U.S. Customary Standards (Diameter in Inches, Area in Square Inches)

	Rod Diameter	**¼**	**½**	**¾**	**1**	**1½**	**2**	**3**
	Rod Area	**0.049**	**0.196**	**0.422**	**0.785**	**1.767**	**3.142**	**7.069**
Piston								
Diameter	**Area**							
1/2	0.196	0.147						
5/8	0.307	0.258	0.111					
3/4	0.442	0.371	0.246					
1	0.785	0.736	0.589	0.363				
1½	1.767		1.571	1.345	0.982			
1¾	2.405			1.983	1.62			
2	3.142			2.719	2.356	1.374		
2½	4.909				4.124	3.142		
3	7.069				6.284	5.302	3.927	
4	12.566				11.781	10.799	9.424	
5	19.635				18.85	17.868	16.493	12.566
6	28.274					26.507	25.132	21.205
8	50.266						47.124	43.197
10	78.540						75.398	56.548
12	113.098						109.956	106.029

Metric Standards (Diameter in Centimeters, Area in Square Centimeters)

Piston								
	Rod Diameter	**0.40**	**0.60**	**1.20**	**2.50**	**3.20**	**4.00**	**5.00**
	Rod Area	**0.126**	**0.283**	**1.131**	**4.909**	**8.043**	**12.566**	**19.635**
Diameter	**Area**							
1.2	1.131	1.005						
2.5	4.909	4.783	4.626					
3.2	8.043	7.917	7.76					
4	12.566	12.44	12.283	11.435				
5	19.635		19.352		18.504	14.726		
6.3	31.173			30.042	26.264			
8	50.266			49.135	45.357	42.223		
10	78.54				73.631	70.497		
12.5	122.719				117.81	114.676	110.153	
16	201.062				196.153	193.019	188.496	
20	314.16				309.251	306.118	301.594	294.525
25	490.875					482.832	478.309	471.24
32	804.25						791.684	684.615

APPENDIX C

Fluid Schematic Symbols Used in This Book

LINES

FLUID CONTAINERS AND CONDITIONERS

Accumulators

Fluid Conditioners

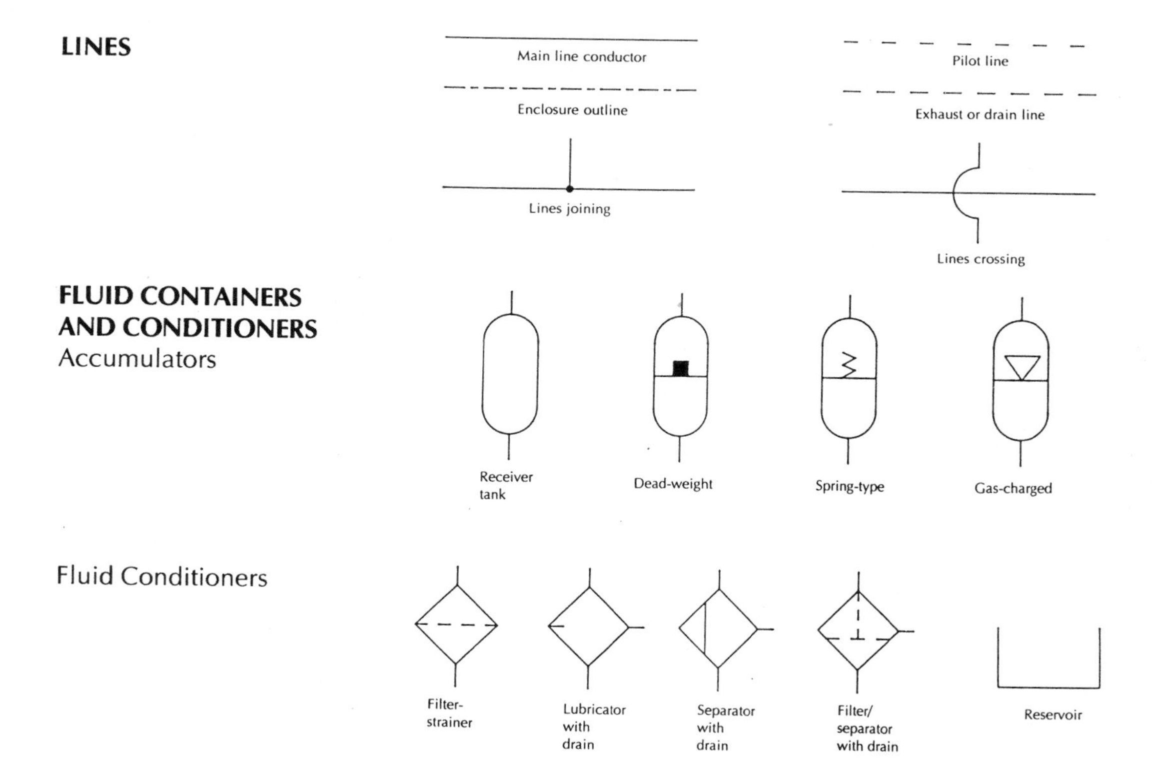

ROTARY DEVICES

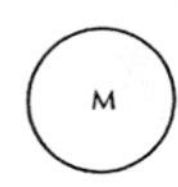

Electric motor

Unidirectional

Variable displacement

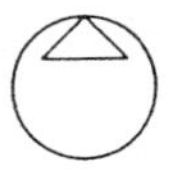
Air compressor

Hydraulic Motors

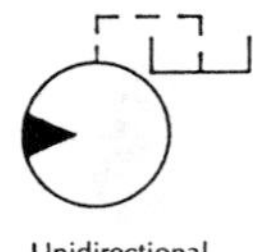
Unidirectional

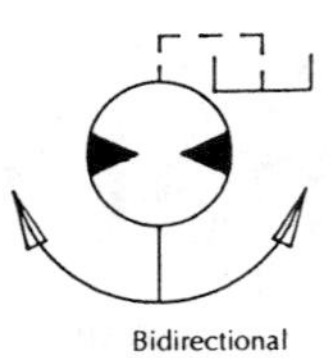
Bidirectional

Variable displacement

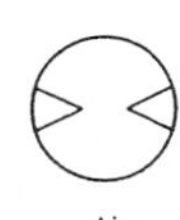
Air motor

PRESSURE CONTROLS

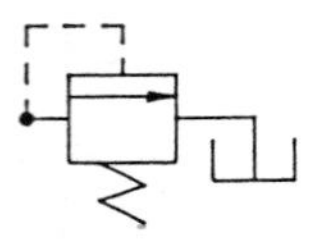
Hydraulic

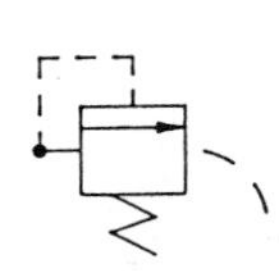
Pneumatic

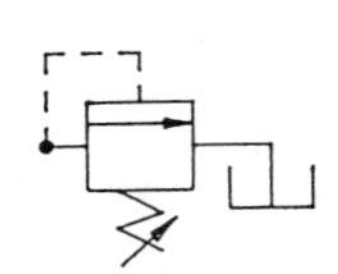
Relief valve

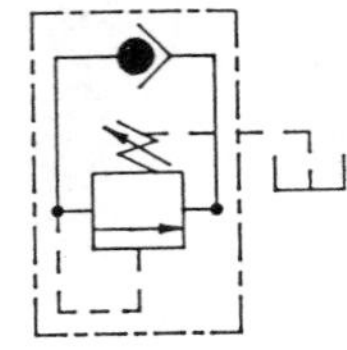
Sequence valve

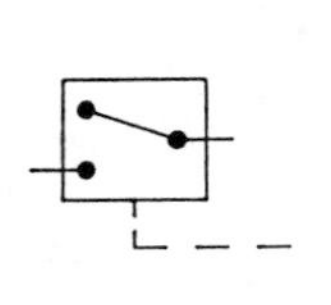
Pressure/electrical switch

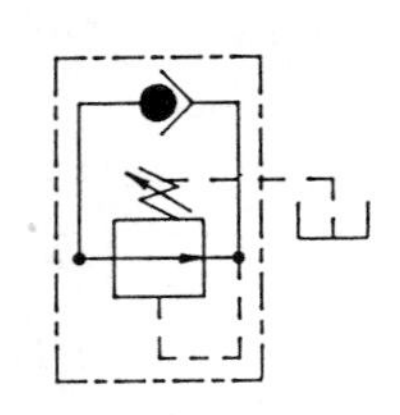
Pressure-reducing valve

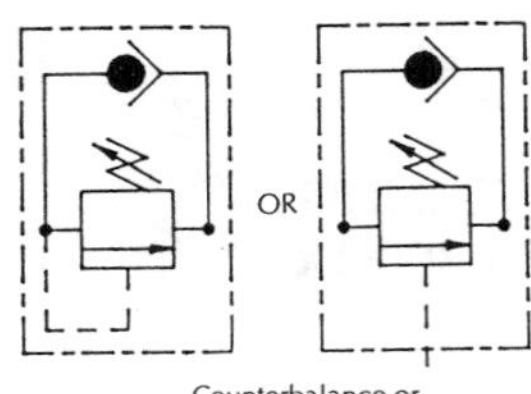

Counterbalance or unloading valve

CHECK VALVES

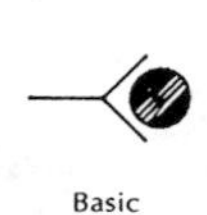
Basic

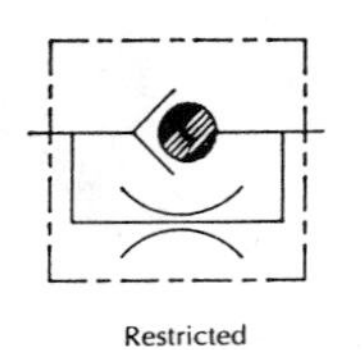
Restricted

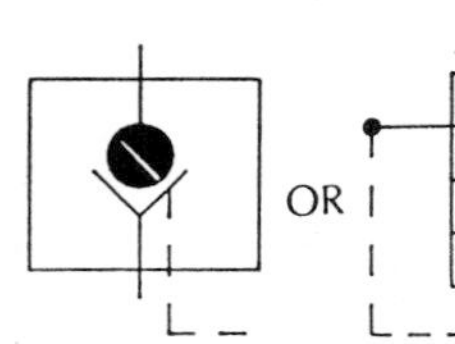

Pilot Override

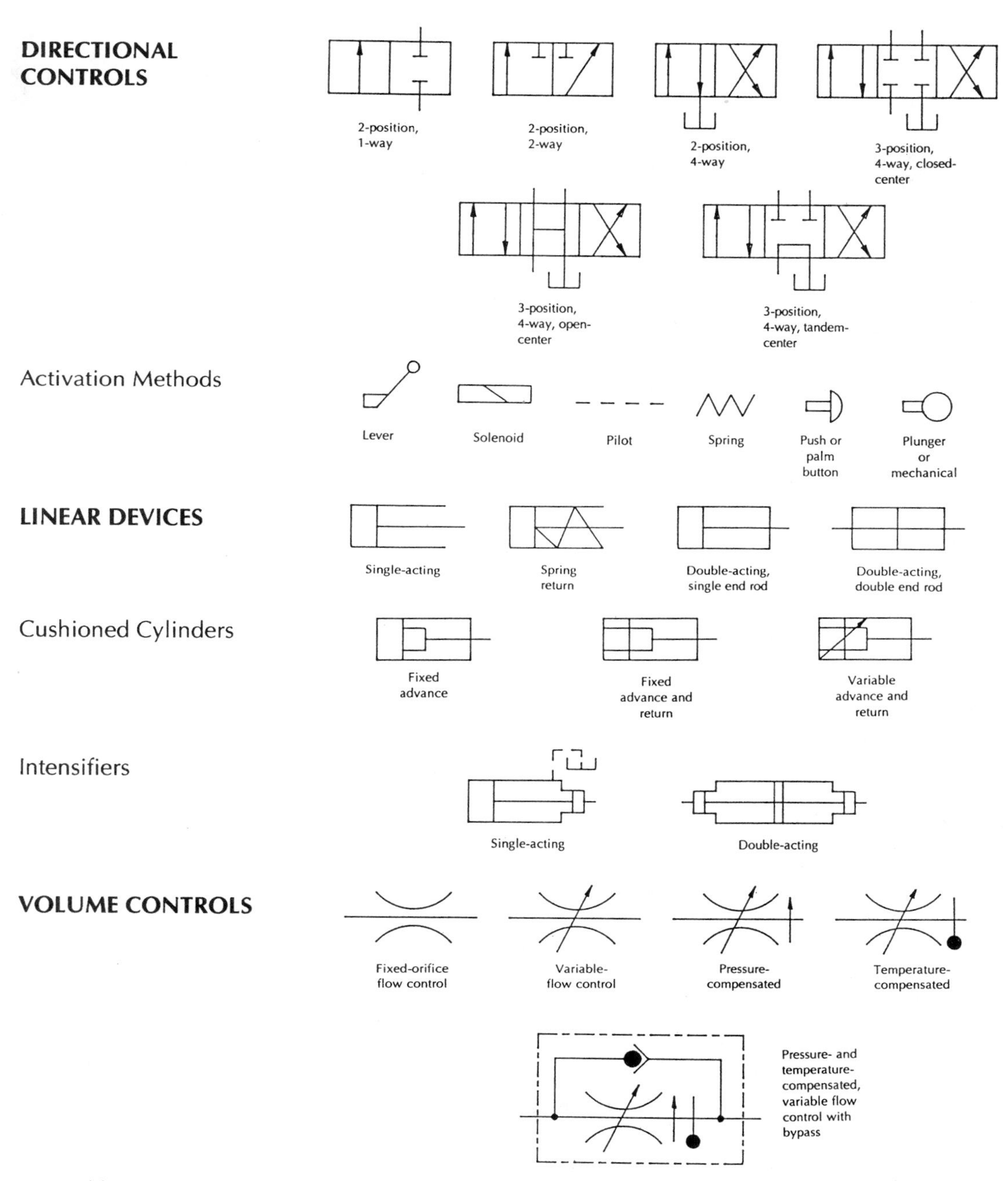

Extracted from USA Standard Graphic Symbols for Fluid Power Diagrams (USASY 32.10-1967) with the permission of the publisher, The American Society of Mechanical Engineers, United Engineering Center, 345 East 47th Street, New York, NY 10017.

APPENDIX D

Answers to Problems

CHAPTER 2

In-Chapter Problems

Problem 2-1 $F1\ (lbs.)\ \{N\} = \dfrac{F2\ (lbs.)\ \{N\} * D2\ (ft.)\ \{m\}}{D1\ (ft.)\ \{m\}}$

Problem 2-2 F1 (lbs.) = 30, or F1 {*N*} = 136.5

Problem 2-3 P (psi) = 15, or P {*kPa*} = 136.5

Problem 2-4 $F2\ (lbs.)\ \{N\} = \dfrac{F1\ (lbs.)\ \{N\} * A2\ (sq.\ in.)\ \{cm^2\}}{A1\ (sq.\ in.)\ \{cm^2\}}$

Problem 2-5 F2 (lbs.) = 60, or F2 {*N*} = 267

Problem 2-6 F2 (lbs.) = 60, or F2 {*N*} = 267

Problem 2-9 W (lbs. ft.) = 3,000, or W {*kJ*} = 4,080

Problem 2-13 Cycles = 24

Practical Design Problem

2.5-1 DIS (cu. in.) = 40, or DIS {cm^3} = 656

2.5-2 Distance (in.) = 2, or Distance {*cm*} = 5.08

2.5-3 Cycles = 2

2.5-4 Time{(sec.)} = 20

2.5-5 P (lbs./in.^{2}) = 200, or P {*kPa*} = 1,379

CHAPTER 3

In-Chapter Problem

Problem 3-1 Specific Gravity (S.G.) = 0.80

CHAPTER 4

In-Chapter Problems

Problem 4-1 C1 (in.) = 474, or C1 {*cm*} = 1,204

Problem 4-2 DIS (cu. in./rev.) = 24.62, or DIS {$cm^3/rev.$} = 698.4

Problem 4-3 V.E. {(%)} = 80

Problem 4-4 DEL (gal./min.) = 127.9, or DEL {*l/min.*} = 484.2

Problem 4-5 V.E. {(%)} = 81

Problem 4-6 W (in. lbs.) = 1,256, or W {*J*} = 143

Problem 4-7 Power (in. lbs./sec.) = 2,301, or Power {*W*} = 128,713

Problem 4-8	hp = 1.09
Problem 4-9	DIS (cu. in./rev.) = 30, or DIS {*cm³/rev.*} = 492
Problem 4-10	DIS (cu. in./rev.) = 7.2, or DIS {*cm³/rev.*} = 118
Problem 4-11	S1 (in.) = 0.776, or S1 {*cm*} = 1.97
Problem 4-13	a. Maximum delivery (gpm) = 60, or Maximum delivery {*lpm*} = 227 b. Minimum delivery (gpm) = 40, or Minimum delivery {*lpm*} = 151
Problem 4-14	P (psi) = 424, or P {*kPa*} = 2,923
Problem 4-15	P (psi) = 471, or P {*kPa*} = 3,248
Problem 4-16	a. Inlet — At lowest velocity, Dia. (in.) = 1.011; or Dia. {*cm*} = 2.568 At highest velocity, Dia. (in.) = 0.639; or Dia. {*cm*} = 1.623 b. Working — At lowest velocity, Dia. (in.) = 0.540; or Dia. {*cm*} = 1.372 At highest velocity, Dia. (in.) = 0.320; or Dia. {*cm*} = 0.813

Practical Design Problem

4.5-1 P (psi) = 438, or P {*kPa*} = 3,010

4.5-2 DEL (gpm) = 7.83, or DEL {*lpm*} = 29.64

4.5-3 P (psi) = 482, or P {*kPa*} = 3,311

4.5-4 DIS (cu. in./rev.) = 4.52, or DIS {*cm³/rev.*} = 74.07

4.5-5 The pressure and delivery for this circuit are not too high; therefore, the piston pumps would not be a wise decision. Either the gear or vane pump would be a good selection. In general, the gear types are less expensive and have good tolerance of the dirty environment encountered with most outdoor portable equipment. Therefore:
Pump type = gear V.E. {(%)} = 70

4.5-6 hp = 3.14, or P {*W*} = 2,341

4.5-7 CAP (gal.) = 15.66 to 23.49, or CAP {*l*} = 58.28 to 88.92

4.5-8 Rating (gal./min.) = 22.37 to 33.56, or Rating {*l/min.*} = 84.69 to 127.03

4.5-9 Inlet lines DIA (in.) = 0.956 to 1.511, or DIA {*cm*} = 1.838 to 2.428
Working lines DIA (in.) = 0.401 to 0.676, or DIA {*cm*} = 1.019 to 1.717

4.5-10 Inlet lines (in.) = 1 ¼, or Inlet lines {*mm*} = 33
Working lines (in.) = ½, or Inlet lines {*mm*} = 14

4.5-11 There is only one deviation from the solutions presented in the previous stepwise procedure. This requires the displacement of the pump to be given in consideration of oversizing to accommodate volumetric efficiency loss. This may be accomplished by dividing the calculated displacement by the volumetric efficiency.

CHAPTER 5

In-Chapter Problems

Problem 5-1
a. P (psi) = 240, or P {*kPa*} =1,655
b. P (psi) = 346, or P {*kPa*} = 2,386

Problem 5-2 Time {(sec.)} = 4.62

Problem 5-3
a. Ext. pressure (psi) = 283, or Ext. pressure {*kPa*} = 1,954
b. Ret. pressure (psi) = 318, or Ret. pressure {*kPa*} = 2,193

Problem 5-4
a. Ext. velocity (ft./sec.) = 0.227, or Ext. velocity {*m/sec.*} = 0.069
b. Ret. velocity (ft./sec.) = 0.255, or Ret. velocity {*m/sec.*) = 0.078

Problem 5-5
a. Ext. force (lbs.) = 4,628, or Ext. force {*kN*} = 20.57
b. Ret. force (lbs.) = 4,545, or Ret. force {*kN*} = 20.20

Problem 5-6
a. Ext. volume (cu. in.) = 30.85, or Ext. volume {cm^3} = 506
b. Ret. volume (cu. in.) = 30.30, or Ret. volume {cm^3} = 497
c. Tot. volume (cu. in.) = 61.15, or Tot. volume {cm^3} = 1,003

Problem 5-7 DEL (gpm) = 3.47, or DEL {*lpm*} = 13.14

Problem 5-8 DEL (gpm) = 0.82, or DEL {*lpm*} = 3.1

Problem 5-9 DIA (in.) = 1.414, or DIA {*cm*} = 3.592

Practical Design Problems

A Single End Rod, Double-Acting Cylinder

PISTON		ROD			
Diameter	Area	Diameter	Area	Annulus Area	Stroke
3 in.	7.07 in.^{2}	1 in.	0.79 in.^{2}	6.29 in.^{2}	60 in.

FORCE		PRESSURE		SUPPLY	
Extension	Retraction	Extension	Retraction	Extension	Retraction
2,000 lbs.	2,000 lbs.	283 psi	318 psi	424.2 in.^{3}	377.4 in.^{3}

			VELOCITY	
Total Supply	Rate	Supply Rate	Extension	Retraction
801.6 in.^{3}	4 cyc/min.	3,206.4 in.^{3}/min.	453.5 in./min.	509.8 in./min.

Port Size	Body Style	Mounting
3/4 in.	Tie-rod	Centerline lug

B Pump

Type	Design	Delivery	Drive Speed	Volumetric Efficiency
Piston	Axial	13.88 gpm	1,750 rpm	90%

	PORT SIZE	
Displacement at V.E.	Inlet	Outlet
2.04 in.^{3}/rev.	1-1/4 in.	3/4 in.

C Directional Control

Positions	Ways	Element	Activation	Port Size
2	4	Sliding spool	Lever	3/4 in.

D Pressure-Relief Valve

Type	Element	Setting	Mounting	Port Size
Simple	Poppet	350 psi	In-line	3/4 in.

E Filtration

		RATING		
Type	Element	Element	Flow	Port Size
Strainer	Wire	60 mesh	30.84 to 46.27 gpm	1-1/4 in.

Description	Mounting
Full Flow	In-line

F Reservoir

CAPACITY	
Minimum	Maximum
27.76 gal.	41.64 gal.

G Electric Motor

Speed	Voltage	Phase	Power at V.E.
1,750 rpm	220	3	3.15hp

H Conductors

Inlet Type	Inlet Description	INLETS (I.D. DECIMAL)		Working Type	Working Description
		Minimum	Maximum		
Pipe	Schedule 40	1.123 in.	1.775 in.	Pipe	Schedule 80

WORKINGS (I.D. DECIMAL)		LINES—NOMINAL	
Minimum	Maximum	Inlets	Workings
0.533 in.	0.900 in.	1-1/4 in.	3/4 in.

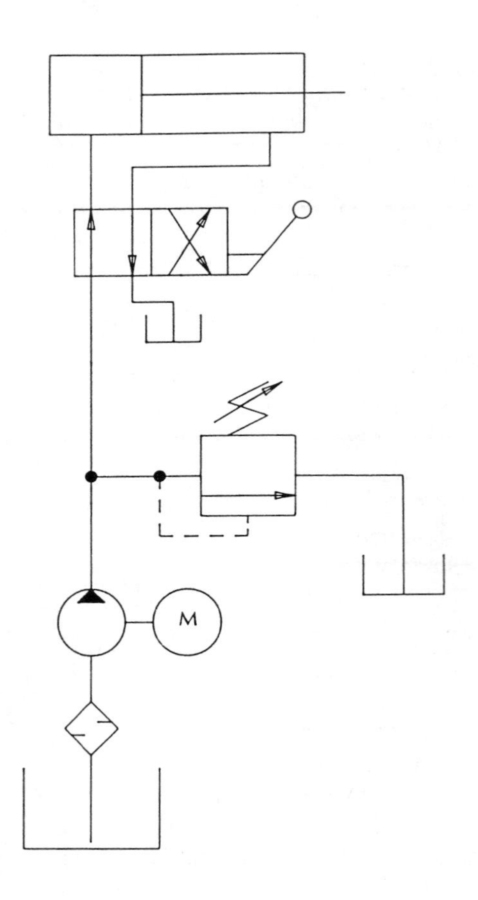

FIGURE D-1
Schematic Linear Piece-Rate Parameter Circuit

A Double End Rod, Double-Acting Cylinder

PISTON		*ROD*			
Diameter	Area	Diameter	Area	Annulus Area	Stroke
10 *cm*	78.5 cm^2	8 *cm*	50.2 cm^2	28.3 cm^2	72 *cm*

FORCE		*PRESSURE*	
Extension	Retraction	Extension	Retraction
6.7 kN	*6.7 kN*	2,367 *kPa*	2,367 *kPa*

A Double End Rod, Double-Acting Cylinder (*continued*)

FEED RATE			*TIME*	
Extension	Retraction	Supply Rate	Extension	Retraction
51 *cm/min.*	51 *cm/min.*	1,443.3 cm^3/*min.*	1.41 sec.	1.41 sec.

Port Size	Body Style	Mounting
6 *mm*	Tie-rod	Threaded end cap

B Pump

Type	Design	Delivery	Drive Speed	Volumetric Efficiency
Vane	Unbalanced	1.44 *lpm*	2,100 rpm	80%

Displacement at V.E.	*PORT SIZE* Inlet	*PORT SIZE* Outlet
0.86 cm^3/*rev.*	6 *mm*	6 *mm*

C Directional Control

Positions	Ways	Element	Activation	Port Size
2	4	Sliding spool	Lever	6 *mm*

D Pressure-Relief Valve

Type	Element	Setting	Mounting	Port Size
Simple	Poppet	2,614 *kPa*	In-line	6 *mm*

E Filtration

		RATING				
Type	Element	Element	Flow	Port Size	Description	Mounting
Strainer	Wire	24 *mesh*	3.6 to 5.4*lpm*	6 *mm*	Full-flow	In-line

F Reservoir

CAPACITY	
Minimum	Maximum
2.88 *l*	4.32 *l*

G Electric Motor

Speed	Voltage	Phase	Power at V.E.
2,100 rpm	220	3	78.44 *W*

H Conductors

Inlet Type	Inlet Description	*INLETS (I.D. DECIMAL)* Minimum	Maximum	Working Type	Working Description
Pipe	STD	0.505 *cm*	0.798 *cm*	Pipe	XS

WORKINGS (I.D. DECIMAL)		*LINES—NOMINAL*	
Minimum	Maximum	Inlets	Workings
0.224 *cm*	0.381 *cm*	6 *mm*	6 *mm*

See Figure D-2.

A Differential Cylinder

PISTON		*ROD*			
Diameter	Area	Diameter	Area	Annulus Area	Stroke
3 in.	7.07 in.^{2}	2.121 in.	3.53 in.^{2}	3.53 in.^{2}	12 in.

FORCE		*PRESSURE*	
Extension	Retraction	Extension	Retraction
3,500 lbs.	1,500 lbs.	992 psi	425 psi

FIGURE D-2
Schematic Linear Feed-Rate Parameter Circuit

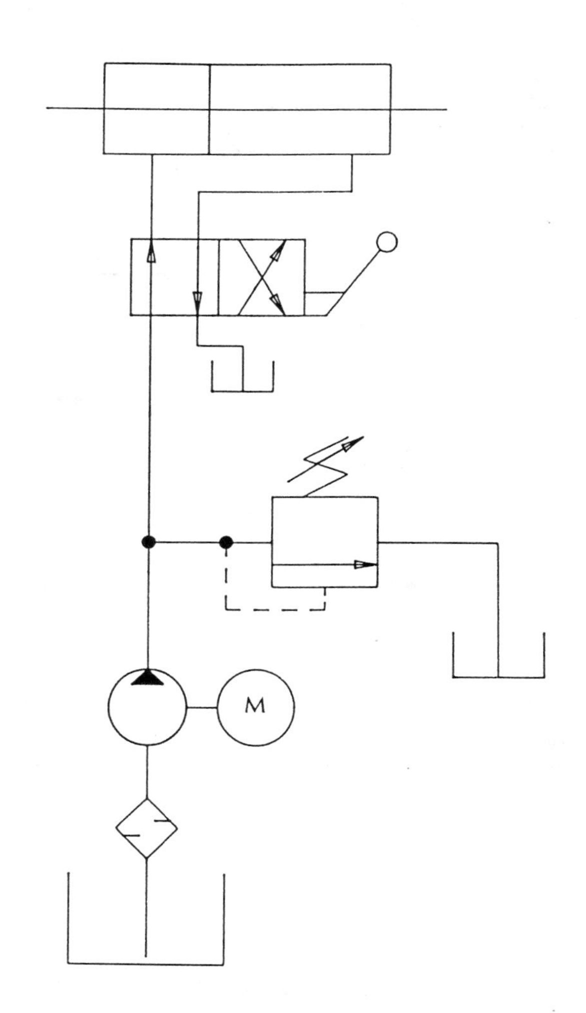

A Differential Cylinder (*continued*)

FEED RATE			*TIME*	
Extension	Retraction	Supply Rate	Extension	Retraction
15 in./sec.	15 in./sec.	3,177 in.^{3}/min.	0.8 sec.	0.8 sec.

Port Size	Body Style	Mounting
3/4 in.	One-piece	Trunion

B Pump

Type	Design	Delivery	Drive Speed	Volumetric Efficiency
Gear	External	13.75 gpm	600 rpm	70%

	PORT SIZE	
Displacement at V.E.	Inlet	Outlet
7.56 in.^{3}/rev.	1-1/4 in.	3/4 in.

C Directional Control

Positions	Ways	Element	Activation	Port Size
2	4	Sliding spool	Lever	3/4 in.

D Pressure-Relief Valve

Type	Element	Setting	Mounting	Port Size
Simple	Poppet	1,091 psi	In-line	3/4 in.

E Filtration

		RATING				
Type	Element	Element	Flow	Port Size	Description	Mounting
Strainer	Wire	60 *mesh*	39.3 to 58.9 gpm	1-1/4 in.	Full flow	In-line

F Reservoir

CAPACITY	
Minimum	Maximum
27.5 gal.	41.25 gal.

G Electric Motor (with gear reduction)

Speed	Voltage	Phase	Power at V.E.
1,750 rpm	220	3	12.49 hp

H Conductors

Inlet Type	Inlet Description	INLETS (I.D. DECIMAL) Minimum	INLETS (I.D. DECIMAL) Maximum	Working Type	Working Description
Pipe	Schedule 40	1.267 in.	2.003 in.	Pipe	Schedule 80

WORKINGS (I.D. DECIMAL) Minimum	WORKINGS (I.D. DECIMAL) Maximum	LINES—NOMINAL Inlets	LINES—NOMINAL Workings
0.530 in.	0.896 in.	1-1/4 in.	3/4 in.

See Figure D-3

CHAPTER 6

In-Chapter Problems

Problem 6-1

P (psi) = 262.5, or P {*kPa*} = 1,810

Problem 6-2

a. Primary Cylinder
Ext. Vol. (cu. in.) = 141.4, or Ext. Vol. {cm^3} = 2,317
Ret. Vol. (cu. in.) = 125.8, or Ext. Vol. {cm^3} = 2,062
b. Secondary Cylinder
Ext. Vol. (cu. in.) = 37.7, or Ext. Vol. {cm^3} = 618
Ret. Vol. (cu. in.) = 28.3, or Ret. Vol. {cm^3} = 464
c. Cycle Volume
Vol. (cu. in.) = 333.2, or Vol. {cm^3} = 5,461
d. Supply (cu. in./min.) = 1,666, or Supply {cm^3/*min.*} = 27,305
e. DEL (gpm) = 7.21, or DEL {*lpm*} = 27.31

Problem 6-3

DEL (gpm) = 16.3, or DEL {*lpm*} = 61.7

Problem 6-4

See Figure D-4.

Problem 6-5

The directional control has an open center. In the center position, pump delivery is directed to the tank through the directional control. Any fluid contained in either end of the cylinder is also free to return to the tank through the directional control.

FIGURE D-3
Schematic Regenerative Circuit

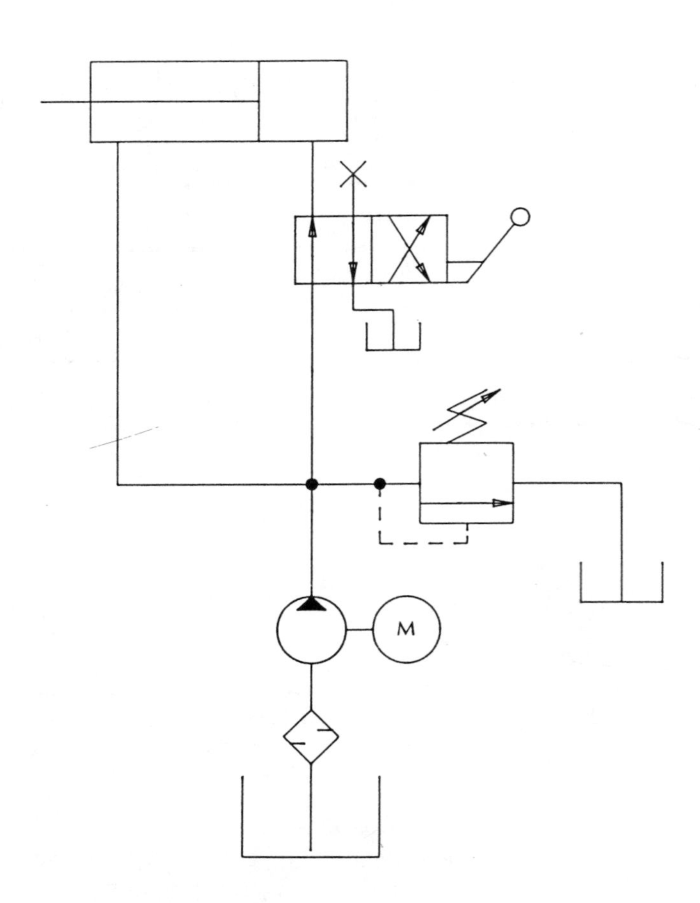

Problem 6-6 See Figure D-5.

Problem 6-7 This directional control has a tandem center. In the center position, pump supply is directed to the tank through the directional control. Any fluid contained in either end of the cylinder is blocked and excluded from flow at the directional control.

Problem 6-8 See Figure D-6.

Problem 6-9 Both directional controls have closed centers. With pump supply blocked at either valve, flow may be directed to the other valve and cycle that cylinder independently. Closed cylinder ports will block all flow from cylinders in the center position.

Problem 6-10 P (psi) = 251, or P {*kPa*} = 1,731

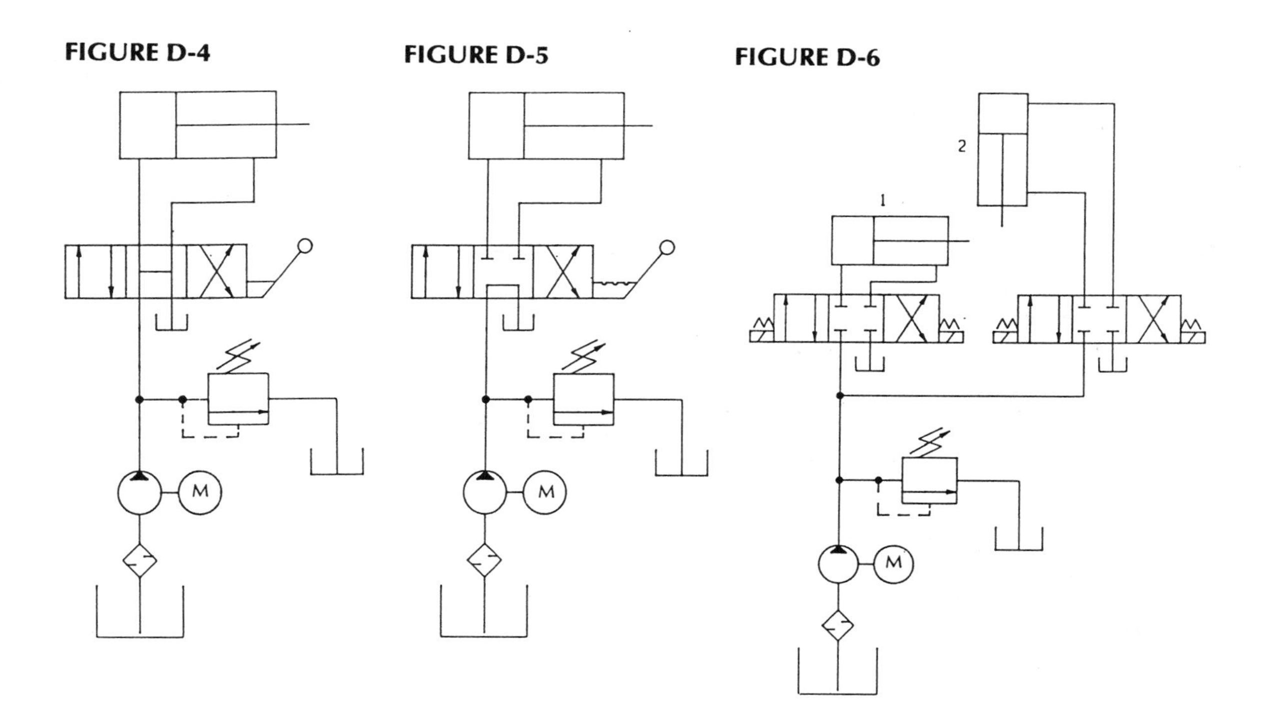

Practical Design Problem

A Single End Rod, Double-Acting Cylinder—Raising Cylinder

PISTON		ROD			
Diameter	Area	Diameter	Area	Annulus Area	Stroke
8*cm*	50.24 cm^2	3.2 *cm*	8.04 cm^2	42.20 cm^2	91.4 *cm*

FORCE		PRESSURE		SUPPLY	
Extension	Retraction	Extension	Retraction	Extension	Retraction
13.3 *kN*	0 *kN*	2,647.3 *kPa*	0 *kPa*	4,591.9 cm^3	3,857.1 cm^3

A Single End Rod, Double-Acting Cylinder—Raising Cylinder (*continued*)

			VELOCITY	
Total Supply	Rate	Supply Rate	Extension	Retraction
8,449 *cm³*	2 cyc/min.	16,898 *cm³/min.*	508.9 *cm/min.*	605.8 *cm/min.*

Port Size	Body Style	Mounting
14 *mm*	Tie-rod	Extended tie-rod

B Single End Rod, Double-Acting Cylinder—Sliding Cylinder

PISTON		ROD			
Diameter	Area	Diameter	Area	Annulus Area	Stroke
5 *cm*	19.63 *cm²*	2 *cm*	3.14 *cm²*	16.49 *cm²*	*120 cm*

FORCE		PRESSURE		SUPPLY	
Extension	Retraction	Extension	Retraction	Extension	Retraction
6.6 kN	0 *kN*	3,362.2 *kPa*	0 *kPa*	2,355.6 *cm³*	1,978.8 *cm³*

			VELOCITY	
Total Supply	Rate	Supply Rate	Extension	Retraction
4,334.4 *cm³*	2 cyc/min.	8,668.8 *cm³/min.*	1,302.4 *cm/min*	1550.4 *cm/min*

Port Size	Body Style	Mounting
14 *mm*	Flange	Flange

C Pump

Type	Design	Delivery	Drive Speed	Volumetric Efficiency
Vane	Balanced	25.57 *lpm*	1,450 rpm	80%

C Pump (*continued*)

	PORT SIZE	
Displacement at V.E.	Inlet	Outlet
22.04 $cm^3/rev.$	24 *mm*	14 *mm*

D Master System Directional Control

Positions	Ways	Element	Activation	Port Size
3	4	Sliding spool	Foot treadle	14 *mm*

E Position Parameter Sequencing Directional Control

Positions	Ways	Element	Activation	Port Size
2	1	Sliding Spool	Plunger	14 *mm*

F Pressure-Relief Valve

Type	Element	Setting	Mounting	Port Size
Simple	Poppet	3,698.4 *kPa*	In-line	14 *mm*

G Filtration

		RATING				
Type	Element	Element	Flow	Port Size	Description	Mounting
Strainer	Wire	24 mesh	63.93 to 95.89 *lpm*	24 *mm*	Full Flow	Flange

H Reservoir

CAPACITY	
Minimum	Maximum
51.14 *l*	76.71 *l*

I Electric Motor

Speed	Voltage	Phase	Power at V.E.
1,450 rpm	220	3	1,791 *W*

J Conductors

Inlet Type	Inlet Description	*INLETS (I.D. DECIMAL)* Minimum	Maximum	Working Type	Working Description
Pipe	STD	2.126 *cm*	3.363 *cm*	Pipe	XS

WORKINGS (I.D. DECIMAL) Minimum	Maximum	*LINES—NOMINAL* Inlets	Workings
0.943 *cm*	1.607 *cm*	24 *mm*	14 *mm*

FIGURE D-7
Schematic Position Parameter Sequence Circuit

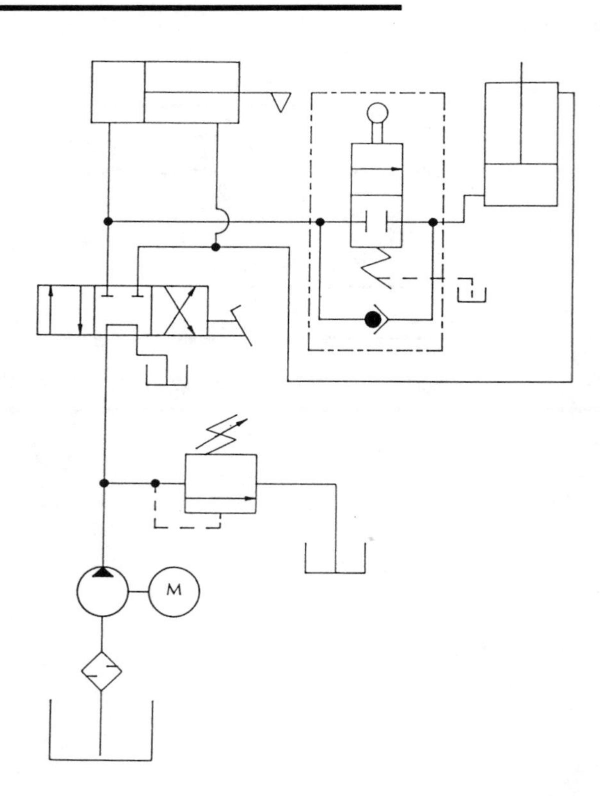

CHAPTER 7

In-Chapter Problems

Problem 7-1

First: 4-in. or *10.2-cm* diameter ram
Second: 3-in. or *7.6-cm* diameter ram
Third: 2-in. or *5.1-cm* diameter ram

Problem 7-2

P (psi) = 350, or P {*kPa*} = 2,413

Problem 7-3

a. P (psi) = 51, or P {*kPa*} = 352
b. P (psi) = 56, or P {*kPa*} = 386
c. P (psi) = 308, or P {*kPa*} = 2,124

Problem 7-4

SV1: P (psi) = 175, or P {*kPa*} = 1,207
SV2: P (psi) = 311, or P {*kPa*} = 2,144
SV3: P (psi) = 234, or P {*kPa*} = 1,613
SV4: P (psi) = 350, or P {*kPa*} = 2,413

Problem 7-5

PRV: P (psi) = 815, or P {*kPa*} = 5,619

Problem 7-6

P (psi) = 382, or P {*kPa*} = 2,634

Problem 7-7

P (psi) = 803, or P {*kPa*} = 5,536

Problem 7-8

P (psi) = 1,751, or P {*kPa*} = 12,073

Problem 7-9

P (psi) = 398, or P {*kPa*} = 2,744

Problem 7-10

P (psi) = 1,926, or P {*kPa*} = 13,279

Practical Design Problem

Note: The sizes of cylinders' piston and rod diameters have been selected to make the schematic (Figure 7-16) correct. Other cylinder sizes may be acceptable. However, this may necessitate the addition, deletion, or relocation of some pressure-reducing valves.

A Single End Rod, Double-Acting Cylinder—Loading Cylinder

PISTON		*ROD*			
Diameter	Area	Diameter	Area	Annulus Area	Stroke
1-1/2 in.	1.77 in.^{2}	1 in.	0.79 in.^{2}	0.98 in.^{2}	12 in.

FORCE		*PRESSURE*		*SUPPLY*	
Extension	Retraction	Extension	Retraction	Extension	Retraction
300 lbs.	300 lbs.	169 psi	306 psi	21.2 in.^{3}	11.8 in.^{3}

			VELOCITY	
Total Supply	Rate	Supply Rate	Extension	Retraction
33.0 in.^{3}	2 cyc/min.	66.0 in.^{3}/min.	391.2 in./min.	706.8 in./min.

Port Size	Body Style	Mounting
1/4 in.	Screw	Flange

B Single End Rod, Double-Acting Cylinder—Clamping Cylinder

PISTON		*ROD*			
Diameter	Area	Diameter	Area	Annulus Area	Stroke
4 in.	12.56 in.^{2}	3-1/2 in.	9.61 in.^{2}	2.95 in.^{2}	6 in.

FORCE		*PRESSURE*		*SUPPLY*	
Extension	Retraction	Extension	Retraction	Extension	Retraction
10,000 lbs.	1,000 lbs.	796 psi	339 psi	75.4 in.^{3}	17.7 in.^{3}

			VELOCITY				
Total Supply	Rate	Supply Rate	Extension	Retraction	Port Size	Body Style	Mounting
93.1 in.^{3}	2 cyc/min.	186.2 in.^{3}/min.	55.2 in./min.	234.6 in./min.	1/4 in.	Flange	Flange

C Single End Rod, Double-Acting Cylinder—Thrust Cylinder

PISTON		*ROD*			
Diameter	Area	Diameter	Area	Annulus Area	Stroke
3 in.	7.07 in.^{2}	2 in.	3.14 in.^{2}	3.93 in.^{2}	20in.

FORCE		*PRESSURE*		*SUPPLY*	
Extension	Retraction	Extension	Retraction	Extension	Retraction
5,000 lbs.	0 lbs.	707 psi	0 psi	141.4 in.^{3}	78.6 in.^{3}

			VELOCITY	
Total Supply	Rate	Supply Rate	Extension	Retraction
220 in.^{3}	2 cyc/min.	440 in.^{3}/min.	97.8 in./min.	176.4 in./min.

Port Size	Body Style	Mounting
1/4 in.	Tie-rod	Centerline lug

D Pump

Type	Design	Delivery	Drive Speed	Volumetric Efficiency
Vane	Unbalanced	3.0 gpm	1,750 rpm	80%

	PORT SIZE	
Displacement at V.E.	Inlet	Outlet
50 in.^{3}/rev.	3/4 in.	1/4 in.

E Reservoir

CAPACITY	
Minimum	Maximum
6 gal.	9 gal.

F Filtration

Type	Element	RATING Element	RATING Flow	Port Size	Description	Mounting
Strainer	Wire	60 mesh	7.5 to 11.25 gpm	3/4 in.	Full Flow	In-line

G Master System Directional Control

Positions	Ways	Element	Activation	Port Size
3	4	Sliding Spool	Solenoid	1/4 in.

H Position Parameter Sequencing Directional Control

Positions	Ways	Element	Activation	Port Size
2	1	Sliding spool	Plunger	1/4 in.

I Pressure Parameter Sequence Valve—1

Type	Element	Setting	Mounting	Port Size
Compound	Spool	876 psi	Subplate	1/4 in.

J Pressure Parameter Sequence Valve—2

Type	Element	Setting	Mounting	Port Size
Compound	Spool	250 psi*	Subplate	1/4 in.

*Minimum Setting

K Pressure Parameter Sequence Valve—3

Type	Element	Setting	Mounting	Port Size
Compound	Spool	373 psi	Subplate	1/4 in.

L Pressure-Reducing Valve—1 (If Required)

Type	Element	Setting	Mounting	Port Size
Direct-acting	Spool	707 psi	Lug	1/4 in.

M Pressure-Reducing Valve—2 (If Required)

Type	Element	Setting	Mounting	Port Size
Direct-acting	Spool	306 psi	Lug	1/4 in.

N Pressure-Relief Valve

Type	Element	Setting	Mounting	Port Size
Compound	Piston	964 psi	Subplate	1/4 in.

O Electric Motor

Speed	Voltage	Phase	Power at V.E.
1,750 rpm	220	3	2.11 hp

P Conductors

Inlet Type	Inlet Description	*INLETS (I.D. DECIMAL)* Minimum	Maximum	Working Type	Working Description
Pipe	Schedule 40	0.554 in.	0.875 in.	Pipe	Schedule 80

WORKINGS (I.D. DECIMAL)		*LINES—NOMINAL*	
Minimum	Maximum	Inlets	Workings
0.248 in.	0.418 in.	3/4 in.	1/4 in.

CHAPTER 8

In-Chapter Problems

Problem 8-1

DEL (gpm) = 6.55, or DEL {*lpm*} = 24.79

Problem 8-2

ΔP (psi) = 49, or ΔP {*kPa*} = 338

Problem 8-3

ΔP (psi) = 400, or ΔP {*kPa*} = 2,758

Problem 8-4

a. Pump Delivery
   Max. VEL: I.D. (in.) = 0.325, or I.D. {*cm*} = 0.894
   Min. VEL: I.D. (in.) = 0.590, or I.D. {*cm*} = 1.450
b. Cylinder Delivery
   Max. VEL: I.D. (in.) = 0.282, or I.D. {*cm*} = 0.716
   Min. VEL: I.D. (in.) = 0.482, or I.D. {*cm*} = 1.224
c. Relief Exhaust Delivery
   Max. VEL: I.D. (in.) = 0.200, or I.D. {*cm*} = 0.508
   Min. VEL: I.D. (in.) = 0.340, or I.D. {*cm*} = 0.867

Problem 8-5

Decimal Size: I.D. (in.) = 0.364, or I.D. {*cm*} = 0.92
Nominal Size (Schedule 40 or STD): I.D. (in.) = ¼, or I.D. {*mm*} = 8

Problem 8-6

a. Max. P: P.T. (psi) = 1,063, or P.T. {*kPa*} = 7,329
b. Min. P: P.T. (psi) = 752, or P.T. {*kPa*} = 5,185

Problem 8-7

Flow contol used is meter-in in both stroke directions.

Problem 8-8

Reduced cylinder speed will occur at any time that no manipulations to the circuit other than shifting of the three-position, four-way directional control occurs.

Problem 8-9

Rapid cylinder movement will occur during either stroke direction if the palm button on the two-position, one-way directional control is depressed.

Practical Design Problem

A Loading, Indexing, and Clamping Cylinder

PISTON		*ROD*			
Diameter	Area	Diameter	Area	Annulus Area	Stroke
10 *cm*	78.5 cm^2	8 *cm*	50.2 cm^2	28.3 cm^2	55.9 *cm*

FORCE		*PRESSURE*	
Extension	Retraction	Extension	Retraction
17.8 *kN*	11.6 *kN*	2,268 *kPa*	4,099 *kPa*

VELOCITY		*TIME*	
Extension	Retraction	Extension	Retraction
5 *cm/sec.*	14 *cm/sec.*	11.18 sec.	4 sec.

Port Size	Body Style	Mounting
11 *mm*	One-piece	Base lug

B Drill-Head Cylinder

PISTON		*ROD*		
Diameter	Area	Diameter	Area	Annulus Area
6.3 *cm*	31.2 cm^2	2 *cm*	3.14 cm^2	28.06 cm^2

FORCE		*PRESSURE*		
Extension	Retraction	Extension	Retraction	Stroke
9.8 *kN* (init.) 12 *kN* (term.)	9.8 *kN*	3,141 *kPa* 3,846 *kPa*	3,493 *kPa*	27.9 *cm* traverse 22.9 *cm* feed 51.8 *cm* total

B Drill-Head Cylinder (*continued*)

FEED RATE		
Extension	Retraction	Supply Rate
12.7 *cm/sec.* (traverse) 457.2 *cm/min.* (feed)	14.1 *cm/sec.*	23,774 cm^3*/min.*

TIME				
Extension	Retraction	Port Size	Body Style	Mounting
5.2 sec.	3.67 sec.	11 *mm*	Tie-rod	Threaded end cap

C Pump

Type	Design	Delivery	Drive Speed	Volumetric Efficiency
Piston	Axial	23.77 *lpm*	1,450 rpm	90%

	PORT SIZE	
Displacement at V.E.	Inlet	Outlet
18.21 cm^3*/rev.*	24 *mm*	11 *mm*

D Flow-Control Valve

Method	Type	Control Setting	Delivery through Valve at Flow-Control Setting
Meter-out	Multirange	457.2 *cm/min.*	12.83 *l/min.*

Allowed Pressure Drop	Pressure-Compensation Design	Delivery Range	Port Size
2,758 *kPa*	Pressure reduction	0 to 25.66 *lpm*	11 *mm*

E Reservoir

CAPACITY	
Minimum	Maximum
47.54 *l*	71.31 *l*

F Filtration

		RATING				
Type	Element	Element	Flow	Port Size	Description	Mounting
Strainer	Wire	24 mesh	52.82 to 79.23 *lpm*	24 *mm*	Full flow	In-line

G Master System Directional Control

Positions	Ways	Element	Activation	Port Size
3	4	Sliding spool	Lever	11 *mm*

H Position Parameter Sequencing Directional Control

Positions	Ways	Element	Activation	Port Size
2	1 (N/C)	Sliding spool	Plunger	11 *mm*

I Feed Shunt Directional Control

Positions	Ways	Element	Activation	Port Size
2	1 (N/O)	Sliding spool	Plunger	11 *mm*

J Pressure Parameter Sequence Valve

Type	Element	Setting	Mounting	Port Size
Compound	Spool	2,495 *kPa*	Subplate	11 *mm*

K Pressure-Relief Valve

Type	Element	Setting	Mounting	Port Size
Compound	Piston	6,314 *kPa*	In-line	11 *mm*

L Electric Motor

Speed	Voltage	Phase	Power at V.E.
1,450 rpm	220	3	2,502 *W*

M Conductors

Inlet Type	Inlet Description	Working Type	Working Description	INLETS (I.D. DECIMAL)	
				Minimum	Maximum
Pipe	STD	Pipe	XS	1.933 *cm*	3.057 *cm*

	WORKINGS (I.D. DECIMAL)		LINES—NOMINAL	
	Minimum	Maximum	Inlets	Workings
Pump	0.909 *cm*	1.550 *cm*	24 *mm*	11*mm*
Cylinder	0.704*cm*	1.201*cm*		
Relief	0.575*cm*	0.980*cm*		

CHAPTER 9

In-Chapter Problems

Problem 9-1 P (psi) = 345, or P {*kPa*} = 2,382

Problem 9-2 T.R. (in. lbs./100 psi) = 159, or T.R. {*N m/1,000 kPa*} = 25.48

Problem 9-3 DEL (gpm) = 8.12, or DEL {*lpm*} = 30.63

Problem 9-4 Power (hp) = 47.4, or Power {*kW*} = 35.3

Problem 9-5 T (in. lbs.) = 1,500, or T {*N cm*} = 16,987

Problem 9-6
a. P (psi) = 2,591, or P {*kPa*} = 17,864
b. P (psi) = 2,850, or P {*kPa*} = 19,650

Problem 9-7 DEL (gpm) = 5.41, or DEL {*lpm*} = 20.5

Problem 9-8
BV-1: P (psi) = 3,800, or P {*kPa*} = 26,200
BV-2: P (psi) = 2,533, or P {*kPa*} = 17,464

Problem 9-9 P (psi) = 1,549, or P {*kPa*} = 10,990

Problem 9-10
Min.: P (psi) = 1,644, or P {*kPa*} = 11,335
Max.: P (psi) = 1,994, or P {*kPa*} =13,478

Practical Design Problem

Problem 9-11 T (in. lbs.) = 11,250, or T {*N m*} = 500

Problem 9-12 Distance (in.) = 47.1, or Distance {*cm*} = 119.6

A Hydraulic Motor

Operating Speed (max.)	Torque Load	Type	Design	Volumetric Efficiency
4.25 rpm	11,250 in. lbs.	Gear	External	70%

Displacement	Required Delivery	*OPERATING PRESSURE* (at Motor) Full Delivery	*OPERATING PRESSURE* (at Relief Valve) Minimum Delivery	Starting Pressure (at Motor)
100 in.^{3}/rev.	2.63 gpm	706.5 psi	1,106.7 psi	777.2 psi

Torque Rating	Mounting	Port Size
1,592.4 in. lbs./100 psi	Flange	1/4 in.

B Pump

Type	Design	Delivery	Drive Speed	Volumetric Efficiency
Vane	Balanced	2.63 gpm	1,750 rpm	80%

Displacement at V.E.	*PORT SIZE* Inlet	*PORT SIZE* Outlet
0.43 in.^{3}/rev	1/2 in.	1/4 in.

C Flow Control Valve

Method	Type	Motor Speed at Flow Control (Lowest Setting)	Delivery through Valve at Flow Control Setting
Meter-in	Single-range	2.34 rpm	1.45 gal./min.

C Flow Control Valve (*continued*)

Allowed Pressure Drop	Pressure Compensation Design	Delivery Range	Port Size
400 psi	Pressure relief	0 to 5.26 gpm	1/4 in.

D Reservoir

CAPACITY	
Minimum	Maximum
5.26 gal.	7.89 gal.

E Filtration

		RATING				
Type	Element	Element	Flow	Port Size	Description	Mounting
Strainer	Wire	60 mesh	6.58 to 9.86 gpm	1/2 in.	Full flow	In-line

F Master System Directional Control

Positions	Ways	Element	Activation	Port Size
3	4	Sliding spool	Solenoid	1/4 in.

G Pressure-Relief Valve

Type	Element	Setting	Mounting	Port Size
Simple	Poppet	1,107 psi	Flange	1/4 in.

H Electric Motor

Speed	Voltage	Phase	Power at V.E.
1,750 rpm	220	3	2.12 hp

I Conductors

Inlet Type	Inlet Description	Working Type	Working Description	*INLETS (I.D. DECIMAL)* Minimum	*INLETS (I.D. DECIMAL)* Maximum
Pipe	Schedule 40	Pipe	Schedule 80	0.158 in.	0.820 in.

	WORKINGS (I.D. DECIMAL) Minimum	*WORKINGS (I.D. DECIMAL)* Maximum	*LINES—NOMINAL* Inlets	*LINES—NOMINAL* Workings
Pump	0.232 in.	0.392 in.	1/2 in.	1/4 in.
Motor	0.172 in.	0.291 in.		
Relief	0.155 in.	0.262 in.		

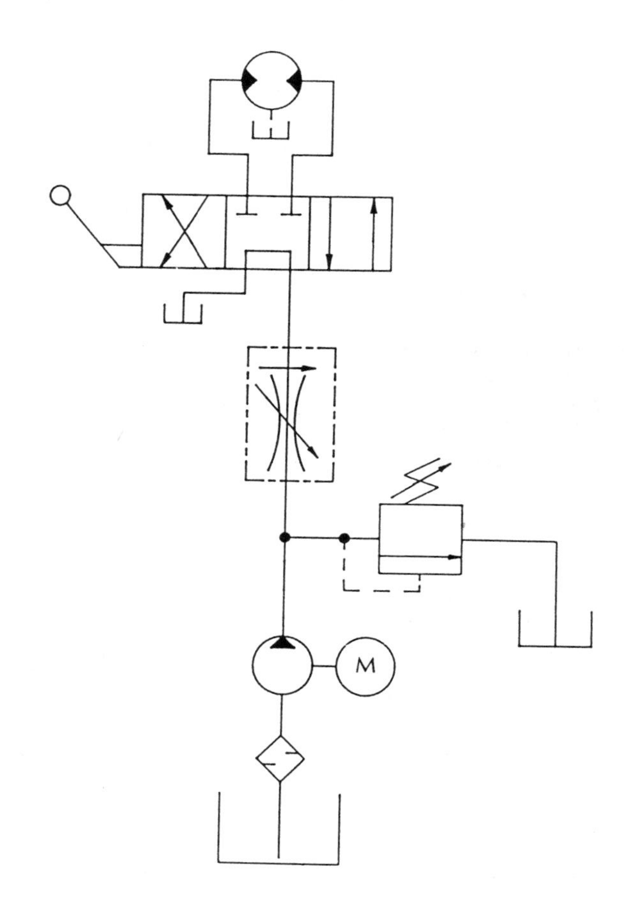

FIGURE D-8
Schematic Rotary with Dependent Meter-In Flow-Control Circuit

CHAPTER 10

In-Chapter Problems

Problem 10-1
a. P (psi) = 750, or P {*kPa*} = 5,171
b. Supply (cu. in.) = 231, or Supply {cm^3} = 3,785

Problem 10-2
a. P (psi) = 841.12, or P {*kPa*} = 5,874
b. Supply (cu. in.) = 255.62, or Supply {cm^3} = 4,188

Problem 10-3
F2 (lbs.) = 125, or F2 {*kN*} = 0.56

Problem 10-4
P (psi) = 250, or P {*kPa*} = 1,724

Problem 10-5
V2 (cu. in.) = 883.23, or V2 {cm^3} = 14,472
V3 (cu. in.) = 345.23, or V3 {cm^3} = 22,042

Problem 10-6
V1 (cu. in.) = 86.28, or V1 {cm^3} = 63,677

Problem 10-7
P (psi) = 12,793, or P {*kPa*} = 88,208

Problem 10-8
S (cu. in.) = 4.74, or S {cm^3} = 77.7

Practical Design Problem

A Single End Rod, Double-Acting Cylinder

PISTON		*ROD*			
Diameter	Area	Diameter	Area	Annulus Area	Stroke
6.3 *cm*	31.17 cm^2	1.6 *cm*	2.01 cm^2	29.16 cm^2	20.3 *cm* [normal advance] 2.5 *cm* [intensified]

FORCE		*PRESSURE*		*SUPPLY*	
Extension	Retraction	Extension	Retraction	Extension	Retraction
126.7 *kN* [intensified] 7.8 *kN* [normal]	7.8 *kN*	40,648.1 *kPa* 2,502.4 *kPa*	2,674.9*kPa*	632.8 cm^3 [normal advance]	664.8 cm^3 [full return]

A Single End Rod, Double-Acting Cylinder *(continued)*

Total Supply	Rate	Supply Rate	*VELOCITY* Extension	Retraction
1,297.6 cm^3	12 cyc/min.	15,572 cm^3/min.	833.17 *cm/min.*	890.60 *cm/min.*

TIME Extension	Retraction	Port Size	Body Style	Mounting
1.46 sec.	1.54 sec.	14 *mm*	One-piece	Flange

B Intensifier

Necessary Cylinder Displacement	*INTENSIFIER PISTON* Diameter	Area	*INTENSIFIER ROD* Diameter	Area
77.9 cm^3	8 *cm*	50.27 cm^2	2.4 *cm*	4.52 cm^2

FORCE Cylinder Piston	Intensifier Rod	*INTENSIFIER PRESSURE* Piston	Rod	Stroke
126.7 *kN*	18.38 *kN*	3,656.3 *kPa*	40,681.1 *kPa*	17.23 *cm*

SUPPLY INPUT FROM PUMP Extension	Retraction	Rate	Supply Rate
866.2 cm^3	0 cm^3	12 cyc/min.	10,394 cm^3/min.

Velocity Extension	Time Extension	*PORT SIZE* Piston End	Rod End	Mounting
516.61 *cm/min.*	2.0 sec.	14 *mm*	6 *mm*	Centerline lug

C Pump

Type	Design	Delivery	Drive Speed	Volumetric Efficiency
Vane	Balanced	25.97 *lpm*	1,450 rpm	80%

C Pump (continued)

Displacement at V.E.	PORT SIZE Inlet	PORT SIZE Outlet
22.39 *cm^3/rev.*	24 *mm*	14 *mm*

D Gas-Charged Accumulator

Necessary Supply (S)	Load Volume (V2)	Discharge Volume (V3)	Precharge Pressure (P1)	Load Pressure (P2)
2,163.7 cm^3	4,797.9 cm^3	6,961.6cm^3	1,320.2 kPa_a	6,115.7 kPa_a

Discharge Pressure (P3)	Precharge Volume (V1)	Size	Element Type	Port Size
3,757.7 *kPa*	22,225.8 cm^3	17.43 *l*	Bladder	14 *mm*

E Reservoir

CAPACITY	
Minimum	Maximum
69.37 *l*	95.34 *l*

F Filtration

Type	Element	RATING Element	RATING Flow	Port Size	Description	Mounting
Strainer	Wire	24 *mesh*	64.93 to 97.39 *lpm*	24 *mm*	Partial flow	In-line

G Pressure-Relief Valve

Type	Element	Setting	Mounting	Port Size
Compound	Piston	6,014.3 kPa_g	Subplate	14 *mm*

H Electric Motor

Speed	Voltage	Phase	Power at V.E.
1,450 rpm	220	3	2,603.7 *W*

I Master System Directional Control

Positions	Ways	Element	Activation	Port Size
3	4	Sliding spool	Solenoid	14 *mm*

J Isolator Directional Control

Positions	Ways	Element	Activation	Port Size
2	2	Sliding spool	Plunger	14 *mm*

K Pilot Override Check Valve

Positions	Ways	Element	Activation	Port Size
2	1 (N/C)	Poppet/piston	Pilot	14 *mm*

L Accumulator Shunt Directional Control—1

Positions	Ways	Element	Activation	Port Size
2	1 (N/C)	Sliding spool	Pilot	14 *mm*

M Accumulator Shunt Directional Control—2

Positions	Ways	Element	Activation	Port Size
2	1 (N/C)	Sliding spool	Lever	14 *mm*

N Check Valves

Positions	Ways	Element	Activation	Port Size
2	1 (N/C)	Poppet	Pilot	14 *mm*

O Inlet and Working Conductors

Inlet Type	Inlet Description	INLETS (I.D. DECIMAL) Minimum	INLETS (I.D. DECIMAL) Maximum	Working Type	Working Description
Pipe	STD	2.143 *cm*	3.389 *cm*	Pipe	XS

WORKINGS (I.D. DECIMAL)		LINES—NOMINAL	
Minimum	Maximum	Inlets	Workings
0.951 *cm*	1.620 *cm*	24 *mm*	14 *mm*

P Intensifier Rod End Conductor

Type	Description	Resolved Delivery from Rod End
Pipe	XS	2.34 *lpm*

WORKINGS (I.D. DECIMAL)		LINES—NOMINAL
Minimum	Maximum	Workings
0.285 *cm*	0.486 *cm*	6 *mm*

CHAPTER 11

In-Chapter Problems

Problem 11-1 Max. P (psi) = 9,240, or Max. P {*kPa*} = 63,708

Problem 11-2 a. Safety Factor P (psi) = 7,600, or Safety Factor P {*kPa*} = 52,400
b. W.T. (in.) = 0.028, or W.T. {*mm*} = 0.715

Practical Design Problem

A Single End Rod, Double-Acting Shear Cylinder

PISTON		*ROD*			
Diameter	Area	Diameter	Area	Annulus Area	Stroke
4 in.	12.56 in.^{2}	3.5 in.	9.62 in.^{2}	1.94 in.^{2}	10 in.

FORCE		*PRESSURE*		*SUPPLY*	
Extension	Retraction	Extension	Retraction	Extension	Retraction
4,500 lbs.	750 lbs.	358.3 psi	386.6 psi	125.6 in.^{3}	19.4 in.^{3}

			VELOCITY	
Total Supply	Rate	Supply Rate	Extension	Retraction
145 in.^{3}	5 cyc/min.	725 in.^{3}/min.	126.2 in./min.	817.2 in./min.

Extension Time	Retraction Time	Port Size	Body Style	Mounting
4.76 sec.	0.73 sec.	P&H, 1/2 in.; T, 5/8 in.	Tie-rod	Centerline lug

B Single End Rod, Double-Acting Clamp Cylinder

PISTON		*ROD*			
Diameter	Area	Diameter	Area	Annulus Area	Stroke
3 in.	7.07 in.^{2}	2.5 in.	4.91 in.^{2}	2.16 in.^{2}	8 in. [normal advance] 1.5 in. [intensified]

FORCE		*PRESSURE*		*SUPPLY*	
Extension	Retraction	Extension	Retraction	Extension	Retraction
19,000 lbs. [intensified]	950 lbs.	2,687.4 psi	439.8 psi	56.56 in.^{3} [normal advance]	20.52 in.^{3} [full return]
950 lbs. [normal]		134.4 psi			

B Single End Rod, Double-Acting Clamp Cylinder (*continued*)

Total Supply	Rate	Supply Rate	*VELOCITY* Extension	Retraction
77.08 in.^{3}	5 cyc/min.	385.4 in.^{3}/min.	224.2 in./min.	734 in./min.

Extension Time	Retraction Time	Port Size	Body Style	Mounting
2.14 sec.	0.78 sec.	P&H, 1/2 in.; T, 5/8 in.	One-piece	Threaded end cap

C Hydraulic Motor

Operating Speed (max.)	Torque Load	Type	Design	Volumetric Efficiency
130 rpm	400 in. lbs.	Gear	External	70%

		OPERATING PRESSURE		
Displacement	Required Delivery	(at Motor) Full Delivery	(at Relief Valve) Minimum Delivery	Starting Pressure (at Motor)
4 in.^{3}/rev.	3.22 gpm	628 psi	1,028 psi	691 psi

Torque Rating	Mounting	Port Size
63.69 in. lbs./100 psi	Foot lug	P, 1/4 in.; H, 5/16 in.; T, 3/8 in.

D Intensifier

	INTENSIFIER PISTON		*INTENSIFIER ROD*	
Necessary Cylinder Displacement	Diameter	Area	Diameter	Area
10.61 in.^{3}	3 in.	7.07 in.^{2}	1 in.	0.79 in.^{2}

FORCE		*INTENSIFIER PISTON*		
Cylinder Piston	Intensifier Rod	Piston	Rod	Stroke
19,000 lbs.	2,123 lbs.	300.3 psi	2,687.4 psi	13.43 in.

D Intensifier *(continued)*

SUPPLY INPUT FROM PUMP			
Extension	Retraction	Rate	Supply Rate
95 in.^{3}	0 in.^{3}	5 cyc/min.	475 in.^{3}/min

Velocity Extension	Time Extension	*PORT SIZE* Piston End	*PORT SIZE* Rod End	Mounting
224.2 in./min.	3.59 sec.	P&H, 1/2 in.; T, 5/8 in.	P, 1/8 in.; H, 3/16 in.; T, 1/4 in.	Extended tie-rod

E Pump—Linear Side

Type	Design	Delivery	Drive Speed	Volumetric Efficiency
Vane	Balanced	6.86 gpm	1,750 rpm	80%

Displacement at V.E.	*PORT SIZE* Inlet	*PORT SIZE* Outlet
1.13 in.^{3}/rev.	P, H, & T, 1-1/4 in.	P & H, 1/2 in.; T, 5/8 in.

F Pump—Rotary Side

Type	Design	Delivery	Drive Speed	Volumetric Efficiency
Vane	Balanced	3.22 gpm	1,750 rpm	80%

Displacement at V.E.	*PORT SIZE* Inlet	*PORT SIZE* Outlet
0.43 in.^{3}/rev.	P & H, 3/4 in.; T, 1 in.	P, 1/4 in.; H, 5/16 in.; T, 3/8 in.

G Gas-Charged Accumulator

Necessary Supply (S)	Load Volume (V2)	Discharge Volume (V3)	Precharge Pressure (P1)	Load Pressure (P2)
316.9 in.^{3}	360.1 in.^{3}	677 in.^{3}	161.3 psi_a	854.5 psi_a

G Gas-Charged Accumulator (*continued*)

Discharge Pressure (P3)	Precharge Volume (V1)	Size	Element Type	Port Size
454.5 psi_a	1,907.7 $in.^3$	8.26 gal.	Piston	P & H, 1/2 in.; T, 5/8 in.

H Reservoir

CAPACITY	
Minimum	Maximum
28.42 gal.	38.50 gal.

I Filtration—Linear Side

		RATING				
Type	Element	Element	Flow	Port Size	Description	Mounting
Strainer	Wire	60 mesh	17.15 to 25.73 gpm	P, H, & T, 1-1/4 in.	Full flow	In-line

J Filtration—Rotary Side

		RATING				
Type	Element	Element	Flow	Port Size	Description	Mounting
Strainer	Wire	60 mesh	8.05 to 12.08 gpm	P & H, 3.4 in.; T, 1 in.	Full flow	In-line

K Master System Directional Control—Linear Side

Positions	Ways	Element	Activation	Port Size
3	4	Sliding spool	Lever	P & H, 1/2 in.; T, 5/8 in.

L Master System Directional Control—Rotary Side

Positions	Ways	Element	Activation	Port Size
3	4	Sliding spool	Solenoid	P, 1/4 in.; H, 5/16 in.; T, 3/8 in.

M Isolator Directional Control

Positions	Ways	Element	Activation	Port Size
2	2	Sliding spool	Plunger	P & H, 1/2 in.; T, 5/8 in.

N Pilot Override Check Valve

Positions	Ways	Element	Activation	Port Size
2	1 (N/C)	Poppet/piston	Pilot	P & H, 1/2 in.; T, 5/8 in.

O Accumulator Shunt Directional Control—1

Positions	Ways	Element	Activation	Port Size
2	1 (N/C)	Sliding spool	Pilot	P & H, 1/2 in.; T, 5/8 in.

P Accumulator Shunt Directional Control—2

Positions	Ways	Element	Activation	Port Size
2	1 (N/C)	Sliding spool	Lever	P & H, 1/2 in.; T, 5/8 in.

Q Check Valves

Positions	Ways	Element	Activation	Port Size
2	1 (N/C)	Poppet	Pilot	P & H, 1/2 in.; T, 5/8 in.

R Flow-Control Valve—Rotary Side

Method	Type	Motor Speed at Flow Control (Lowest Setting)	Delivery through Valve at Flow-Control Setting
Meter-in	Single-range	60 rpm	1.49 gpm

Allowed Pressure Drop	Pressure-Compensation Design	Delivery Range	Port Size
400 psi	Pressure-relieving	0 to 6.44 gpm	P, 1/4 in.; H, 5/16 in.; T, 3/8 in.

S Sequence Valve—1

Type	Element	Setting	Mounting	Port Size
Compound	Spool	330 psi_g	Subplate	P & H, 1/2 in.; T, 5/8 in.

T Sequence Valve—2

Type	Element	Setting	Mounting	Port Size
Compound	Spool	425.3 psi_g	Subplate	P & H, 1/2 in.; T, 5/8 in.

U Pressure-Relief Valve—Linear Side

Type	Element	Setting	Mounting	Port Size
Compound	Piston	839.8 psi_g	In-line	P & H, 1/2 in.; T, 5/8 in.

V Pressure-Relief Valve—Rotary Side

Type	Element	Setting	Mounting	Port Size
Compound	Piston	1,028 psi_g	In-line	P, 1/4 in.; H, 5/16 in.; T, 3/8 in.

W Electric Motor

Speed	Voltage	Phase	Power at V.E.
1,750 rpm	220	3	6.61 hp

X Inlet and Working Conductors—Linear Side

Type	Relief Valve Pressure	Safety Factor	Safety Factor Pressure	Pipe Material	Pipe Tensile Rating
Pipe, Tube, and Hose	839.8 psi	8 : 1	6,718.4 psi	AIS1 1010	55,000 psi

INLETS (I.D. DECIMAL)		*WORKINGS (I.D. DECIMAL)*			
Minimum	Maximum	Minimum	Maximum	Pipe Schedule	Pipe Wall Thickness
0.837 in.	1.342 in.	0.374 in.	0.633 in.	40	0.051 in.

X Inlet and Working Conductors—Linear Side *(continued)*

Tube Material	Tube Tensile Rating	Tube Wall Thickness	*PIPE—NOMINAL*		*TUBE—NOMINAL*	
			Inlets	Workings	Inlets	Workings
5020 Alum.	38,000 psi	0.057 in.	1-1/4 in.	1/2 in.	1-1/4 in.	5/8 in.

Hose Description	Hose Operating Pressure Rating	*HOSE—NOMINAL*	
		Inlets	Workings
SAE 100R1	2,000 psi	1-1/4 in. or –20*	1/2 in. or –8

*SAE 100R4 suction hose

Y Working Conductors—Intensifier Rod End

Type	Intensifier Pressure	Safety Factor	Safety Factor Pressure	Pipe Material	Pipe Tensile Rating
Pipe, Tube, and Hose	2,687.4 psi	4 : 1	10,750 psi	AIS1 1010	55,000 psi

Derived Delivery	*WORKINGS (I.D. DECIMAL)*		Pipe Schedule	Pipe Wall Thickness	Pipe Nominal Size
	Minimum	Maximum			
0.767 gpm	0.125 in.	0.212 in.	80	0.040	1/8 in.

Tube Material	Tube Tensile Rating	Tube Wall Thickness	Tube Nominal Size	Hose Description	Hose Operating Pressure Rating	Hose Nominal Size
6061 Alum.	45,000 psi	0.030 in.	1/4 in.	SAE 100R2B	5,000 psi	3/16 or –3

Z Conductors—Rotary Side

Type	Intensifier Pressure	Safety Factor	Safety Factor Pressure	Pipe Material	Pipe Tensile Rating
Pipe, Tube, and Hose	1,028 psi	6 : 1	6,168 psi	AISI 1010	55,000 psi

	WORKINGS (I.D. DECIMAL)		*INLETS (I.D. DECIMAL)*	
	Minimum	Maximum	Minimum	Maximum
Pump	0.257 in.	0.434 in.	0.574 in.	0.907 in.
Motor	0.174 in.	0.295 in.		
Relief	0.188 in.	0.318 in.		

Z Conductors—Rotary Side *(continued)*

Pipe Schedule	Pipe Wall Thickness	*PIPE—NOMINAL* Inlets	Workings	Tube Material	Tube Tensile Rating	Tube Wall Thickness
40	0.030 in.	3/4 in.	1/4 in.	5020 Alum.	38,000 psi	0.031 in.

TUBE—NOMINAL Inlets	Workings	Hose Description	Hose Operating Pressure Rating	*HOSE—NOMINAL* Inlets	Workings
1 in.	3/8 in.	100R1	2,500 psi	3/4 in. or –12*	5/16 in. or –5

*SAE 100R4 suction hose

CHAPTER 12

In-Chapter Problems

Problem 12-1 P (psi) = 200, or P {*kPa*} = 1,379

Problem 12-2 TE2 (°F) = 5,454, or TE2 {*°C*} = 3,012

Problem 12-3 ΔP (psi) = 11.64, or ΔP {*kPa*} = 80.93

Problem 12-4 P2 (psi_g) = 133.4, or P2 {kPa_g} = 920

Problem 12-5 St (in.) = 5.83, or St {*cm*} = 14.8

Problem 12-6 P2 (psi_g) = 131.56, or P2 {kPa_g} = 907

Problem 12-7 P2 (psi_g) = 122.54, or P2 {kPa_g} = 845

Problem 12-8 VE1 (in../min.) = 1,596, or VE1 {*cm/min.*} = 4,054

Problem 12-9 T (ft. lbs.) = 2.63, or T {*N m*} =3.5

Problem 12-10 DIS (cu. in./rev.) = 1.59, or DIS {cm^3/*rev.*} = 25.5

Problem 12-11 DEL (scfm) = 3.04, or DEL {*scmm*} = 0.08

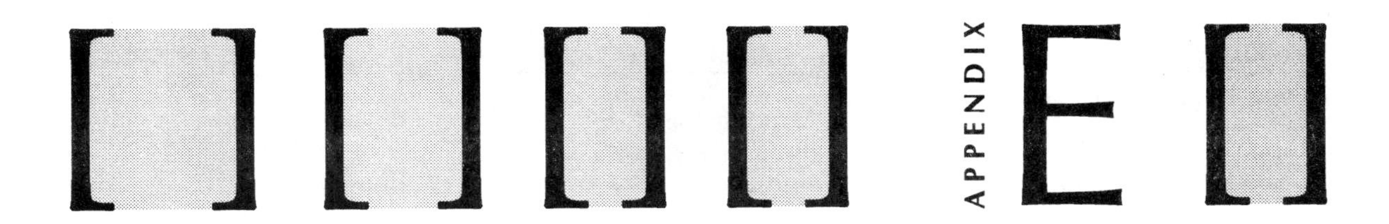

Conversions between U.S. Customary and Metric Units

Quantity	U.S. Unit	U.S. Symbol	SI Unit	SI Symbol
Length	1 inch	in.	20.54 millimeters	mm
	1 inch	in.	2.54 centimeters	cm
	1 inch	in.	0.0254 meter	m
	1 foot	ft.	0.3048 meter	m
	1 yard	yd.	0.9144 meter	m
Area	1 square inch	in.^{2} or sq. in.	6.4516 square centimeters	cm^2
	1 square inch	in.^{2} or sq. in.	0.00064516 square meter	m^2
	1 square foot	ft.^{2} or sq. ft.	0.028317 square meter	m^2
Volume	1 cubic inch	in.^{3} or cu. in.	16.387064 cubic centimeters	cm^3
	1 cubic foot	ft.^{3} or cu. ft.	0.028317 cubic meter	m^3
Velocity	1 inch per second	in./sec.	2.54 centimeters per second	cm/sec.
	1 inch per minute	in./min.	2.54 centimeters per minute	cm/min.
	1 foot per second	ft./sec.	30.48 centimeters per second	cm/sec.
	1 foot per minute	ft./min.	0.3048 meters per minute	m/min.

Quantity	U.S. Unit	U.S. Symbol	SI Unit	SI Symbol
Supply Rate	1 cubic inch per second	in.^{3}/sec. or cu. in./sec.	16.3871 cubic centimeters per second	cm^3/sec.
	1 cubic inch per minute	in.^{3}/min. or or cu. in./min.	16.3871 cubic centimeters per minute	cm^3/min.
	1 cubic foot per minute	ft.^{3}/min. or cu. ft./min.	0.028317 cubic meter per minute	m^3/min.
Delivery	1 gallon per minute	gal./min. or gpm	227.1247 liters per minute	l/min. or lpm
	1 standard cubic foot per minute	scfm	0.028317 standard cubic meter per minute	scmm
Force/Mass	1 pound weight	lb.$_w$	0.453592 kilogram	kg
	1 pound force	lb.$_f$	4.448222 Newtons	N
	1 pound force	lb.$_f$	0.00448222 kiloNewton	kN
Energy/Torque	1 pound inch	lb. in.	0.112985 Newton meter	N m
	1 pound foot	lb. ft.	1.355818 Newton meters	N m
	1 pound foot	lb. ft.	1.355818 Joules	J
Pressure	1 pound per square inch	lb./in.^{2} or psi	0.006894757 Newtons per square meter	N/m^2
	1 pound per square inch	lb./in.^{2} or psi	6.894757 kiloPascals	kPa
	1 atmosphere	14.7 psi	101.4 kiloPascals	kPa
Power	1 pound foot per second	lb. ft./sec.	1.355818 Newton meters per second	N m/sec.
	1 pound foot per second	lb. ft./sec.	1.355818 Joules per second	J/sec.
	1 pound foot per second	lb. ft/sec.	1.355818 Watts	W
	1 pound foot per minute	lb. ft./min.	0.022597 Watt	W
	1 horsepower	hp	745.7 Watts	W
Temperature	1 degree Farenheit	°F	1.8 degrees Celsius + 32	°C
	1 degree Rankine	°R	1.8 degrees Kelvin	°K

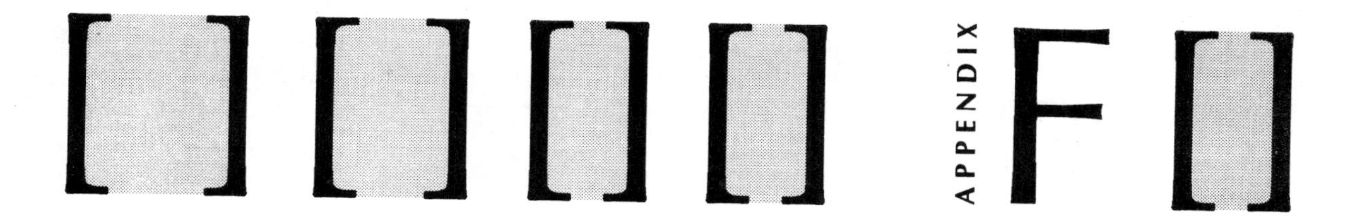

Selected Formulation Used in This Book

Area

Area of a Circle

$\pi r^2 = 0.7854D^2$

Annulus Area of Cylinder

A_{ann} (sq. in.) {*cm²*} = A_{pis} (sq. in.) {*cm²*} – A_{rod} (sq. in.) {*cm²*}

Area of Inside Diameter of Conductor

$$A \text{ (sq. in.) } \{cm^2\} = \frac{\text{DEL (gpm) } \{lpm\} * 0.3208 \ \{0.1667\}}{\text{VEL (ft./sec.)} \{m/sec.\}}$$

Displacement

Of a Cylinder during Extension

VOL_{ext} (cu. in.) {*cm³*} = A_{pis} (sq. in.) {*cm²*} * St (in.){*cm*}

Of a Cylinder during Retraction or of Double-End Rod Cylinder

VOL_{ret} (cu. in.) {*cm³*} = A_{ann} (sq. in.) {*cm²*} * St (in.) {*cm*}

Of a Motor or Pump

$$\text{DIS (cu. in./rev.)} \{cm^3/rev.\} = \frac{\text{DEL (gpm)} \{lpm\} * 231 \text{ (cu. in./gal.)} \{1{,}000\ cm^3/l\}}{\text{Speed} \{\text{(rev./min.)}\}}$$

Supply Rate

Linear Actuation—Piece-Rate Parameter

$$\text{S.R. (cu. in./min.)} \{cm^3/min.\} = S_{tot} \text{ (cu. in./cyc)} \{cm^3/cyc\} * \text{C.R.} \{\text{(cyc/min.)}\}$$

Linear Actuation—Feed-Rate Parameter

$$\text{S.R. (cu. in./min.)} \{cm^3/min.\} = \text{VEL (in./min.)} \{cm/min.\} * \text{A (sq. in.)} \{cm^2\}$$

Rotary Actuation

$$\text{S.R. (cu. in./min.)} \{cm^3/min.\} = \text{DIS (cu. in./rev.)} \{cm^3/rev.\} * \text{SP} \{\text{(rev./min.)}\}$$

Pump Delivery Demand

Of a Linear Circuit—Piece-Rate Parameter

$$\text{DEL (gpm)} \{lpm\} = \frac{\text{S.R. (cu. in./min.)} \{cm^3/min.\}}{231 \text{ (cu. in./gal.)} \{1{,}000\ cm^3/l\}}$$

Of a Linear Circuit—Feed-Rate Parameter

$$\text{DEL (gpm)} \{lpm\} = \frac{\text{VEL (ft./sec.)} \{m/sec.\} * \text{A (sq. in.)} \{cm^2\}}{0.3208 \{0.1667\}}$$

Of a Rotary Circuit

$$\text{DEL (gpm)} \{lpm\} = \frac{\text{DIS (cu. in./rev.)} \{cm^3/rev.\} * \text{SP} (\{\text{rev./min.}\})}{231 \text{ (cu. in./gal.)} \{1{,}000\ cm^3/l\}}$$

Speed

Of a Cylinder during Extension

$$\text{VEL (in./min.)} \{cm/min.\} = \frac{\text{DEL (gpm)} \{lpm\} * 231 \text{ (cu. in./gal.)} \{1{,}000\ cm^3/l\}}{A_{pis} \text{ (sq. in.)} \{cm^2\}}$$

$$\text{VEL (in./min.) } \{\textit{cm/min.}\} = \frac{\text{S.R. (cu. in./min.) } \{\textit{cm}^3\textit{/min.}\}}{A_{pis} \text{ (sq. in.) } \{\textit{cm}^2\}}$$

$$\text{VEL (ft./sec.) } \{\textit{m/sec.}\} = \frac{\text{DEL (gpm) } \{\textit{lpm}\} * 0.3208 \ \{\textit{0.1667}\}}{A_{pis} \text{ (sq. in.) } \{\textit{cm}^2\}}$$

Of a Cylinder during Retraction or of Double-End Rod Cylinder

$$\text{VEL (in./min.) } \{\textit{cm/min.}\} = \frac{\text{DEL (gpm) } \{\textit{lpm}\} * 231 \text{ (cu. in./gal.) } \{\textit{1,000 cm}^3\textit{/l}\}}{A_{ann} \text{ (sq. in.) } \{\textit{cm}^2\}}$$

$$\text{VEL (in./min.) } \{\textit{cm/min.}\} = \frac{\text{S.R. (cu. in./min.)}\{\textit{cm}^3\textit{/min.}\}}{A_{ann} \text{ (sq. in.) } \{\textit{cm}^2\}}$$

$$\text{VEL (ft./sec.) } \{\textit{m/sec.}\} = \frac{\text{DEL (gpm) } \{\textit{lpm}\} * 0.3208 \ \{\textit{0.1667}\}}{A_{ann} \text{ (sq. in.) } \{\textit{cm}^2\}}$$

Of a Motor

$$\text{SP } \{\text{(rev./min.)}\} = \frac{\text{DEL (gpm) } \{\textit{lpm}\} * 231 \text{ (cu. in./gal.) } \{\textit{1,000 cm}^3\textit{/l}\}}{\text{DIS (cu. in./rev.) } \{\textit{cm}^3\textit{/rev.}\}}$$

Energy

Work of Cylinder Movement

W (lb. in.) {*N cm*} = F (lbs.) {*N*} * St (in.) {*cm*}

W (lb. in.) {*J*} = P (lbs./sq. in.) {*Pa*} * Q (cu. in.) {m^3}

Work of Motor Rotation

T (lb. in.) {*N cm*} = F (lbs.) {*N*} * Radius (in.) {*cm*}

$$\text{T (lb. in.) } \{\textit{N m}\} = \frac{\text{DIS (cu. in./rev.) } \{\textit{cm}^3\textit{/rev.}\} * \text{P (lbs./sq. in.) } \{\textit{kPa}\}}{2\pi \ \{\textit{2,000 } \pi\}}$$

Pressure

Of Cylinder during Extension

$$P\ (\text{psi})\ \{kPa\} = \frac{F\ (\text{lbs.})\ \{kN\}}{A_{pis}\ (\text{sq. in.})\ \{m^2\ or\ 10{,}000\ cm^2\}}$$

Of Cylinder during Retraction or of Double-End Rod Cylinder

$$P\ (\text{psi})\ \{kPa\} = \frac{F\ (\text{lbs.})\ \{kN\}}{A_{ann}\ (\text{sq. in.})\ \{m^2\ or\ 10{,}000\ cm^2\}}$$

Of Motor

$$P\ (\text{psi})\ \{kPa\} = \frac{T\ (\text{lb. in.})\ \{N\ m\} * 2\pi\ \{2{,}000\pi\}}{DIS\ (\text{cu. in./rev.})\ \{cm^3/rev.\}}$$

Power

Power (lb. in./sec.) {*W*} = P (psi) {*kPa*} * DEL (gpm) {*lpm*} * 3.85 {*0.01667*}

Power (hp) = P (psi) * DEL (gpm) * 0.000583

In Linear or Rotary Circuit

$$\text{Power (lb. in./sec.)}\ \{W\} = \frac{P_{rel}\ (\text{psi})\ \{kPa\} * DEL\ (\text{gpm})\ \{lpm\} * 3.85\ \{0.01667\}}{V.E.\ \{(\%)\}}$$

$$\text{Power (hp)} = \frac{P_{rel}\ (\text{psi}) * DEL\ (\text{gpm}) * 0.000583}{V.E.\ (\%)}$$

Of Motor

$$\text{Power (hp)} = \frac{T\ (\text{lb. in.}) * SP\ (\text{rev./min.}) * 0.000158}{V.E.\ (\%)}$$

Index

Note: Numbers in *italics* denote figures or problems.

W